D0501316

*f*P

ALSO BY KEN CROSWELL

The Alchemy of the Heavens (1995)

PLANET QUEST

The Epic Discovery of Alien Solar Systems

KEN CROSWELL

Illustrations by [sic]

THE FREE PRESS
New York London Toronto Sydney Singapore

fP

THE FREE PRESS
A Division of Simon & Schuster Inc.
1230 Avenue of the Americas
New York, NY 10020

Designed by Carla Bolte

Manufactured in the United States of America

10 9 8 7 6 5 4 3 2 1

Library of Congress Cataloging-in-Publication Data

Croswell, Ken.
Planet quest : the epic discovery of alien solar systems / Ken Croswell.
p. cm.
Includes bibliographical references and index.
1. Extrasolar planets. I. Title.
QB820.C76 1997
523—dc21 97-9473
CIP

ISBN 0-684-83252-6

Contents

About the Author

KEN CROSWELL is an astronomer in Berkeley, California, who earned his Ph.D. from Harvard University. His first book, *The Alchemy of the Heavens,* explored the Milky Way Galaxy and was a finalist for a 1995 *Los Angeles Times* Book Prize. His work has appeared in *Astronomy, New Scientist, The New York Times,* and *Sky and Telescope.* He also writes for the *Star Date* radio program, which airs on two hundred stations.

Introduction: Bruno's Execution

> To believe that [suns illuminate other worlds], it would be necessary to have the mind of Giordano Bruno, who was burned by the judgment of the Holy Inquisition for having maintained such an impertinence.
>
> —French cleric Marc Antoine Guigues, 1700

IT WAS Thursday morning, February 17, 1600, and the crowd in Rome was jeering. A fifty-one-year-old former priest, condemned as a heretic by the Inquisition, was being led, in chains, to his death.

Giordano Bruno had spent the last eight years of his life in the dark dungeons of the Inquisition, where he was repeatedly interrogated and probably tortured. He refused to recant, however, hoping instead to convince the Inquisitors, and even the Pope himself, that his beliefs were correct. Of these beliefs, one of the most radical was that the stars were other suns, circled by planets like the Earth.

Bruno was born in 1548, just five years after Copernicus published his revolutionary claim that the Sun, not the Earth, constituted the center of the universe around which all else revolved. The idea was so daring that the publisher prefaced Copernicus's work with a statement that the model was only a mathematical device to calculate planetary positions and did not reflect reality. Bruno, though, gladly adopted Copernicus's heliocentric model and took it even further. Copernicus thought the stars were merely part of the firmament, a finite sphere that encased the solar system, but Bruno proclaimed that they were actually distant suns scattered throughout a universe infinite in size. Around these suns circled planets. Bruno also predicted that additional planets, too distant to be seen, revolved around the Sun beyond the orbit of Saturn, the farthest planet then known.

Because Bruno advanced these ideas, he might seem to have been a link between Copernicus and Galileo, the great Italian astronomer whose telescopic observations in the early 1600s would support Copernicus's conception of the solar system. But Copernicus and Galileo were scientists, whereas Bruno was a philosopher steeped in medieval mysticism and magic; he eschewed observations and disdained mathematics. Bruno believed the Earth circled the Sun not because observations said so but because he thought the Earth was literally alive, and like all other living creatures, it must move.

Similarly, Bruno's belief in an infinite universe, filled with innumerable planets circling other stars, stemmed not from science but from religion. God is infinite, said Bruno, so his universe must be, too. "Thus is the excellence of God magnified and the greatness of his kingdom made manifest; he is glorified not in one, but in countless suns; not in a single earth, a single world, but in a thousand thousand, I say in an infinity of worlds," wrote Bruno in his 1584 work, *On the Infinite Universe and Worlds*. Elsewhere Bruno explained that these planets could not be seen because they were fainter than their stars. He imagined seeing a large ship at the harbor, surrounded by small boats. If we then see a large ship in the distance, said Bruno, it should also be surrounded by small boats, even though we cannot see them: the near and the far obey the same rule. "We are a celestial body for the Moon and for every other celestial body," wrote Bruno, "and we are the firmament just as much as they are for us." He thus contradicted Aristotle, whom the Church esteemed. Aristotle had held that the Earth, flawed and imperfect, was separate from the heavens, which were perfect and changeless.

Neither Copernicus nor Bruno was entirely original in his beliefs. Long before Copernicus, Greek astronomer Aristarchus had proposed that the Earth circled the Sun. And long before Bruno, Greek philosopher Epicurus had suggested that other "worlds"—by which he meant other solar systems with an Earthlike planet, rather than a star, at their centers—existed elsewhere. Both ideas challenged the teachings of the Church, but Bruno's were far more radical. Copernicus had yanked the Earth away from the privileged center of the universe; now Bruno was saying that the Earth was not even unique. If there were other planets with intelligent beings, then where did that leave Jesus Christ? Did those other planets need their own Jesus? Might they even have religions different from the Catholic faith?

Back in the 1570s, while Bruno was only in his twenties, Church authorities began to accuse him of heretical thinking. He was then a monk in Italy, and he fled his native land to wander for over ten years through Switzerland, France, England, and Germany. In 1591, however, he made the fatal mistake of accepting an invitation from a professed admirer who offered him free lodging in Venice. The next year, this person betrayed him to the Inquisition, and Bruno's eight-year-long ordeal began.

It is not clear today exactly why the Church condemned Bruno. In the end, it found him guilty of eight heresies, but precisely what those heresies were is unknown, because the documents relating to Bruno's final trial perished in the 1800s, before any historians examined them. However, older documents do survive, and they accuse Bruno of both religious and scientific errors. Two of his alleged scientific errors were his belief in an infinite universe and in planets circling other stars.

The final sentence, handed down by the Inquisition in early 1600, mentioned Bruno's eight heresies and then said: "We hereby, in these documents, publish, announce, pronounce, sentence, and declare thee the aforesaid Brother Giordano Bruno to be an impenitent and pertinacious heretic, and therefore to have incurred all the ecclesiastical censures and pains of the Holy Canon, the laws and the constitutions, both general and particular, imposed on such confessed impenitent pertinacious and obstinate heretics. . . . We ordain and command that thou must be delivered to the Secular Court . . . that thou mayest be punished with the punishment deserved. . . . Furthermore, we condemn, we reprobate, and we prohibit all thine aforesaid and thy other books and writings as heretical and erroneous, containing many heresies and errors, and we ordain that all of them which have come or may in future come into the hands of the Holy Office shall be publicly destroyed and burned in the square of St. Peter before the steps and that they shall be placed upon the Index of Forbidden Books, and as we have commanded, so shall it be done. . . . Thus pronounce we, the undermentioned Cardinal General Inquisitors."

Bruno's response was eloquent: "Perchance you who pronounce my sentence are in greater fear than I who receive it."

At the time, most heretics condemned to the stake were not actually burned. They were executed in prison and burned only in effigy, and for centuries afterward certain Catholic apologists claimed that Bruno met such a fate. Some even claimed there never was a Bruno.

But eyewitness accounts recorded what actually happened on that Thursday morning, four hundred years ago: Giordano Bruno, poet, philosopher, and proponent of alien worlds, was bound to a stake in the ironically named Field of Flowers, a fire was lit, and he was burned alive.

PLANETS ARE vital for life: without them, intelligent life would never have arisen on Earth, for we owe our lives to the planet beneath our feet. But planets are so difficult to see that only in the 1990s, four centuries after Bruno's death, did astronomers succeed in fulfilling his dream by detecting genuine planets around other stars.

This book is the story of these recent discoveries and the people who made them. *Planet Quest* opens with an investigation of the one solar system scientists know best, our own. The narrative then shifts back in time, to an era when Saturn marked the known edge of the solar system, and joins astronomers as they look for and discover distant worlds orbiting the Sun.

Then the book sets off for other stars, culminating with the stunning discoveries of the 1990s: the astonishing 1991 find of planets orbiting a dead star known as a pulsar, and the rich harvest of planet discoveries around Sunlike stars that began in 1995. These discoveries took astronomers closer than they had ever been to their goal of finding life among the stars, and this book's penultimate chapter explores new techniques that promise to divulge smaller planets, like Earth, circling stars like the Sun, the very planets that can harbor life. The last chapter of the book explores the challenging and controversial possibility that starships might someday journey to the planets of other stars.

As we explore the solar system and the Galaxy, studying old planets and searching for new ones, we will see just how hostile all other known planets are—from scorched Mercury to frigid Pluto, and even to the new worlds that astronomers have found around other stars. These forbidding worlds should make us appreciate more fully the special nature of the Earth, a warm, wet world that still remains amazingly unique and whose ultimate fate is in our own hands.

I THANK the astronomers who talked with me: Roger Angel, Hartmut Aumann, Alan Boss, Adam Burrows, Paul Butler, Heinrich Eichhorn, Dale Frail, George Gatewood, Fred Gillett, Robert Harrington, Shrinivas Kulkarni, Shiv Kumar, David Latham, Jack Lissauer, Andrew Lyne, Geoffrey Marcy, Brian Marsden, Michel Mayor, Tadashi Nakajima, Robert Noyes, Bohdan Paczyński, Didier Queloz, Kenneth Seidelmann, Michael Shao, Bradford Smith, Myles Standish, David Stevenson, Richard Terrile, Clyde Tombaugh, Peter van de Kamp, Paul Weissman, George Wetherill, and Alex Wolszczan.

I especially thank those who read the manuscript and offered comments: Robert Havlen, George Musser, Richard Pogge, and Rob Radick. Of course, these people are not responsible for any errors I have made.

I thank Stefano Arcella of [sic] for the many illustrations that accompany the text.

For their support of this project, I thank my agent, Lew Grimes, and my editor, Stephen Morrow.

1

A Good Planet Is Hard to Find

NINE MAJESTIC planets race around the Sun, trumpeting the same wealth of diversity that pervades all of nature. From the cream-colored clouds of Venus and the orange sands of Mars to the green and blue atmospheres of Uranus and Neptune; from the great red spot of Jupiter to the bright white rings that bedeck golden Saturn; from the gray craters that puncture Mercury to the icy plains of far-off Pluto: though only nine planets strong, our solar system teems with a treasure of planetary possibilities.

But the most stunning of the nine planets is the water-covered one that is third from the Sun, the only world in the entire cosmos known to support intelligent life. Enveloped by a warm, moist, benevolent atmosphere, the Earth thrives with millions of different species that nestle in numerous niches. This vibrant planet stands out all the more vividly when compared with its eight siblings, for even the harshest parts of Earth—such as frigid Antarctica or the blistering desert of the Sahara—would actually be havens of gentleness if they appeared on any other planet.

Astronomers did not always know this. At one time, the two nearest planets, Venus and Mars, seemed sufficiently Earthlike that it was thought they too might harbor advanced life. The clouds of Venus, which cloaked the planet's surface, suggested abundant water and evoked visions of oceans or tropical forests adorning this neighbor world. Today, astronomers know that Venus is hellishly hostile, its surface smothered beneath an atmosphere ninety times thicker than Earth's and searing at a temperature that would melt lead. A century ago, the red planet Mars also beckoned: its "canals" and polar caps meant water, and green or blue areas were surely plants or seas that modulated with the rhythm of the changing seasons. Today, astronomers know that Mars is a frozen desert whose thin atmosphere cannot even protect the planet from the Sun's lethal ultraviolet rays. The canals were mere illusions, as were the green-blue colors, triggered in the eye by a little gray sprinkled among the planet's predominant red and orange, the opposites of green and blue; and the seasonal changes in these colors result from huge dust storms that occasionally sweep over the planet.

These distressing discoveries pushed possible abodes of alien life far from Earth and entirely out of the solar system, to unseen planets orbiting other stars like the Sun. This simple fact inspires a modern astronomical adventure: the quest for new and possibly Earthlike planets around other stars. If ours were the only solar system in the Galaxy, then ours would also be the only civilization in the Galaxy. Yet if planets abound throughout the Milky Way, then so might life itself.

A second consideration also drives the search for alien worlds. The discovery of other solar systems would enable astronomers to compare planets elsewhere with the nine that orbit the Sun and thus examine the general properties of planetary systems and how they form and evolve. In a similar way, astronomers have long exploited the stars and the galaxies. Most of what is known about the evolution of the Sun has come from observing other stars, and much of what is known about our Galaxy has been obtained by examining others beyond.

Unfortunately, planets compare poorly with stars and galaxies, because so few planets are known. Stars and galaxies litter the sky in vast quantities, but until 1991 known planets numbered a mere nine. Nine. That is far too small and local a sample from which to draw firm conclusions about planets throughout the universe, just as nine autobiographies of people from the same small town could hardly begin to fathom the entire human race. The discovery of other planets would help reveal the rules that govern planetary systems and allow astronomers to see whether our own solar system is typical or instead highly unusual—two possibilities that carry radically different implications for the existence of extraterrestrial intelligence.

The Four Astronomical Ingredients of Life

From a broad astronomical view, four crucial actors work together to create life: stars, galaxies, planets, and stars again.

Stars are the first prerequisite for life, because they are master alchemists, transforming hydrogen and helium, the two lightest and most common elements, into heavier elements like carbon, nitrogen, oxygen, and iron—elements that life requires. Stars perform this celestial magic by fusing light elements together, in nuclear reactions that power the stars with the energy they need to live and shine. At the end of their lives, stars toss these newly minted elements into space, either by violently exploding as supernovae or by gently casting off their outer atmospheres. Every star, in its own way, contributes to the Galaxy. Massive stars, such as the blue supergiant Rigel and the red supergiant Betelgeuse, both in the constellation Orion, will churn out large amounts of oxygen and magnesium, ejecting these elements when the stars die; less massive stars, which do not explode but instead eject their atmospheres into space, produce most of the carbon and nearly all of the nitrogen in the cosmos.

These precious elements do no good unless they are recycled back into new stars and planets. This is why the second prerequisite for life is a galaxy. Not just any galaxy will work, however. If the galaxy is too small, as most galaxies are, then a dying star's harvest of heavy elements escapes the galaxy's weak gravitational grasp and drifts into intergalactic space, where their potential is completely wasted. Fortunately, a few galaxies, such as our own Milky Way, are giants that dwarf the rest. These galaxies harbor hundreds of billions of stars whose collective gravity retains stellar ejecta. Brimming with heavy elements, this life-giving debris can then enrich the beautiful interstellar clouds of gas and dust that give birth to new stars and planets. The most famous such stellar nursery is the Orion Nebula, whose glow is visible to the naked eye in the sword of Orion. This and all the other stellar spawning grounds that dot the Milky Way's spiral

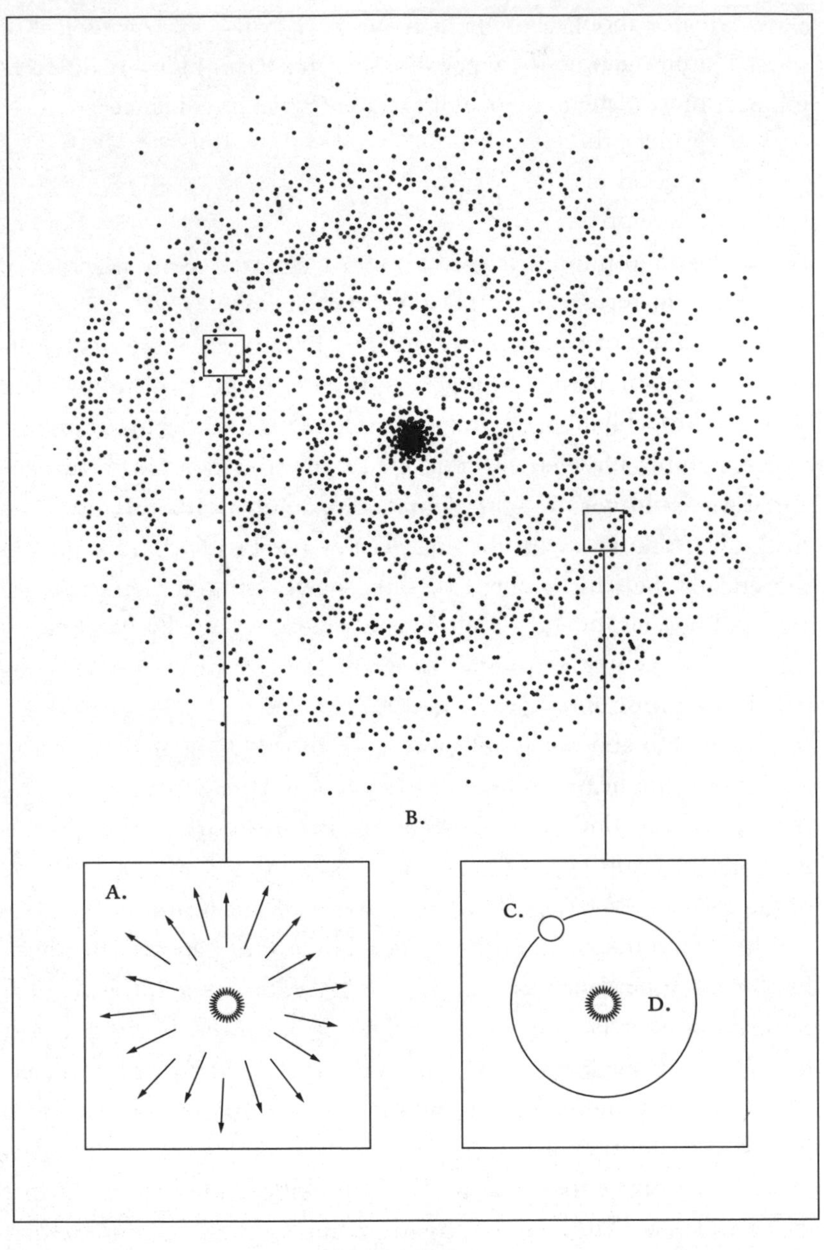

FIGURE 1

Life requires four key astronomical objects: stars (A) to create heavy, life-giving elements; a giant galaxy (B) to retain those heavy elements and recycle them into new stars and planets; a planet (C) on which life can arise; and a star (D) to light and warm that planet.

arms produce about ten new stars each year. If each star has about ten planets orbiting it, then every year the Milky Way Galaxy creates a hundred new planets.

Such planets, the subject of this book, are the third key ingredient for life. A good planet provides a stable platform on which life can arise and develop, aided by a solid surface, seas of liquid water, and a nurturing atmosphere. Planets like Earth are dense concentrations of life-giving material, oases of heavy elements like oxygen and iron in a cosmic desert of lifeless hydrogen and helium, which make up over 99.8 percent of all atoms in the universe. Vital though planets are, most are probably lifeless. Even in our own solar system, hostile planets outnumber life-supporting planets eight to one, and good planets may be still more scarce in the cosmos at large. The easiest planets for astronomers to detect around other stars are gargantuan worlds like Jupiter and Saturn, gas giants that probably do not have life. So the ultimate and far more difficult goal of planet hunters is the discovery of small, rocky worlds like Earth, for at least one such planet has fulfilled its promise and produced actual life.

The fourth and final requirement for life is a star again—not, this time, to create heavy elements but instead to shower its planets with light and heat. Sunlight warms the world, keeping most of Earth's water liquid, and liquid water is vital for life. Furthermore, energy from the Sun powers most life on Earth. Bright yellow stars like the Sun supply abundant light and heat. And contrary to statements that frequently appear in popular science books, the Sun *is* bright, outperforming 95 percent of the stars in the Galaxy. The few stars much brighter than the Sun will not do, because they burn themselves out so fast that advanced life never has time to evolve on their planets; stars much dimmer than the Sun will not work either, because in order to receive enough warmth, a planet would have to huddle so close to its star that the star's gravity would force one planetary hemisphere to face the star and the other to turn away, thereby frying one side and freezing the other. The Sun is therefore a perfect star, without which the Earth's oceans would freeze and its plants wither.

In this chain of four prerequisites, only one link is weak: planets. The first two requirements have already been met, for stars long since dead have enriched the Milky Way and other giant galaxies with plenty of heavy, life-giving elements. And the fourth requirement has also been met, because about 4 percent of stars are bright and yellow like the Sun, giving the Milky Way billions of good stars.

But planets? No other world like the Earth has yet been seen around any other star like the Sun. Hence, though extremely difficult, the search for such planets will completely determine whether intelligent life can exist elsewhere in the cosmos.

All Planets Bright and Dim

Although modern astronomers face a great challenge simply in finding other stars' planets, five of the Sun's planets glow so brightly that they were known to the ancients. Venus, aptly named for the goddess of love and beauty, sparkles in the eastern sky before sunrise or the western sky after sunset. The planet outshines every other celestial object except the Sun and the Moon, and at its greatest brightness, Venus appears twenty times brighter than the brightest star in the night sky, brilliant Sirius. The planet Mars can also dazzle. At its best, Mars also outdoes every star in the night sky, and it bears such a bloody hue that the ancients named it for the god of war. Mars is indeed red for the same reason that blood is. Blood turns red when the iron-bearing compound hemoglobin joins with oxygen, and the Martian surface is red because the same two elements combine to form iron oxide, or rust. Beyond the orbit of Mars lies a third brilliant planet, Jupiter, named for the king of the gods. Jupiter's immense size compensates for its greater distance so that the planet always outshines every star in the night sky.

On their night, all three planets—Venus, Mars, and Jupiter—command the attention of even a casual observer. The ancients also knew two less prominent planets. One, though bright, clings so tightly to the Sun that it appears only briefly, darting from the eastern sky just before sunrise to the western sky just after sunset. Because of the planet's swiftness, the ancients named it for the messenger of the gods, Mercury. Fainter but less elusive is Saturn, which is brighter than all but a handful of stars. In Roman mythology, Saturn was the father of Jupiter.

Together with the Sun and the Moon, the five planets conferred

TABLE 1–1
The Daily Planet

Day	Object
Sunday	Sun
Monday	Moon
Tuesday	Mars
Wednesday	Mercury
Thursday	Jupiter
Friday	Venus
Saturday	Saturn

their names on the seven days of the week, in ways both obvious (Sunday, Monday, Saturday) and not (Tuesday, Wednesday, Thursday, Friday). If ancient people had recognized that the Earth was also a planet, perhaps the week would have eight days, one of them Earthday.

All five planets stood out from the stars because they moved, and the word *planet* derives from the Greek word *wanderer*. As astronomers know today, the stars also move, and often faster than planets, but they are so far away that they look stationary. By contrast, the planets are so close that they appear to move against this starry backdrop, both because of their own motion and because the Earth is moving around the Sun. In the same way, a speeding automobile driver sees nearby road signs rush past, while distant mountains seem fixed.

Even the brightest and most spectacular planets depend on the Sun for their beauty, in that they do not generate their own visible light but simply reflect the Sun's. If the Sun stopped shining, Mercury, Venus, Mars, Jupiter, and Saturn would turn black, depriving observers of the brightest objects in the sky. As one ventures beyond Saturn, one appreciates just how dim planets can be, even though the Sun tries to illuminate them. Because of their greater distance and smaller size, Uranus is just barely visible to the naked eye, Neptune is visible only through binoculars, and a view of Pluto requires a telescope, for Pluto is over a thousand times fainter than the faintest star the naked eye can see. All three planets lie more than a billion miles from the Sun, so the light they receive is weak and the little reflected toward Earth is further attenuated by the great distance between them and us.

The dimness of these far-off worlds is bad, but much worse awaits those astronomers who seek extrasolar planets—planets beyond the Sun's. Part of the problem stems from the enormous distances of other stars. Even the nearest star system to the Sun, Alpha Centauri, is so remote that astronomers resort to a new unit, the light-year, to measure its distance. Alpha Centauri is 4.3 light-years from the Earth, which means that the star's light requires 4.3 years to reach us. This works out to a distance of 25 trillion miles, thousands of times greater than dim Pluto's. To put this immensity into perspective, imagine shrinking the universe so that the distance between the Sun and the Earth were 1 inch. Then all the Sun's planets would be within easy reach, even far-off Pluto, which on average would lie just 40 inches away from the Sun. But on the same scale, a light-year equals 1 mile, so Alpha Centauri would be 4.3 miles from the Sun. And every other star in the night sky would be farther still.

Furthermore, as viewed from Earth, an extrasolar planet would hug its star, which would shine millions or billions of times more brightly and wash out the planet's faint light. Seeing a planet around another star is like trying to glimpse the flicker of a candle amid a raging forest fire.

Planets, then, confront astronomy with a fundamental paradox: despite their supreme importance to life and thus to what many view as the spiritual side of the universe, they barely affect the *physical* universe at all. They emit no light of their own, and their mass is so minuscule that their gravity disturbs their stars only slightly. Even giant Jupiter, the largest planet in the solar system, has just a thousandth the mass of the Sun; therefore, the planet tugs on the Sun little and shifts its position so subtly that an alien astronomer would have to employ sophisticated instruments in order to detect the planet's presence.

President John Kennedy once said that humanity must go to the Moon not because doing so is easy but because it is hard. He could just as well have been talking about the search for extrasolar planets. Discerning worlds around other stars is so difficult that successful planet hunters must arm themselves with every possible weapon that might aid them in their attack. Perhaps the most powerful of these is an understanding of the one solar system astronomers know best: the nine planets that circle the Sun, one of which gave rise to the people who now seek new worlds among the stars.

2

A Living Solar System

Before brave explorers sail the seas in search of new and exotic lands, they ought first to prepare themselves by surveying their own country. Its properties may reappear elsewhere and help them make sense of new discoveries. One's native land might have hills and valleys, forests and deserts, rivers and lakes—features that a foreign land might also possess. Whether at home or abroad, these features would probably obey the same basic laws, such as the tendency of rivers to flow downhill from the mountains to the ocean. Similarly, the general properties of our solar system and the laws it follows might be echoed in the planetary systems of other stars. For this reason, before venturing across the cosmic abyss that separates the Sun's planets from worlds circling other stars, astronomers should first examine the solar system they know best.

One Star, Nine Planets

Denigrated, dismissed, belittled, and bashed by ignorant science writers who persist in calling it an "average" star, the Sun constitutes the centerpiece of the solar system, harboring over 99 percent of its mass and anchoring all the planets with a gravitational grip so strong that it can snare objects trillions of miles distant. The Sun is anything but average, for it outshines 95 percent of the Galaxy's stars. Deep within the Sun's core, nuclear reactions convert 600 million tons of hydrogen into helium each second, thereby generating the intense energy that lights and warms the Earth. Even from a distance of 93 million miles, the Sun is so powerful that it can melt telescopes and blind those foolish enough to stare at it.

The first of the Sun's planets, Mercury, is a small world that huddles close to the solar flame and is scorched by the same. The planet rotates slowly, so the days and nights are long, and the temperature range is enormous. During the day, Mercury hits a sizzling 800 degrees Fahrenheit, but at night the temperature plunges to 300 degrees below zero. Only Venus is hotter; only Uranus, Neptune, and Pluto are colder. Equally forbidding is Mercury's lunarlike landscape—gray with hills and countless craters.

Although Venus lies twice as far from the Sun as Mercury, it is even hotter, boiling at a temperature, day and night, of 860 degrees Fahrenheit. Blame the planet's thick, carbon dioxide atmosphere, which produces an extreme greenhouse effect that traps solar energy.

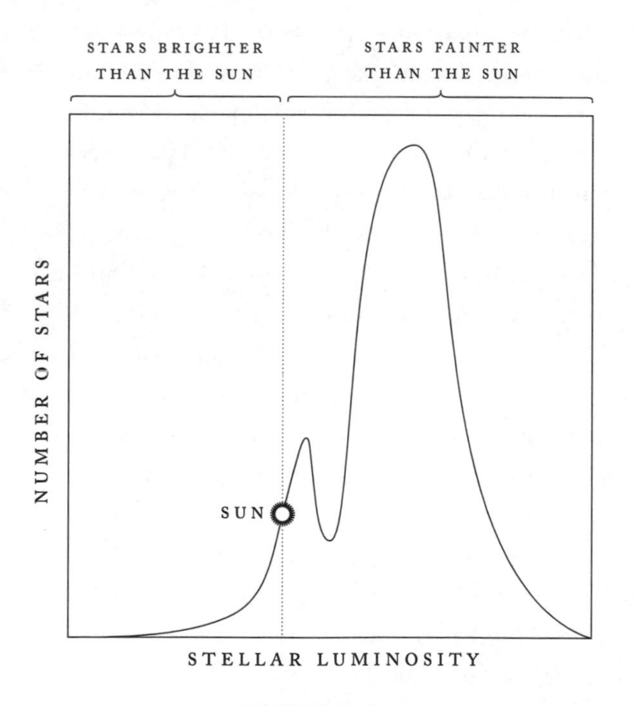

FIGURE 2

The Sun is no average star. It outshines most other stars in the Galaxy.

In terms of size and mass, though, this hostile planet so resembles our own that it was once nicknamed the Earth's twin.

The third planet from the Sun needs no introduction. It is a world of water: seawater covers 71 percent of its surface, rainstorms nourish its plentiful plants and trees, and immense sheets of ice cloak its south pole. The Earth's atmosphere is a mix of volatile oxygen (21 percent) buffered by tranquil nitrogen (78 percent). Conveniently for the many life forms that inhabit this world, some of the oxygen forms an ozone (O_3) layer that shields the planet's surface from deadly ultraviolet radiation. Oxygen was originally injected into the atmosphere billions of years ago by primitive life that developed in the oceans; ancient life therefore paved the way for modern. Earth is the only small planet that has a large satellite, which is simply called "the Moon."

Alas, none of Earth's benefits extend to the next planet out, the red planet Mars. A world half the size of Earth, Mars is a desert whose thin atmosphere has essentially no oxygen or ozone. In the past, though, Mars may have been warmer and wetter, perhaps giving rise to some form of life, whose fossils may still be preserved among the Martian sands. In 1996, a team of scientists claimed to have discovered fossils of Martian life in a meteorite that came from the red planet. Although other scientists sharply attacked this claim, the search for fossils will be a chief objective of a manned mission to Mars, because such fossils would allow scientists to investigate how life arose and evolved on another planet. Two tiny moons, which may be captured asteroids, whirl around Mars and bear the names Phobos ("fear") and Deimos ("terror"), appropriate companions for the god of war.

Beyond Mars lies a belt dotted with thousands of known asteroids, and beyond them dwells the solar system's planetary behemoth, giant Jupiter. The mightiest planet in the solar system, Jupiter is eleven times larger than the Earth and contains over twice as much mass as all the other planets put together. It is a gas giant, an enormous ball of hydrogen and helium entombing a much smaller core of rock and ice. Under Jupiter's supreme command are sixteen satellites, four of them large like the Earth's Moon.

Next out from the Sun is the solar system's artistic masterpiece: Saturn, a golden world encircled by bright white rings. Like Jupiter, Saturn is a gas giant consisting mostly of hydrogen and helium, but

Saturn has only 30 percent of Jupiter's mass. Nevertheless, Saturn rules more moons than any other planet in the solar system. Saturn's largest moon, Titan, has a nitrogen atmosphere thicker than the Earth's.

Lying twice as far from the Sun as Saturn is the dim world Uranus, whose green color gives it the look of a cosmic pea. Both Uranus and its even fainter neighbor, blue Neptune, are giants that are considerably smaller than Jupiter and Saturn. Uranus and Neptune are not gas giants like Jupiter and Saturn, for their hydrogen-helium atmospheres form only envelopes around dominant cores of water and rock.

The final world on this reconnaissance of the solar system is odd little Pluto, the smallest planet, which lies so far from the Sun's warmth that the temperature hovers around -400 degrees Fahrenheit. Unlike the four giant planets, Pluto has a tenuous atmosphere and a solid surface that a spacecraft could easily land on. Between 1979 and 1999, Pluto's highly elliptical orbit carried the little world closer to the Sun than Neptune.

Planetary Patterns

Each individual planet fascinates astronomers, but as a group these worlds exhibit patterns that may also prevail elsewhere. Of particular importance is a planet's distance from the Sun, for this largely dictates the planet's temperature, orbital period, and even size. Although planetary distances can be measured in miles or kilometers, astronomers prefer to use the astronomical unit, or AU: the mean distance between the Sun and the Earth, about 93 million miles. The Earth, then, is 1 AU from the Sun, Mercury and Venus lie less than 1 AU from the Sun, and the planets from Mars through Pluto lie more than 1 AU from the Sun.

Contrary to the belief of both ordinary folks and even some astronomers, the planets are not uniformly spaced from Mercury to Pluto. Instead, the inner planets bunch together, and the outer ones spread apart. If the six planets from Mercury through Saturn retained their spacing, they could all fit between the present orbits of Saturn and Uranus. Planetary distances display a general pattern, called the Titius-Bode law, which states that one planet lies roughly twice as far from the Sun as the planet next in. For example, Venus is about twice as far as

TABLE 2–1
Planetary Distances from the Sun

	Mean Distance from the Sun			
Planet	astronomical units	miles	kilometers	light-time
Mercury	0.3871	36,000,000	57,900,000	3.22 light-minutes
Venus	0.7233	67,200,000	108,200,000	6.02 light-minutes
Earth	1.0000	92,960,000	149,600,000	8.32 light-minutes
Mars	1.5237	141,600,000	227,900,000	12.67 light-minutes
Jupiter	5.203	483,600,000	778,300,000	0.72 light-hours
Saturn	9.54	886,700,000	1,427,000,000	1.32 light-hours
Uranus	19.2	1,780,000,000	2,870,000,000	2.66 light-hours
Neptune	30.1	2,790,000,000	4,500,000,000	4.17 light-hours
Pluto	39.5	3,670,000,000	5,910,000,000	5.48 light-hours

Mercury, Saturn is about twice as far as Jupiter, and Uranus is about twice as far as Saturn. Other solar systems may follow a similar rule.

A planet's distance strongly affects its temperature, since the inner worlds receive strong sunlight that keeps them much warmer than the cold worlds in the outer reaches of the solar system. The right temperature is crucial to life, which needs a temperature where water exists as a liquid. In the solar system, only Earth is lucky enough to reside in the narrow temperature range that allows liquid water. For other stars, astronomers can use the distance-temperature relation to predict the approximate temperatures of orbiting planets, even if these planets cannot be seen. To estimate temperatures, astronomers must also know how luminous the star itself is, because a planet 1 AU from a bright star like the Sun is warmer than a planet that same distance from an average star.

Other factors also influence temperature. If a planet's atmosphere contains greenhouse gases, such as carbon dioxide, water vapor, and

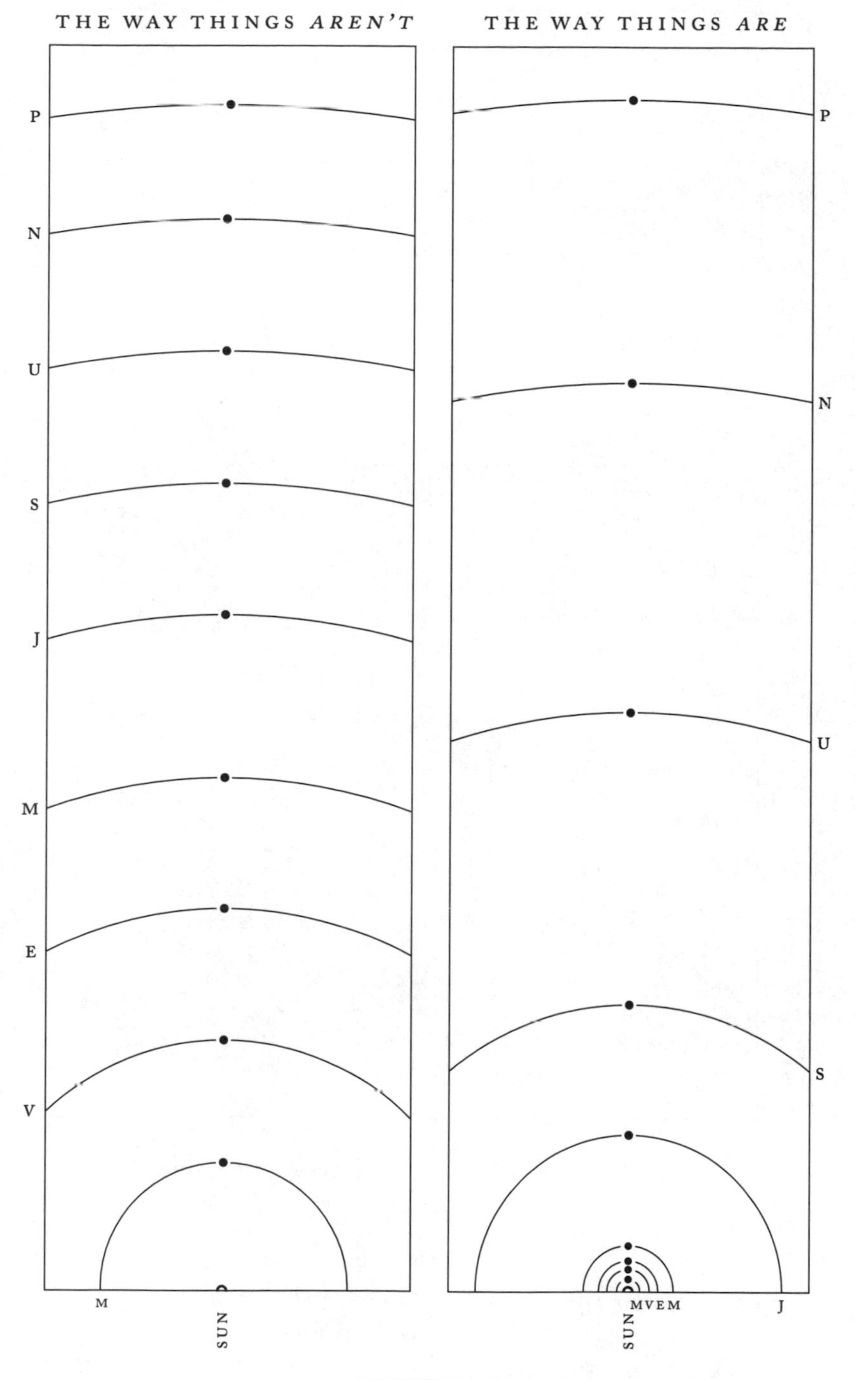

FIGURE 3

Planets are not spaced uniformly (left); instead, the inner planets bunch together, and the outer ones spread apart (right).

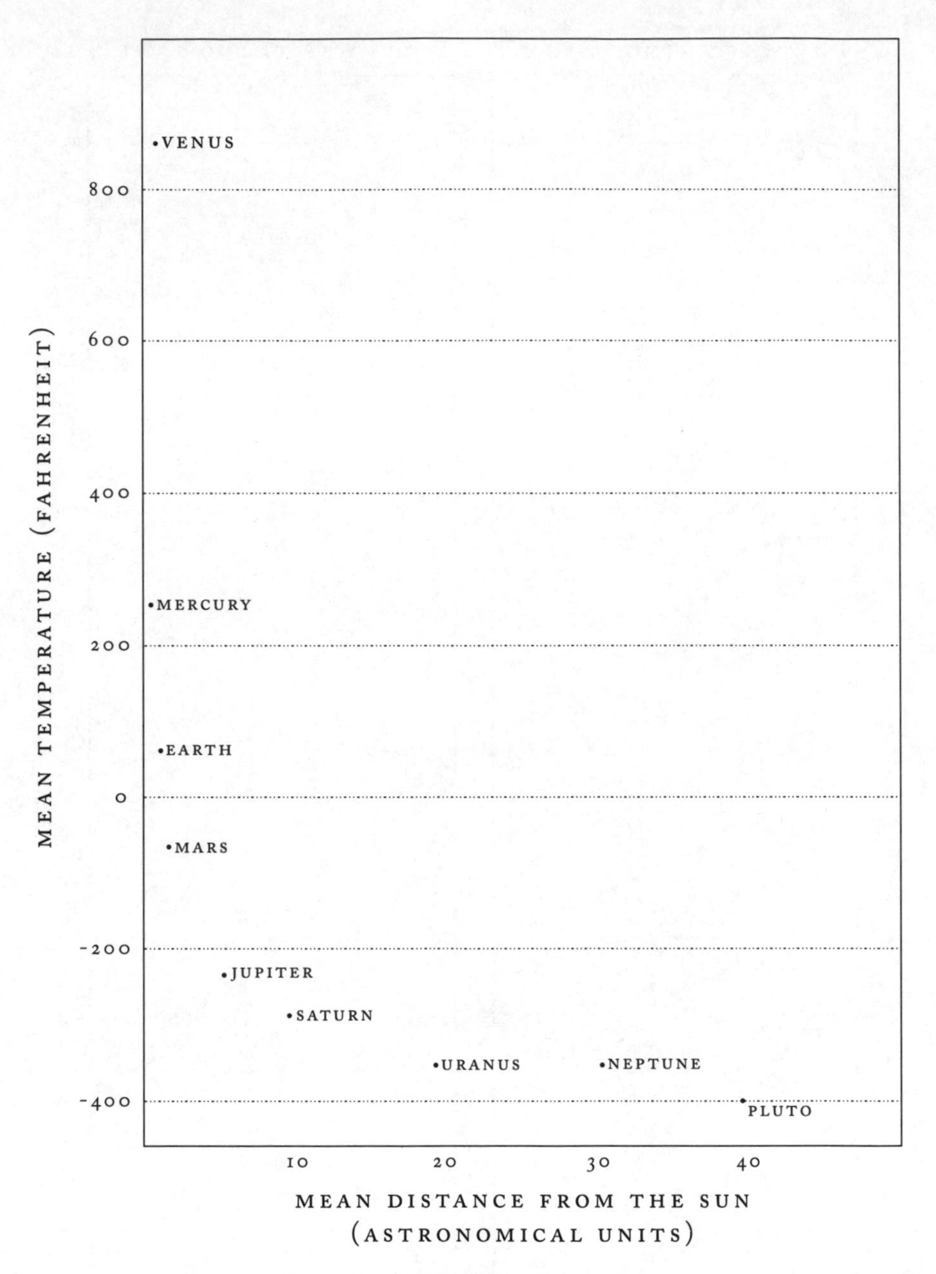

FIGURE 4

In general, the farther a planet is from the Sun, the colder it is; however, an intense greenhouse effect keeps Venus extremely hot.

TABLE 2–2
Planetary Temperature

Planet	Sunlight Intensity (Earth = 1.00)	Mean Temperature		
		Fahrenheit	Celsius	Kelvin
Mercury	6.67	+250	+120	390
Venus	1.91	+860	+460	730
Earth	1.00	+60	+16	289
Mars	0.43	–67	–55	218
Jupiter	0.037	–236	–149	124
Saturn	0.011	–289	–178	95
Uranus	0.0027	–353	–214	59
Neptune	0.0011	–353	–214	59
Pluto	0.00064	–400?	–240?	30?

ozone, the planet traps warmth from its star. This explains why Venus is hotter than Mercury even though the former lies nearly twice as far from the Sun. Earth and Mars also harbor greenhouse gases. If it weren't for the greenhouse effect, Earth would be an ice-covered world that might never have spawned life, because greenhouse gases boost the temperature by 60 degrees Fahrenheit. The Martian atmosphere consists almost entirely of carbon dioxide but is so thin that it raises the red planet's temperature by only about 10 degrees Fahrenheit.

Three of the four giant planets also elevate their temperatures, but not by the greenhouse effect; rather, they somehow generate heat themselves. Astronomers know this because they observe that Jupiter, Saturn, and Neptune radiate more heat than they receive from the Sun. This excess heat keeps Neptune as warm as Uranus, even though Neptune is a billion miles farther from the Sun. Because of these factors, temperature estimates for planets around other stars are necessarily approximate.

Perhaps the most striking pattern to the solar system is the enormous variation in planetary size from one part of the solar system to

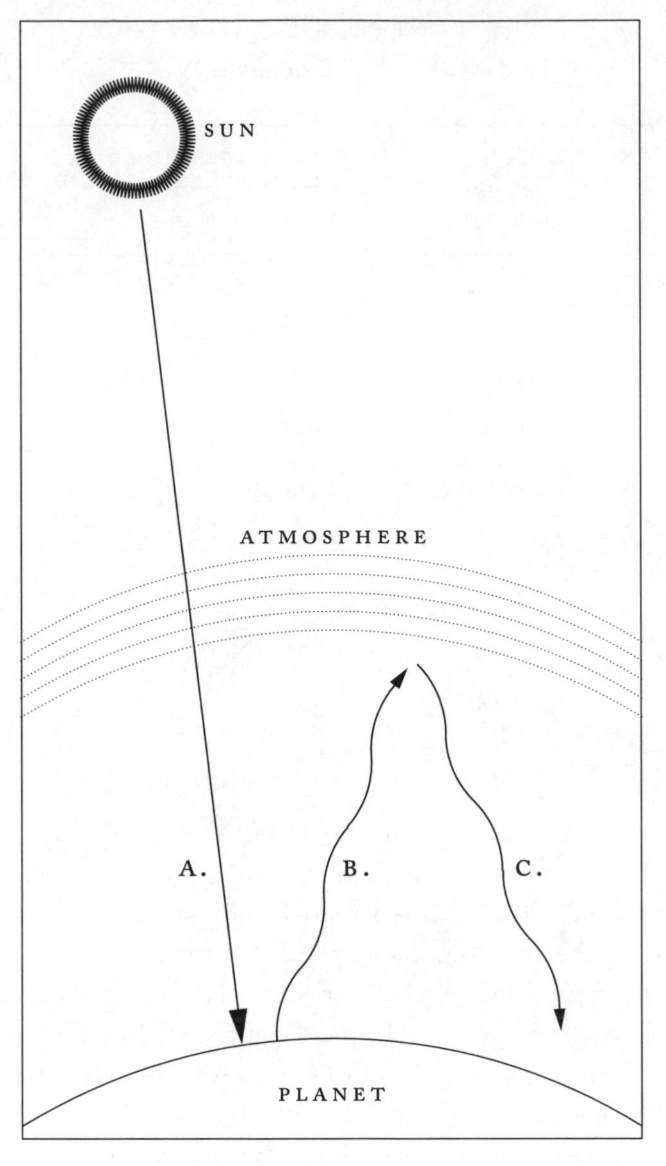

FIGURE 5

The greenhouse effect. The Sun's visible light (A) heats a planet's surface. This heat is radiated at infrared wavelengths (B), which are too long for the human eye to see, and can be bounced back to the surface by atmospheric gases (C).

another: the inner planets—Mercury, Venus, Earth, and Mars—are all small, and all the large planets—Jupiter, Saturn, Uranus, and Neptune—lie far from the Sun. Astronomers expect most other solar systems to follow the same rule.

TABLE 2–3
Planetary Size

Planet	Mass (Earth = 1.00)	Equatorial Diameter		Mean Density (water = 1.00)
		miles	kilometers	
Mercury	0.055	3,032	4,879	5.43
Venus	0.815	7,521	12,104	5.24
Earth	1.000	7,926	12,756	5.52
Mars	0.107	4,222	6,794	3.93
Jupiter	317.83	88,846	142,984	1.33
Saturn	95.16	74,898	120,536	0.69
Uranus	14.54	31,763	51,118	1.27
Neptune	17.15	30,775	49,528	1.64
Pluto	0.002	1,430	2,300	2.0

The four inner planets' combined mass adds up to just under twice the Earth's. Earth is the largest inner planet, Venus runs a close second, and Mars and Mercury come in last. All four planets are tough worlds of rock and iron, some five times denser than water. All have solid surfaces on which life could have evolved if the temperature had been right. Because this actually happened on one of these worlds, small rocky planets are the most promising abodes for extraterrestrial life. Unfortunately, their low masses also make these so-called terrestrial planets the hardest to detect, since their weak gravity perturbs their stars the least.

Beyond the asteroid belt lies an entirely different region, the realm of the giant planets, which have thick hydrogen-helium atmospheres surrounding dense cores. Jupiter and Saturn have 318 and 95 times the mass of the Earth, and Uranus and Neptune contain 15 and 17 Earth masses, respectively. Yet each giant has a core of roughly 10 Earth masses. The difference is the topping of hydrogen and helium, which is enormous on Jupiter and Saturn and merely thick on Uranus and Neptune. This is why only Jupiter and Saturn are called gas giants—the hydrogen and helium dominate their mass—whereas for Uranus and Neptune these gases make up only a fraction.

On the fringe of the solar system is tiny Pluto, a dwarf among giants. Many astronomers today do not consider Pluto a full-fledged planet but instead the largest member of the Kuiper belt, a zone of small icy bodies just beyond Neptune's orbit. The Kuiper belt supplies short-period comets, those having orbital periods of less than two hundred years. Every now and then, a gravitational disturbance tosses a Kuiper belt object inward, and a planet's gravity may trap the object in a short-period orbit. Whenever the object nears the Sun's heat, its ice vaporizes and it sports a cometary tail. Far beyond the Kuiper belt lies another swarm of icy bodies, the Oort cloud, from which long-period comets emerge. These comets may take many thousands of years to revolve once around the Sun. Objects in the Oort cloud are too far away to be seen from Earth, but since 1992 astronomers have detected many objects in the nearer Kuiper belt.

A planet's distance from the Sun determines how much time the planet needs to complete an orbit, the planet's year. Earth circles the Sun once a year; planets closer to the Sun require less time and planets farther more. There are two reasons for this. First, the farther a planet is from the Sun, the larger is its orbit and so the longer it takes to complete it. Second, the farther a planet is, the more slowly it moves, because the Sun's gravitational pull on it is weaker. Mercury and Pluto illustrate both factors. Pluto lies a hundred times farther from the Sun than does Mercury and so has an orbit a hundred times larger; in addition, Pluto moves ten times more slowly. As a result, Pluto takes a thousand times longer to complete an orbit: Mercury revolves every quarter year, whereas distant Pluto requires two and a half centuries to do the same.

This relation between planetary distance and orbital period is quite useful in exploring other solar systems. Suppose an astronomer sees a Sunlike star being tugged by an invisible planet circling the star every twelve years, the orbital period of Jupiter. Even though the planet cannot be seen, the astronomer can deduce that the planet is Jupiter's distance from its star and that the planet is probably too cold for life. Or suppose astronomers do manage to spot a planet at Jupiter's distance from a Sunlike star. Then they can conclude that the newfound

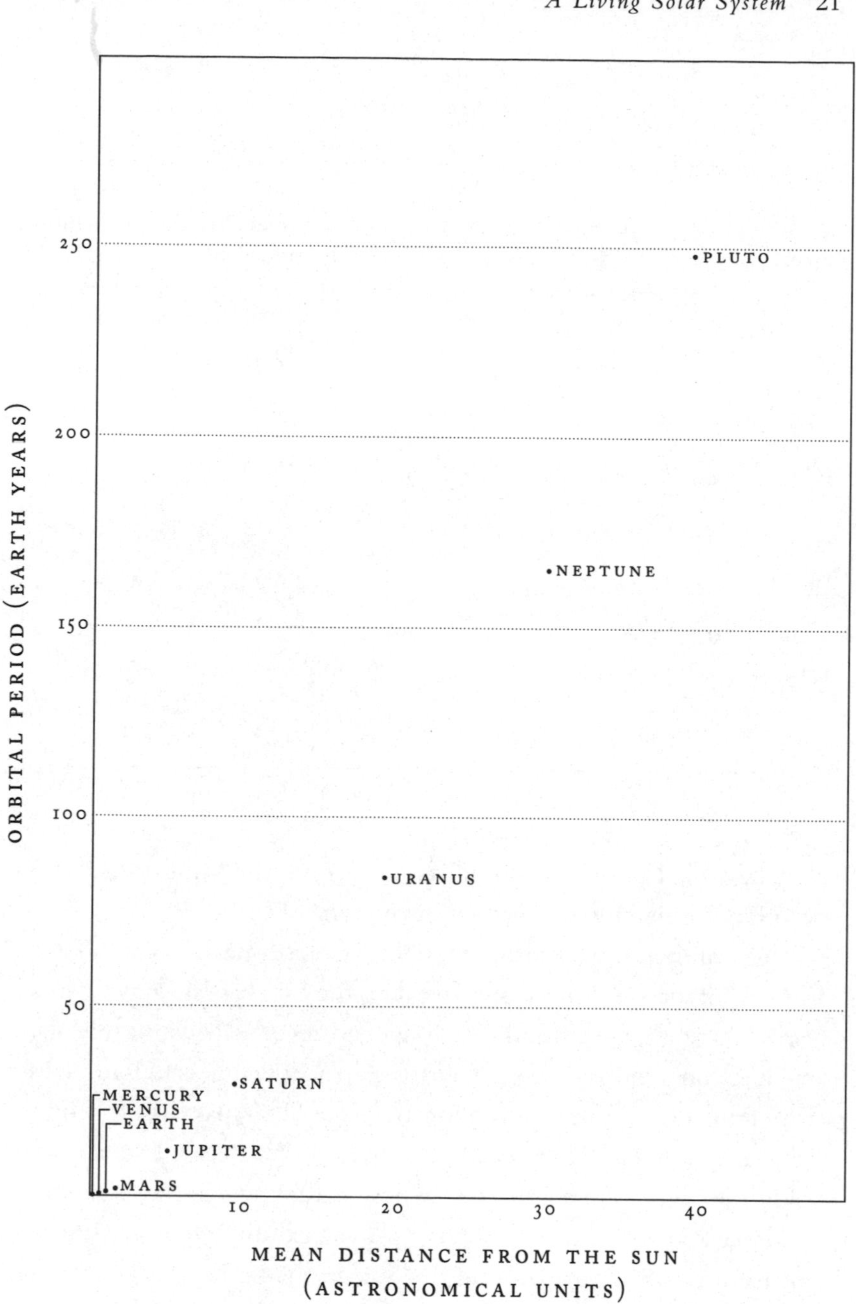

FIGURE 6

The farther a planet is from the Sun, the longer the planet takes to revolve around the Sun, because the planet has a larger orbit and moves more slowly.

TABLE 2–4
Planetary Paths

Planet	Length of Year	Mean Speed (km/sec.)	Orbital Eccentricity	Orbital Inclination (degrees)
Mercury	88 days	47.9	0.206	7.0
Venus	225 days	35.0	0.007	3.4
Earth	365.256 days	29.8	0.017	0.0
Mars	687 days	24.1	0.093	1.8
Jupiter	11.9 years	13.1	0.048	1.3
Saturn	29.5 years	9.6	0.054	2.5
Uranus	84.0 years	6.8	0.047	0.8
Neptune	165 years	5.4	0.009	1.8
Pluto	248 years	4.7	0.25	17.1

planet's orbital period is around twelve years, the same as Jupiter's, even if no orbital motion has yet been seen.

One complication exists, though, because stars have different amounts of mass. A less massive star than the Sun has less gravity, so its planets move sluggishly and take longer to revolve. Most nearby stars are faint, cool objects called red dwarfs. A planet that lies at Jupiter's distance from a red dwarf having a fourth of the Sun's mass would require twice as long to complete an orbit, or twenty-four years. The opposite holds true for a more massive star, whose planets would revolve faster.

The motions of the solar system's planets exhibit two other important features: all revolve around the Sun in the same direction as the Sun spins, counterclockwise as viewed from above the Earth's north pole, and their orbits lie close to a single plane. This is because the planets were born in a flat disk of material circling the newborn Sun; hence, the planets revolve in the same direction as the Sun rotates, and they never stray far from the original plane of that primordial disk. The orbital inclination, relative to the plane of the Earth's orbit, is an angle expressed in degrees. Except for Pluto, all the planets have orbital inclinations of less than eight degrees. In other solar systems, planetary orbits also probably line up, more or less, in a single plane. Strangely, though, the Sun itself is slightly tilted, for its equator makes an angle of seven degrees with the mean plane of the planets' orbits.

Especially useful in future journeys will be one of the most remarkable features of planetary orbits, their circularity. Technically, planetary orbits are elliptical, but most are so close to being perfect circles that the eye can hardly tell the difference. Astronomers express how round or elliptical an orbit is by using orbital eccentricity, which ranges from 0.00 for a perfect circle to just under 1.00 for the most elongated orbit that an object bound to a star can follow. All but two planets have orbital eccentricities of less than 0.10. Furthermore, the three planets with the most eccentric orbits are also the least massive: Pluto (orbital eccentricity 0.25), Mercury (orbital eccentricity 0.21), and Mars (orbital eccentricity 0.09). All other planets have quite circular orbits. In particular, the giant planets have orbital eccentricities of less than 0.06, and giants will be the easiest planets to detect around other Sunlike stars.

Orbital eccentricity will be most helpful as astronomers discover invisible objects around other stars. The dark object might really be a planet—or simply a star too faint to see. Most double stars have *high* orbital eccentricities, so orbital eccentricity provides a good diagnostic to distinguish stars from planets.

Astronomers expect that planets around other stars spin, as the Sun's planets do. For most planets, and for all of the giants, the spin

TABLE 2-5

Bright Nearby Double Stars

Stars	Distance from Sun (light-years)	Orbital Period (years)	Orbital Eccentricity
Alpha Centauri A and B	4.35	80	0.52
Sirius A and B	8.5	50	0.59
Procyon A and B	11.4	40	0.36

period is short, a day or less. Planetary spin axes are not exactly perpendicular to their orbits. Earth's axis tilts 23.4 degrees, which causes the seasons, because when one hemisphere tilts sunward, it warms up and soon experiences summer. Some planets, though, are much more radical: Uranus, with an axial tilt of 97.9 degrees, lies on its side as it rotates, perhaps because a large object smashed into it and knocked it over; and Venus spins backward.

Most planets have moons revolving around them. Although over five dozen moons exist in the solar system, only seven are large, and Earth is blessed with one of them, whose tides stir the seas and may have pushed life from the ocean onto the land. Jupiter has four large moons, Saturn one, and Neptune one. Of the giant planets, only Uranus lacks a large moon. And though Pluto's one moon, Charon,

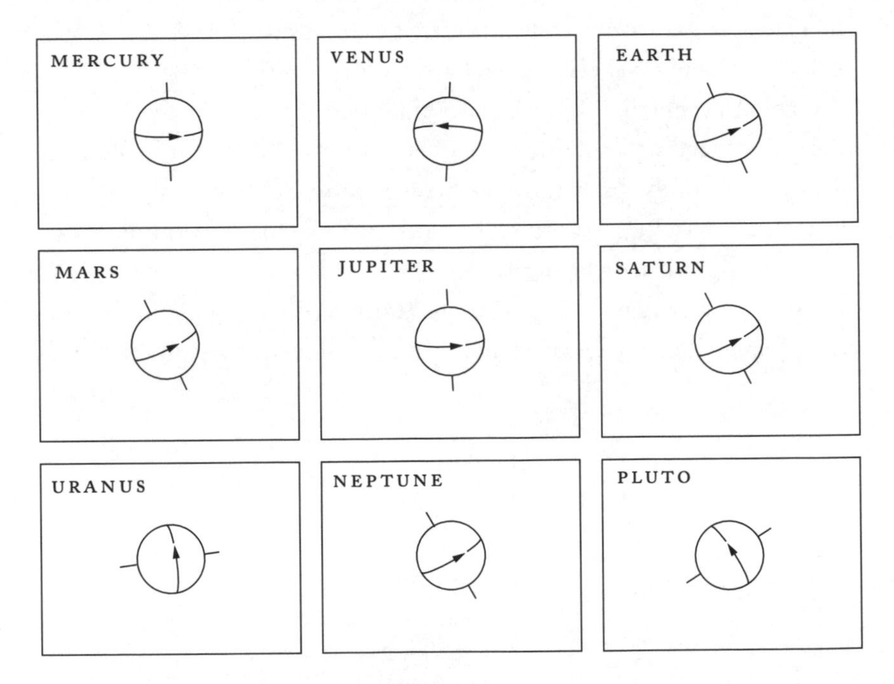

FIGURE 7

Six of the nine planets, including Earth, stand more or less upright and rotate in the same direction as the Sun; if viewed from north of the solar system, they would turn in the counterclockwise direction. However, Venus spins backward, and Uranus and Pluto lie on their sides as they spin.

TABLE 2–6
A Day in a Planet's Life

Planet	Length of Day	Axial Tilt (degrees)
Mercury	58.65 days	2
Venus	243.0 days	177.3
Earth	23 hours 56 minutes	23.4
Mars	24 hours 37 minutes	25.2
Jupiter	9 hours 55 minutes	3.1
Saturn	10 hours 39 minutes	26.7
Uranus	17 hours 14 minutes	97.9
Neptune	16 hours 7 minutes	29.6
Pluto	6.39 days	122.5

TABLE 2–7
Moons

Planet	Number of Moons	Big Moons
Mercury	0	—
Venus	0	—
Earth	1	Moon
Mars	2	—
Jupiter	16	Io, Europa, Ganymede, Callisto
Saturn	18	Titan
Uranus	15	—
Neptune	8	Triton
Pluto	1	—

TABLE 2–8
Planetary Encounters

Planet	First Flyby	First Orbiter	First Lander
Mercury	1974, U.S.A.		
Venus	1962, U.S.A.	1975, U.S.S.R.	1970, U.S.S.R.
Earth	—	—	—
Mars	1965, U.S.A.	1971, U.S.A.	1976, U.S.A.
Jupiter	1973, U.S.A.	1995, U.S.A.	
Saturn	1979, U.S.A.	2004? U.S.A.	
Uranus	1986, U.S.A.		
Neptune	1989, U.S.A.		
Pluto			

is not large, it is enormous compared with the planet's small size: Charon is half Pluto's diameter.

Finally, what astronomers know about a planet depends on its distance from Earth and the number and quality of spacecraft it has received. The most primitive type of spacecraft encounter is a flyby, in which a spacecraft passes the planet for a quick look. A more thorough mission puts a spacecraft into orbit around the planet for prolonged study, and the most difficult actually lands a craft on the planet's surface.

In a nutshell, then, searchers of new solar systems can take the following expectations with them as they set off on their voyage to other stars: planets close to a star are hot or warm, and planets far away are cold; the inner planets are small and rocky, while the outer planets are large and partially gaseous; the inner planets revolve faster than the outer ones; all planets in a solar system revolve in the same direction as the star spins; all planets, and especially the giant ones, have nearly circular orbits that lie close to a single plane; and, if our solar system is typical, other stars have about ten planets.

But is the solar system typical? Can astronomers even expect other

TABLE 2–9
Features of Our Solar System

Planets revolve around their star in the same direction.
Planets lie near a single plane.
Planets have nearly circular orbits.
Small planets lie near their star, big ones far from it.
The solar system has nine planets.

stars to have planets? Just because the Sun has planets does not guarantee their existence elsewhere. After all, if planets had not formed here, neither would we; thus, our very existence forces the Sun to have planets. Although ongoing searches will ultimately reveal the prevalence or rarity of extrasolar planets, astronomers can get a sneak preview by considering how the solar system formed, 4.6 billion years ago.

Primal Clues

Theories for the solar system's origin can either motivate or discourage the search for planets around other stars. If the planets formed as a natural consequence of the Sun's birth, then other newborn stars should also form planets, and planets should revolve around many of the Galaxy's stars. But if the Sun's planets arose in a freak accident, then most stars must be planetless.

In 1755, Prussian philosopher Immanuel Kant offered a theory that should have inspired planet seekers, but few if any read it, because his publisher went bankrupt and most of Kant's books were impounded to pay off the publisher's debts. Kant pictured the solar system as arising from a diffuse cloud of particles that settled into a plane and conglomerated into the six planets then known. The inner planets were denser than the outer ones because the heavier particles had sunk closer to the Sun. The outer planets were larger, though, since regions farther from the Sun had larger volumes, allowing Jupiter and Saturn to sweep up more particles. Because this process should also have occurred elsewhere, Kant's theory implied that other stars had planets.

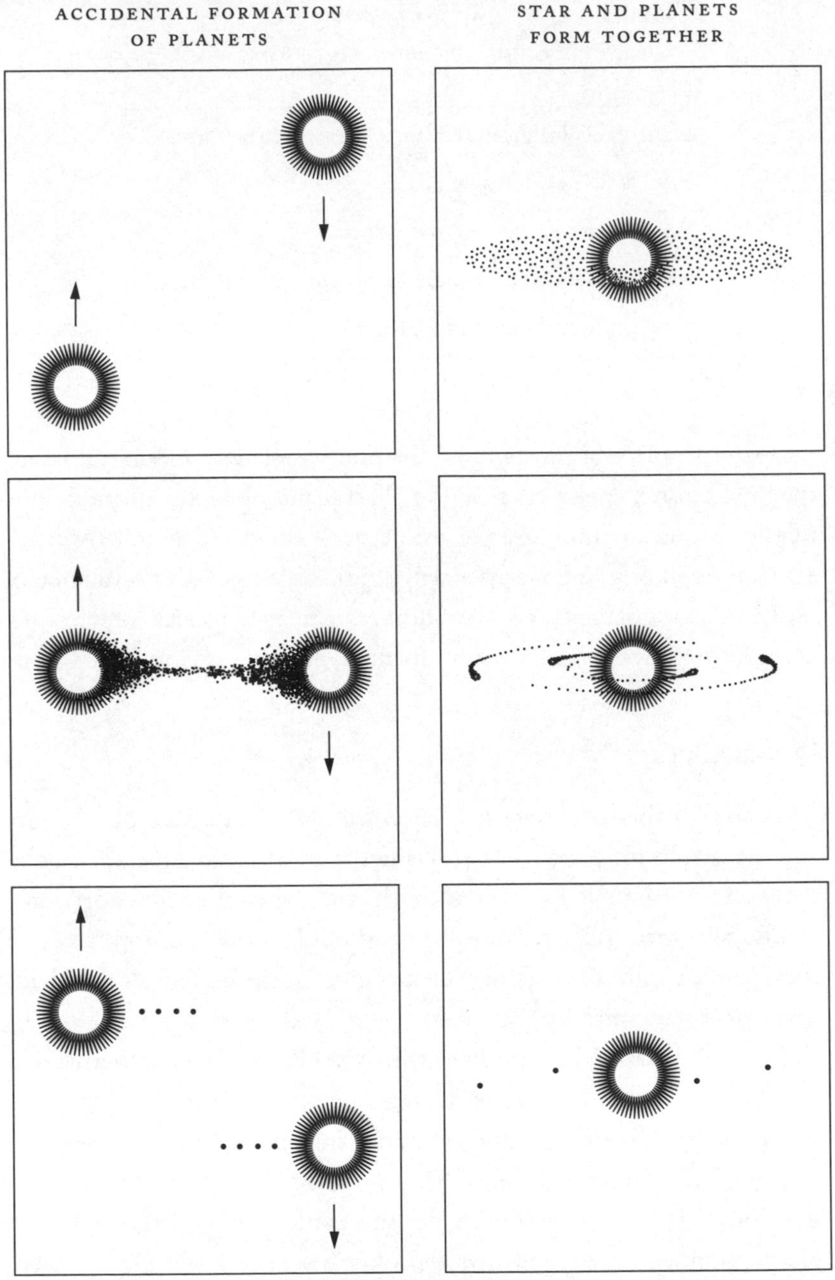

FIGURE 8

If planets form only in freak accidents, such as when stars collide (left), then planets should be rare. If instead planets form when stars do (right), then planets should be plentiful.

In 1796, French astronomer Pierre Simon Laplace published a theory that, like Kant's, held that the planets were a natural consequence of the Sun's evolution. According to Laplace, the Sun's atmosphere once extended over billions of miles of space. Rotation caused it to flatten. The atmosphere then shrank, and as it did so, it rotated faster, like a spinning ice skater drawing in his arms, and the Sun cast out rings of material that revolved around it. These condensed into the planets. Laplace's theory explained the general features of the solar system, such as why the planets revolve around the Sun on circular orbits, in the same plane, and in the same direction. Laplace even saw evidence for his theory in Saturn's rings, which he said might someday condense into satellites just as the rings around the Sun had condensed into planets. But none other than Napoleon Bonaparte complained that Laplace never mentioned God, to which Laplace replied: "Sir, I have no need of that hypothesis." During the 1800s, astronomers generally supported Laplace's theory, but it eventually ran into problems. For example, as the Sun's atmosphere contracted, the Sun should have spun more and more quickly, yet the real Sun spun slowly and had much less "spin momentum"—what physicists call angular momentum—than the planets. This and other concerns led to the theory's demise in 1900.

The theory's main critics, geologist Thomas Chrowder Chamberlin and astronomer Forest Ray Moulton, worked at the University of Chicago and would formulate a radically different model. Chamberlin was examining eruptions on the Sun, which he thought might give birth to planets. Eruptions large enough to produce whole planets, he thought, might be triggered by the gravitational pull of a passing object. In February 1901, this nascent idea received apparent confirmation when a spectacular nova flared up in the constellation Perseus and outshone almost every star in the sky. Today astronomers know that a nova is an exploding star, but at the time many thought a nova occurred when a star collided into another object. Later that year, astronomers detected glowing material around Nova Persei, which suggested that the collision had spewn debris into space.

The Chamberlin-Moulton theory, published in 1905, said that long ago another star had skirted near the Sun and ripped material out of it. Material from the Sun's top layer, which was presumably the lightest, was thrown out the farthest and formed the low-density giant planets—Jupiter, Saturn, Uranus, and Neptune—while denser material from deeper inside the Sun remained close in and formed the rocky inner planets—Mercury, Venus, Earth, and Mars. Years later, British scientists James Jeans and Harold Jeffreys offered a similar theory.

Chamberlin was nothing if not confident. In late 1905, he wrote:

> I believe it is the practice of celestial astronomers . . . to notify their colleagues of important events impending in any part of the universe. This is therefore to inform you that on and after January 1st proximo, the solar system will be run on the new hypothesis. It is not expected that the transition will be attended by any jar or other perceptible perturbation or that the change . . . will occasion any nausea. Everything is expected to work smoothly. . . . The inclination of the Sun's axis will not be regarded as a moral obliquity but merely as a frank confession that once on a time he flirted with a passing star and got twisted a bit, as was natural. The solar family will not appear as prim and precise as heretofore, and at first the neighbors may cast some reflections on it, but it is thought that in time they will come to rather like the new regime and will take kindly to the notion of new families springing up here and there. . . .
>
> Please post notice according to the rules of the game.

This theory for the solar system's origin spelled doom for future planet hunters. It said that few other stars had planets, because stars in our part of the Galaxy rarely collide or even come close to one another. The Chamberlin-Moulton theory therefore meant that few other planets exist and that our civilization might be alone in the Galaxy.

Eventually, however, this theory ran into trouble itself. For one thing, because of the Sun's temperature, the extremely hot material torn from it by a passing star would probably not condense into planets but instead would dissipate into space. In addition, the material would not have enough angular momentum to account for the orbital motion of the planets, especially the outer ones. By 1935, as a result of these objections, the theory was dead.

In the early 1940s, astronomers reported the discovery of planets around two nearby stars, 70 Ophiuchi and 61 Cygni. Although these claims would later be refuted, they demanded a new theory to explain the seeming prevalence of planets. In 1944, German physicist

Carl Friedrich von Weizsäcker, who was helping the Nazis develop nuclear weapons, revived the view that planets form naturally, when a star does, and so should be common. Von Weizsäcker apparently did not know of the recently discovered "planets," but they probably helped his theory get a favorable reception.

According to modern theory, the planets originated in a disk of gas and dust that revolved around the newborn Sun. The inner part of the Sun's disk was rapidly rotating and thus hot, and within this hot region only hardy substances with high melting temperatures—rock and iron—could condense. These substances gathered into small asteroidlike objects called planetesimals, which smashed together and formed the four inner planets. This is why Mercury, Venus, Earth, and Mars are dense worlds of rock and iron. In the cool outer reaches of the solar nebula, though, ice condensed and became part of the planets. Because of this extra material, and because of the large area of the outer disk, the planets grew large. As they did so, their gravity grabbed gaseous material from the disk, mostly hydrogen and helium. Jupiter and Saturn attracted so much that these two gases made up most of each planet's mass, whereas Uranus and Neptune got far less. Pluto and the icy bodies now in the Kuiper belt escaped collision with the eight main planets and thus survived.

This theory explains the overall properties of the solar system. The planets all lie in a single plane because the primordial solar disk was flat, and they all revolve around the Sun in the same direction because the disk itself did. The theory also explains why the inner planets are small and rocky and the outer ones large and partially gaseous. Asteroids and comets are simply leftover debris from the solar system's messy formation.

From the perspective of modern planet hunters, this theory promises plenty of planets, because other stars should have gone through the same process. There are new worlds to discover and new worlds to explore—worlds that will shed light on our own and may just harbor life. But the search for new worlds actually began closer to home, with the very first new planets ever found. These were not unseen worlds circling other stars but instead full-fledged members of our own solar system that were so dim and distant they eluded the ancients: Uranus, Neptune, and Pluto.

3

The Sun's Distant Outposts

ALTHOUGH THE prospect of new worlds circling distant stars beckons modern astronomers, the first new planets appeared around the very star that lights the Earth. These worlds underscore just how little most planets herald themselves, because for millennia three outer planets eluded all observers, and another world went unrecognized even though it lay right beneath astronomers' feet.

In a sense, planetary discovery began with the Earth, which was once thought to be the immobile center of the universe. During the 1500s, Nicolaus Copernicus recognized that the Earth was a planet racing around the Sun between the orbits of Venus and Mars. Nearly two thousand years earlier, Greek astronomer Aristarchus had also proposed a heliocentric solar system. Crude estimates of the distance to the Sun indicated that the Sun was larger than the Earth, so Aristarchus thought the Sun would make a more logical center. The discovery of the Earth as a planet gave the Sun six planetary companions: Mercury, Venus, Earth, Mars, Jupiter, and Saturn.

During the early 1600s, after astronomers started to use telescopes, planets were proposed inside the orbit of Mercury. These had never been seen directly because they were so close to the Sun's glare, which challenged even observers who sought Mercury. When these purported planets passed between the Earth and the Sun, the theory went, they blocked its light and produced the dark spots that Galileo Galilei and others had observed on the Sun. Even before the telescope's invention, spots on the Sun could occasionally be seen, and astronomers attributed them, incorrectly, to the passage of Mercury or Venus in front of the solar disk. The telescope revealed dozens of dark spots, which Galileo thought were actual sunspots. Such blemishes, however, contradicted Aristotle—and the Church—which held that the Sun and other heavenly bodies were perfect and thus spotless. Hence, if the spots were really intervening planets, the Sun could remain unmarred.

The sunspot dispute was soon resolved by the Sun itself, for from 1645 to 1715 nearly all spots vanished from its face. This period, now called the Maunder minimum, corresponded to cold weather on Earth and was part of the "little ice age," when the Sun probably faded. The Maunder minimum more than dimmed the prospects of the planet theory; after all, if the planets really existed, they should have continued to skirt across the Sun.

The first actual discovery of a new planet beyond the Earth took place in the opposite direction, nearly a billion miles past the orbit of ringed Saturn. This stunning discovery, made by a professional musician, marked a rich new movement in the symphony of planetary discovery.

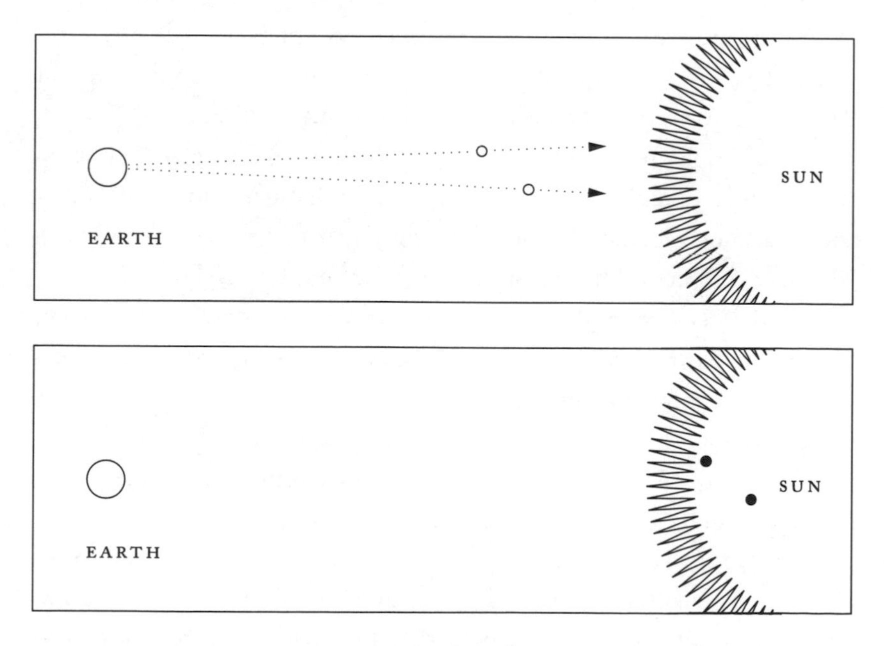

FIGURE 9

Planets between the Sun and Earth (top) can mimic sunspots (bottom), and vice versa.

Uranus

Of the planets visible to the naked eye, Uranus is unique: it had to be discovered. No ancient people recorded the wanderings of this dim green world, and the few astronomers who glimpsed it prior to its discovery thought it was just a faint and insignificant star.

Ironically, the keen-eyed man who would fish this distant world out of the celestial sea had little interest in planets. German-born English astronomer William Herschel had set for himself what he saw as a far greater goal, understanding the stars and the structure of the Milky Way. In contrast, most astronomers of his time concentrated on the solar system. Herschel's training also lay outside the mainstream of astronomy, for he was a musician who did not delve into science until his thirties. He taught himself astronomy and built large telescopes that he deployed to probe the heavens.

Herschel discovered Uranus as he was searching the sky for double stars, which were one step in his grand scheme to fathom the Galaxy. On March 13, 1781, between 10 and 11 P.M., he noticed a peculiar object in Taurus the Bull, near its border with Gemini the Twins. Because both constellations belong to the zodiac, they periodically give refuge to the Sun's planets. The one that caught Herschel's attention lay just beyond the tips of the Bull's horns. Herschel could see that the object was extended rather than sharp and starlike.

He did not think it was a planet. For millennia, the solar system's planetary boundary had been firmly fixed at Saturn, and the idea that other planets circled the Sun beyond Saturn's orbit was as alien a thought as a second moon orbiting the Earth. Herschel therefore believed the peculiar object was either a new comet or a nebulous star. Four nights later, he reobserved the object. "It is a comet," he wrote, "for it has changed its place."

No comet like Herschel's had ever been seen. Wrote French comet hunter Charles Messier, "I am constantly astonished at this comet, which has none of the distinctive characters of comets." The comet had no tail, and it refused to follow the orbit computed for it. Unlike planets, most comets have extremely elliptical orbits around the Sun, so astronomers assumed that Herschel's did, too. Just days after calculating an orbit, however, astronomers found the comet darting from the predicted path. Eventually, they realized the reason: this was no comet pursuing an elliptical orbit but rather a new planet lumbering along a circular orbit in the outermost reaches of the solar system. The new planet lay twice as far from the Sun as Saturn; Herschel's discovery had doubled the size of the known solar system.

Strangely, most astronomers recognized this long before Herschel himself, and months passed until he grew convinced that his comet was really a planet. Nevertheless, the discovery showered Herschel with fame, and he named the new planet after the king of England, George III, who had lost the American colonies, but thanks to Herschel had gained a whole new world. Unfortunately for the king, the name did not stick, and the planet is now called Uranus, the god of the sky, father of Saturn, and grandfather of Jupiter.

Herschel's discovery had been serendipitous, because he found the planet while looking for something else. The feat of this amateur astronomer became all the more remarkable when professional astronomers realized that their peers had seen Uranus over twenty times before the discovery but mistook the planet in every instance for a mere star. The first was John Flamsteed, the astronomer royal in Greenwich. In 1690, over ninety years before Herschel's discovery, Flamsteed observed Uranus in Taurus. He even named the "star" 34 Tauri.

Uranus suggested that even farther and fainter planets might exist unseen in the outer solar system. Indeed, the old, prediscovery observations from Flamsteed and others would play a pivotal role in the discovery of the still more distant Neptune. Long before Neptune's discovery, however, Uranus spurred a hunt for an unseen world much closer to home.

The Asteroids

The discovery that brought fame to an amateur astronomer in England also brought vindication to a professional astronomer in Germany. Johann Elert Bode first suggested the new planet be named Uranus, realized that prediscovery observations might exist, and retrieved Flamsteed's 1690 sighting of the far-off world. Today, though, Bode is best known for advocating a law of planetary distances that predicted an unseen planet between the orbits of Mars and Jupiter.

TABLE 3–1

The Titius-Bode Law of Planetary Distances

	Distance from Sun (astronomical units)	
Planet	Titius-Bode Law	Actual Distance
Mercury	0.40	0.39
Venus	0.70	0.72
Earth	1.00	1.00
Mars	1.60	1.52
?	2.80	—
Jupiter	5.20	5.20
Saturn	10.00	9.54

The law works like this. Take the number 3 and double it, then double that, and double that, and so on:

0, 3, 6, 12, 24, 48, 96.

Now add 4 to each number:

4, 7, 10, 16, 28, 52, 100.

Dividing each number by 10 then yields the numbers in Table 3-1. With one exception—the gap between Mars and Jupiter—these numbers match the distances of the known planets in astronomical units, with 1 astronomical unit being the Earth's distance from the Sun.

This rule is sometimes called Bode's law, or, more correctly, the Titius-Bode law, because in 1766 it was first mentioned by Johann Daniel Titius, who anonymously inserted it into a book he was translating. Titius stressed the gap between Mars and Jupiter, where the law said a planet should lie, 2.8 astronomical units from the Sun. "Why should the Lord Architect have left this space empty?" asked Titius, who thought that it might be filled by unseen moons of Mars and Jupiter. Six years later, Bode read Titius's note and echoed his question: "Can you believe that the Founder of the universe had left this space empty?" Unlike Titius, Bode proposed that an actual planet hid in the gap, an idea that Johannes Kepler had suggested back in 1596.

Most astronomers, though, thought Bode was talking nonsense and dismissed his law as little more than a numerical coincidence. No additional planets circled the Sun, they said, and any planet as close as Bode's would have been found long ago. For these reasons, the idea of an undiscovered planet between the orbits of Mars and Jupiter struck most astronomers as absurd—and one beyond Saturn never even occurred to them.

But the new planet Uranus obeyed the Titius-Bode law almost perfectly. If the numerical sequence was extended, it predicted that the next planet beyond Saturn was 19.6 astronomical units from the Sun. Uranus's actual distance is 19.2 astronomical units, just 2 percent shy of the prediction. Moreover, the planet's discovery proved that additional worlds could indeed exist right in our own solar system. As a result, Uranus convinced many astronomers that Bode was right, stimulating a search for a new planet—not beyond Uranus, but between the orbits of Mars and Jupiter.

In September 1800, six astronomers gathered in Lilienthal, Germany, to lay a trap that would snare the elusive world. They believed the hunt far exceeded the capability of any one astronomer, so they divided the zodiac into twenty-four zones, each to be monitored by a different person. These "celestial police" would enforce the law—the Titius-Bode law, that is—by bringing the errant planet to justice. One astronomer they planned to invite was Sicilian Giuseppe Piazzi, then hard at work on a new star catalogue. Before Piazzi even heard of their plans, he found what they were looking for: on January 1, 1801, he spotted what seemed to be a new star in Taurus, site of Herschel's discovery of Uranus. The object was fainter, however, and whereas Uranus had alerted its discoverer by its unusual appearance, Piazzi's object looked just like a star. Over the following nights, though, he watched as it moved, like a planet.

The strange object, which Piazzi would later name Ceres, quickly escaped. He tracked it into February, when illness forced him to stop, and by the time the news reached Bode and others, Ceres had dodged behind the Sun. Astronomers tried to calculate the orbit of the newfound world, to predict its future position; but the few positions from Piazzi did not suffice to establish an accurate orbit, and the frantic searches that were mounted for Ceres after it emerged from behind the Sun turned up nothing. Ceres seemed lost forever.

Fortunately, the problem intrigued a brilliant German mathematician, Carl Friedrich Gauss. Devising new mathematical methods, Gauss was able to take the few observations from Piazzi and predict an accurate future path for Ceres. In December 1801, Gauss's orbit helped astronomers recover Ceres among the stars of Virgo the Virgin.

Like Uranus, Ceres was a triumph for Bode. Not only did the gap between Mars and Jupiter harbor an apparent planet, as the Titius-Bode law predicted, but its distance from the Sun matched almost exactly that laid down by the law: 2.77 astronomical units, versus a prediction of 2.80 astronomical units.

Then came a bombshell. One of the original planet hunters who had gathered in Lilienthal was Heinrich Olbers, a medical doctor and amateur astronomer. In helping to recover Ceres, Olbers had learned the stars of Virgo, and in March 1802 he noticed a "star" in that constellation which seemed out of place. When he observed it again, he discovered that, like Ceres, it moved. This object, which he named Pallas, lay as far from the Sun as Ceres, so that two planets now existed where the Titius-Bode law had predicted only one. William Herschel called them asteroids, since they were so small that they looked like stars rather than planets. As astronomers know today, Ceres is the largest asteroid, but it is only a quarter the diameter of the Moon.

Olbers proposed that Ceres and Pallas were shrapnel from a planet that had exploded between the orbits of Mars and Jupiter. This implied that other asteroids should exist, and in 1804 and 1807 astronomers found Juno and Vesta. Olbers's theory has not survived, though, because astronomers now think the asteroids are the remains of a planet that never formed. Jupiter's gravity so stirred up the objects inside its orbit that they could not conglomerate into a full-fledged planet.

Today known asteroids number in the thousands, but after Vesta was sighted, several decades elapsed with no further discoveries. The next asteroid was finally spotted in 1845, by an ex-postman who had

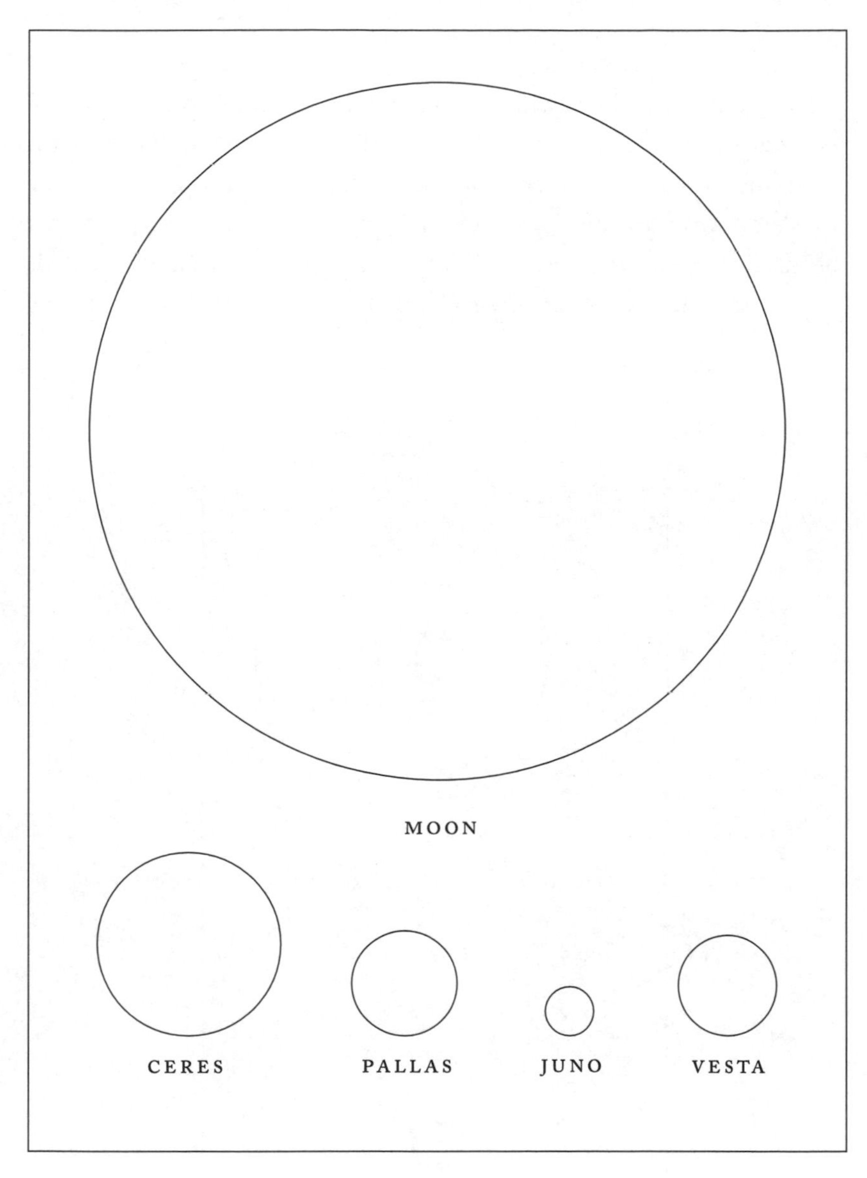

FIGURE 10

The Moon dwarfs all asteroids, including the first four discovered. In fact, the Moon has much more mass than all the asteroids put together.

been looking without success for fifteen years. A much greater discovery would occur the very next year, a discovery that was also spurred on by Uranus, for the planet that Herschel had discovered was being perturbed by a new and still more distant world.

Neptune

Gravity is invisible, but it was through gravity that astronomers first saw Neptune. As a giant planet with seventeen times the mass of the Earth, Neptune exerts a gravitational force on Uranus that speeds it up or slows it down. For this reason, Uranus caused astronomers a continual headache, repeatedly deviating from the orbit that they had calculated.

To compute an accurate orbit for a planet and thus predict its

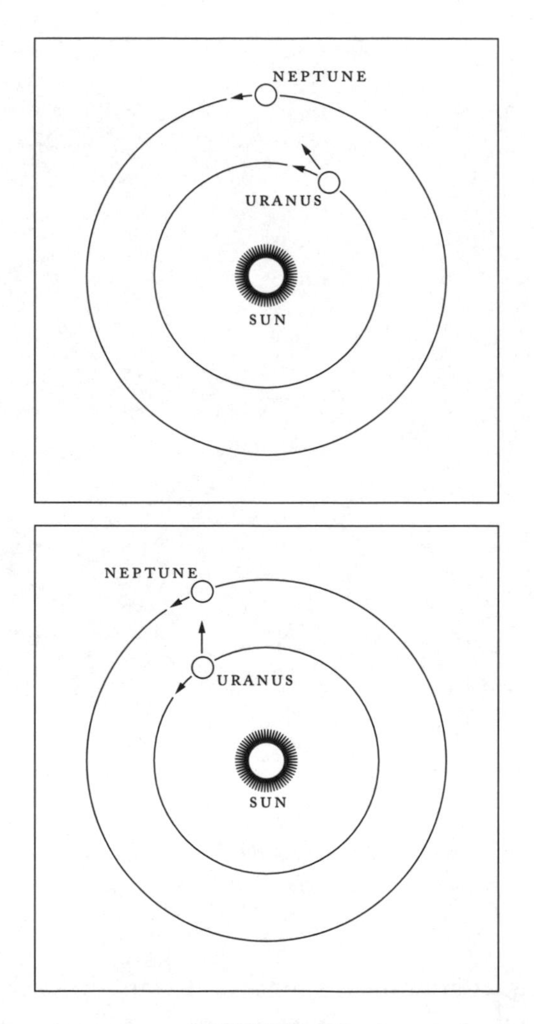

FIGURE 11

Neptune's gravity perturbs Uranus, at times pulling it forward faster (top) and at other times slowing it down (bottom).

future position, astronomers must have observed the planet over at least a complete revolution. Because distant Uranus revolves only once every eighty-four years, the planet, spotted in 1781, would not return to its place of discovery until 1865. Fortunately, the prediscovery observations of Uranus from John Flamsteed and others instantly gave astronomers more than a full cycle. These should have secured the planet's orbit.

Uranus, though, seemed to have a mind of its own. It refused to adhere to its orbit: sometimes the planet moved too quickly, other times too slowly. By the 1830s the problem had grown so acute that astronomers began speculating that an unseen planet lay beyond Uranus and tugged at it through its gravity.

The first to attack this problem with full force was a young English mathematician named John Couch Adams. Adams was mild and modest, traits that would conspire with less desirable ones in others to cost his country a new planet. "Look without envy on the success of others," he once wrote to himself, "and without pride on my own." The son of a farmer, Adams displayed his mathematical talents early, studying algebra when he was ten. After he entered the University of Cambridge, he learned of the problem with Uranus while visiting a bookstore, where he found an old report mentioning the planet's cantankerous nature. That was in 1841, but the press of studies prevented him from investigating Uranus until 1843.

"You see, Uranus is a long way out of his course," Adams wrote to a fellow student. "I mean to find out why. *I think I know*." Adams postulated that an unseen planet lay beyond the orbit of Uranus, and he calculated the properties the planet must have to pull Uranus from its path. One of these properties was the planet's mass, from which the planet derived its gravitational strength and perturbed Uranus. The most crucial property was the planet's celestial position, which would allow astronomers to discover the planet.

Adams completed his work in 1845, predicting a position for the planet that would coincide almost perfectly with Neptune's. He took his result to the astronomer royal, George Biddell Airy, who had written the report that had initially alerted Adams to the problem with Uranus. Adams could not have picked a worse astronomer: Airy was unimaginative and stubborn; he had little patience for young people, especially those, like Adams, he had never heard of; and as for Uranus, Airy had stated his opinion long ago. Whatever the cause of Uranus's trouble, it was probably not another planet. Besides, wrote the master in 1837, "If it be the effect of any unseen body, it will be nearly impossible to ever find out its place."

Yet Adams now bore the position of just that unseen body. After trying and failing several times to meet Airy, Adams left a report that included the new planet's position. Airy never even thought to look for such a planet. For one thing, he distrusted theoretical work. For another, any search would ruin the routine of the Royal Observatory in Greenwich, and above all Airy valued order. As one commentator said, Airy liked his field because, in Airy's words, "astronomy is preeminently the science of order."

Although Airy did reply to Adams's message, he made an objection that Adams considered trivial and to which he did not respond. Perhaps had Adams been bolder and more ambitious, he would have insisted on the validity of his calculation and pressed his case. Thus, the landmark discovery of a planet that the British could have easily found in 1845 was forfeited to foreign astronomers.

Meanwhile, in England's occasional enemy, France, another brilliant scientist was working on the same problem. Urbain Jean Joseph Leverrier did not suffer from Adams's virtues: he was rude and ambitious, and in June 1846 he too predicted the location of an eighth planet. But aside from a brief search by astronomers at Paris Observatory, Leverrier had no luck stirring his countrymen to action, either.

When Leverrier's report arrived in England, Airy saw that the predicted position matched the one that Adams had earlier calculated. This, plus Leverrier's excellent scientific reputation, finally convinced Airy that a new planet might indeed be out there. He still did not want his own observatory to search for it; after all, that would disrupt the orderly functions of the Royal Observatory. So Airy asked James Challis, a Cambridge astronomer, to hunt for the planet.

Challis was the third actor in the British tragedy of Neptune. He had already been approached by Adams three years earlier and had, unlike Airy, encouraged Adams in his attempt to solve the Uranus problem. Challis agreed to Airy's request and on July 29 began to look for the unseen planet.

The search was comic, not cosmic. Rather than observe the region near the predicted planet, Airy instructed Challis to examine a large zone that bristled with thousands of stars. Furthermore, Challis chose to study stars much fainter than Adams had said the planet would be, thereby vastly upping the number of stars to be investigated. And instead of looking for a planetary disk, which would stand out from the stars, Challis laboriously recorded the position of each star and hoped to find one that moved. Following this strategy, Challis actually saw the sought-after planet twice in August but failed to notice it.

Meanwhile, back in France, Leverrier was frustrated. Other than that quick and unsuccessful search from Paris Observatory, no one had bothered to look for "his" planet—and as far as he was concerned, it *was* his planet, since he knew nothing of Adams's work or the hunt then underway at Cambridge. In late August, Leverrier published a second paper in which he again urged astronomers to search for the planet. This time he pointed out that the best approach would be to look for a planetary disk.

Leverrier's new plea elicited no response, so in desperation he sent a letter to another of France's occasional enemies. On September 23, 1846, Johann Gottfried Galle of Berlin Observatory received the letter and was immediately excited. That night, over the objections of his boss, Galle and an equally eager student, Heinrich d'Arrest, began to search for the solar system's eighth planet.

It took only a few minutes. Armed with Leverrier's coordinates, Galle swung the observatory's 9-inch telescope toward the border between the constellations Capricornus and Aquarius, and looked for a planetary disk. Initially he saw nothing unusual, so d'Arrest suggested they use a map. If a planet was really trespassing through the region, the sky would have a "star" that the map did not. Galle doubted the plan would work, because he knew how bad the star maps were, but new maps had just become available, and Galle agreed to try. With Galle at the telescope and d'Arrest at the map, the former described the brightness and position of the stars he saw and the latter checked them on the map. The first star Galle called out, d'Arrest reported, appeared on the map. So did the second. Then, only a few stars later, Galle described the position and brightness of another object, and d'Arrest exclaimed, "That star is not on the map!" From among the faint stars of Capricornus, the Sea Goat, and Aquarius, the Water Carrier, the blue planet that would be named for the god of the sea had finally emerged from the dark celestial ocean. The next night the astronomers located the object again and found that its position had shifted slightly with respect to the background stars, confirming its planetary nature.

Galle wrote to Leverrier: "The planet whose position you have pointed out *actually exists*." Galle's boss, Johann Franz Encke, who had initially opposed the search, also wrote to Leverrier: "Allow me, Sir, to congratulate you most sincerely on the brilliant discovery with which you have enriched astronomy. Your name will be forever linked with the most outstanding conceivable proof of the validity of universal gravitation, and I believe that these few words sum up all that the ambition of a scientist can wish for." It was Leverrier who first suggested the new world be named Neptune, but in early October he changed his mind and wanted the planet named after himself. None too subtle, he then began referring to Uranus as "Herschel."

Meanwhile, back in England, Challis was still plodding along, unaware of the German discovery. On September 29, he observed Neptune again and even noticed that it might have a disk, but he did not pursue the hint. Two days later, he learned that Neptune had been captured and, with horror, that the world could have been his had he only studied his own notes.

On the Continent, a firestorm was brewing, and the discovery of the new planet would soon divide this one. Because of Leverrier's correct prediction, France considered Neptune to be its planet, so when the English revealed Adams's earlier prediction, the French accused England of trying to steal their newfound world. One called Adams's work "clandestine" and went on to say, "No, no! The friends of science will not permit the perpetuation of such a flagrant injustice! . . . Mr. Adams has no right to figure in the history of the discovery of the planet Le Verrier." Leverrier was outraged: "Why has Mr. Adams kept silent for four months? Why had he not spoken since June if he had good reasons to give? Why did he wait until the planet was seen in the telescope?" The French press entered the fray and attacked the British with such vehemence that even Leverrier and his allies disowned them.

The British were equally incensed—not at the French, but at the arrogant Airy and bumbling Challis. "I mourn over the loss to England and to Cambridge of a discovery which ought to be theirs every inch of it," wrote John Herschel, son of William. Challis admitted his mistakes, but Airy did not. "Attacks which would have agonized Flamsteed's every nerve . . . were absolutely and entirely disregarded by Airy," wrote one commentator, who added, "He was perfectly satisfied with himself, and what other people thought or said about him influenced him no more than the opinions of the inhabitants of Saturn."

Ever mild and polite, the first inhabitant of Earth to pinpoint Neptune's position never battled his French rival. In 1847, Adams met Leverrier for the first time, and the two became lifelong friends. Adams would later award Leverrier a medal from the Royal Astronomical Society for his work on the planets.

Although the discovery of Neptune brought fame to Adams and Leverrier, one of the biggest winners was Isaac Newton, whose law of gravity had predicted the existence and place of an entirely new world. Yet Newton's law of gravity, as well as Leverrier's reputation, would soon be tarnished by another planetary problem: the amusing search for Vulcan, a purported world inside the orbit of Mercury.

The Phantom Planet

Mercury, closest and swiftest of the Sun's planets, exasperated astronomers. Whirling around the Sun in eighty-eight days, the planet occasionally passed in front of the solar disk. Astronomers should have been able to predict the times of these transits, but Mercury refused to cooperate. It missed a transit in 1707 by a full day and another in 1753 by several hours—even though the great astronomer Edmond Halley, of comet fame, had predicted it.

Leverrier began to investigate Mercury before his triumph at Neptune, but even he failed to predict the exact timing of an 1848 transit. He delved into the problem more deeply and in 1859 blamed an unseen mass whose gravity pulled Mercury off course, just as Neptune had Uranus. This mass, said Leverrier, might be a belt of asteroids or even an entire planet inside Mercury's orbit.

In late December, Leverrier received a startling letter. It came from a rural doctor and amateur astronomer named Lescarbault, who wrote that on March 26 he had watched as a planet crossed the Sun's disk. The planet, he said, looked like a small, black dot. An excited Leverrier raced to Lescarbault's home, reaching him unannounced on New Year's Eve. Leverrier bombarded the good doctor with numerous questions and checked his reputation with neighbors. This convinced Leverrier that the doctor was honest and the planet real. He called it Vulcan, after the god of fire and husband of Venus, and calculated that it was only about a third of Mercury's distance from the Sun, whose glare hid the world from direct view. Leverrier put its mass at only a quarter of the Moon's. This was too little to perturb Mercury by the observed amount, so he said that other planets like Vulcan must also exist.

In England, the *Monthly Notices of the Royal Astronomical Society* heaped praise on both Lescarbault and Leverrier: "The singular merit of M. Lescarbault's observation will be recognised by all who examine the attendant circumstances; and astronomers of all countries will unite in applauding this second triumphant conclusion to the theoretical inquiries of M. Le Verrier." However, a French astronomer who had observed the Sun when Lescarbault did had seen no planet, and during the 1860s deliberate searches for Vulcan turned up mostly nothing.

Nevertheless, sporadic reports of Vulcan continued to appear, and in 1876 Leverrier predicted that the planet would cross the Sun's disk around March 22, 1877. Hoping to confirm Vulcan's existence once and for all, astronomers scrutinized the Sun, but saw not a thing. Leverrier died later that year, but his planet lived on and soon enjoyed renewed fame. The occasion was a total solar eclipse in 1878. While the Sun's disk was blotted out, two American observers glimpsed a dim object southwest of the Sun. No one else saw it, but that hardly stopped proclamations hailing the discovery of the long-sought Vulcan. Declared *The New York Times:* "The planet Vulcan, after so long eluding the hunters, showing them from time to time only uncertain tracks and signs, appears at last to have been fairly run down and captured." Discrepancies between the two alleged sightings fueled doubts, however, and Vulcan failed to materialize for a predicted transit in 1879. By the 1880s, belief in Vulcan was fading fast, and unsuccessful searches during solar eclipses in the first decade of the 1900s finally killed the planet once and for all. Vulcan would reappear only as Mr. Spock's home world on the American television series *Star Trek*.

Yet the problems with Mercury persisted. The true solution came in 1915, when Albert Einstein formulated his general theory of relativity, a more sophisticated law of gravity than the simple formula written down centuries earlier by Isaac Newton. In weak gravity, such as on Earth, Einsteinian and Newtonian gravity yield the same results, but in extreme cases, Einsteinian gravity differs from Newtonian. Of the solar system's planets, Mercury feels the strongest gravitational pull from the Sun. Furthermore, Mercury's orbit is fairly elliptical, which also accentuates the difference between Einsteinian and Newtonian gravity. Einstein's general theory of relativity finally allowed astronomers to understand Mercury's path. Named for the messenger of the gods, the planet Mercury had brought astronomers word of a new theory of gravity.

The search for Vulcan detracted from a more promising hunting ground, the enormous region of space beyond the orbit of Neptune. Nevertheless, a few astronomers did try to predict the location of a distant planet, often from continuing differences between the observed and predicted positions of Uranus. Among them were David Peck Todd in America, George Forbes in Scotland, and Hans-Emil Lau in Denmark, but nothing came of their searches. Undoubtedly the most notorious planetary prognosticator was the hated American astronomer Thomas Jefferson Jackson See. "Personally I have never had such an aversion to a man or beast or reptile or anything disgusting as I have had to him," wrote one astronomer who witnessed first-hand See's arrogance, deceitfulness, and theft of others' ideas. "The moment he leaves town will be one of vast and intense relief and I never want to see him again under any circumstances. If he comes back, I will have him kicked out of town." See postulated three planets beyond the orbit of Neptune, naming one Cronus. Largely because of See, when a new world was finally found, it was *not* named Cronus.

For a time, before he got fired, See worked at a new observatory in Flagstaff, Arizona. From here, a new world would indeed be seen—the result of a quiet, quarter-century-long quest inspired by the poetic, flamboyant, and controversial Percival Lowell.

Pluto

Percival Lowell was under attack. The wealthy Bostonian, who founded the Flagstaff observatory in 1894, had dedicated his career not to dim, distant planets but to a bright, nearby one: the red planet Mars, which Lowell believed had intelligent beings. But the Martians were in trouble, he said, for Mars was drying up. To carry the planet's remaining water from the polar caps to the equator, the Martians had built canals that Lowell and his assistants observed night after night from atop a hill named for the planet.

Lowell's Martian theory stemmed from more than just science, for he was a fervent pacifist. He hated war and hoped that the inhabitants of Mars would inspire peace on Earth. The Martians did not fight, he said, because their dying planet was so fragile that an attack by one nation would disrupt the water supply and doom them all. "War is a survival among us from savage times," wrote Lowell, adding, "It is something people outgrow. But whether they consciously practice peace or not, nature in its evolution eventually practices it for them, and after enough of the inhabitants of a globe have killed each other off, the remainder must find it more advantageous to work together for the common good." Blood-red Mars, named for the god of war, was actually a world of peace.

Lowell gave lectures and wrote books and articles to promote his Martian theory, and newspapers around the world reported on his research. Although his theories excited the public, they angered astronomers, who attacked him and ostracized his observatory. In 1905, Lowell devised a secret plan to silence his critics, one involving an unseen world he called Planet X. If he could discover the planet, he would at long last earn the credibility that had eluded him in all the years he had studied Mars.

The search began as assistants took photographic plates that Lowell inspected for a moving object. Lowell also tried to predict the planet's location, as Adams and Leverrier had Neptune's, by examining irregularities in the motion of Uranus. However, the task was far worse than the one that had faced Adams and Leverrier. The irregularities in Uranus's motion were much smaller than those which Adams and Leverrier had used. In addition, for Adams and Leverrier,

Neptune had been the solution; for Lowell, it was part of the problem. Not only did Neptune perturb Uranus, but Neptune's orbit was not well determined, because the planet takes 165 years to revolve and had not yet been observed over a full cycle. Therefore, Lowell could not know whether Neptune was following its proper path. This forced Lowell to use only Uranus, which lay farther from and so would be less perturbed by Planet X.

Lowell's calculations were long and tedious, and at one point he fell ill in part because of them. Year after year, he sent possible positions of the planet to the observers in Arizona, finally completing his calculations in 1914. For nearly ten years, he had mounted the most ambitious planetary search in history. He had spent years calculating Planet X's position and deployed an armada of telescopes and observers to capture the distant world. In May, after sending the predicted positions, Lowell wrote to one of the astronomers: "Don't hesitate to startle me with a telegram—FOUND!"

Despite the huge amount of work, though, no planet appeared, and Lowell began to grow discouraged as plate after photographic plate failed to reveal the elusive body that would have brought him the vindication he had sought for so long. By 1915, even as his observers continued to photograph the sky, Lowell's enthusiasm for the project had vanished entirely. The final plate in the Planet X search was taken July 2, 1916, and one of Lowell's last telegrams to Flagstaff read: "Please see that wild flowers on [Mars] hill are never picked. Put up notices."

Lowell was a defeated and exhausted man. He felt great pain over the outbreak of war in Europe, and his observatory was just as ostracized as ever: his Mars work was still ridiculed, and his Planet X remained unfound. In November 1916, he suffered a stroke and died. His brother wrote that Lowell's failure to find Planet X "was the sharpest disappointment of his life." A lifelong friend said it "virtually killed him."

Yet the final telescope that Lowell employed in his search for Planet X actually captured a faint world on the fringe of the solar system, a billion miles more distant than Neptune. On two different nights—March 19, 1915, and April 7, 1915—the photographic plates of Lowell Observatory's 9-inch telescope faithfully recorded tiny Pluto as it hid among the stars. No one noticed the trespasser, and it silently slipped back into the celestial sea.

With Lowell's death, the observatory's search for Planet X ended, not to resume until 1929. Although Lowell had left the observatory over a million dollars, his widow, Constance, challenged the agreement and sued, initiating a dispute that lasted for over ten years and greatly impoverished the observatory.

Constance Lowell's lawsuit almost handed a planet to one of Percival Lowell's rivals, Harvard astronomer William Pickering. Pickering had helped Lowell establish the Flagstaff observatory, but he later criticized Lowell's Martian theory. Over the years, Pickering had predicted the existence of several planets, including one called Planet O. Lowell had not been impressed; after reading one of Pickering's papers, he wrote, "This planet is very properly designated O [and] is nothing at all." In 1919, however, Pickering convinced Mount Wilson Observatory in California to conduct a brief search for Planet O, and four separate photographic plates captured a tiny world beyond the orbit of Neptune. No one noticed.

Meanwhile, a farm boy in the Midwest was exploring the sky for himself, concentrating on planets rather than stars. "Planets are more personal," said Clyde Tombaugh, whose interest in astronomy was sparked in sixth grade when he wondered what sort of geology and geography might exist on other worlds. "The reason I like planets better is that they're a possible place for life to inhabit." Tombaugh built telescopes and used them to observe the planets. Because Lowell Observatory did planetary astronomy, Tombaugh sent Flagstaff drawings he had made of Mars and Jupiter. The drawings impressed the observatory's director, Vesto Melvin Slipher, who was looking for an assistant to renew the fight for Planet X.

But Slipher said nothing about Planet X. "I didn't know until I got there," said Tombaugh. "They never mentioned about looking for a new planet. They just told me I was to operate a new photographic telescope." The new telescope, a 13-inch, was ideal for a planet hunt because it captured a large area of the sky with a single shot.

Before the twenty-two-year-old Tombaugh boarded the train from Kansas to Arizona, his father told him, "Clyde, make yourself useful, and beware of easy women." To someone who had never left the Midwest, Flagstaff was exotic: in place of endless fields of golden wheat and big open skies stood snow-capped mountains and tall pine trees.

Tombaugh began taking photographic plates with the new telescope in April 1929. His tenth plate, of Gemini, actually captured the world he would later discover, but again no one noticed. It was one thing to take the plates; it was quite another to examine them, because a single plate contained tens or hundreds of thousands of stars. To help find the planet, Lowell Observatory had a blink comparator, a device that held two plates taken of the same region of the sky on different nights. An astronomer used the blink comparator to flip the view from one plate to the other. That way, the stationary stars held steady, but a moving object, such as a planet, blinked back and forth. Even with the blink comparator, however, inspecting plates was tedious, and Tombaugh was glad he did not have to do it.

In May, Slipher and his brother, also an astronomer, used the blink comparator to rifle through Tombaugh's plates. Slipher was under the gun: the new telescope had been expensive, and it had been built specifically to find Planet X. Now Slipher had better deliver. But the hasty search turned up nothing.

After this failure, Slipher ordered Tombaugh to blink the plates himself. "I shuddered with the responsibility," said Tombaugh. "I felt overwhelmed, because I thought it'd be a very difficult and tedious task, and frankly at the time I was not very happy about that assignment. But it was a job for me, which was better than pitching hay back on the farm." As Tombaugh began to blink the plates, his spirit sunk even lower. The plates picked up numerous asteroids, whose movement masqueraded as that of the planet he was seeking.

A visiting astronomer, Frank Ross of Yerkes Observatory in Wisconsin, did not help matters. "Young man, I am afraid you are wasting your time," Ross told Tombaugh. "If there were any more planets to be found, they would have been found long before this." Over the summer, when rains in Arizona prevent astronomical observations, Tombaugh devised a strategy that would ultimately capture a world beyond the orbit of Neptune. From now on, Tombaugh would take three plates of each region of the sky—two to blink and the third as a check. In addition, he would always photograph the part of the sky opposite the Sun. Then, as the Earth raced around the Sun, nearby asteroids would appear to move swiftly from one night to the next, whereas a distant planet would appear to move slowly, just as a distant mountain does to a speeding driver.

In September, Tombaugh put his plan into action, taking photographic plates on nights when the Moon was dark and blinking the plates on days when the Moon was near full. The first constellations he photographed using his new strategy were Aquarius and Pisces, which lie far from the star-rich Milky Way and so were fairly easy to examine for planets. Later in the fall, Tombaugh moved into Aries and then Taurus, all the time getting closer and closer to Gemini. Both Taurus and Gemini are in the Milky Way, whose many stars slowed the search for the planet.

Tombaugh reached Gemini in winter. On January 21, he pointed the telescope at the star Delta Geminorum and began to take a photograph. Although the night was clear, a fierce wind suddenly erupted from the northeast and distorted the images of the stars. Tombaugh knew his photograph would be terrible; he even thought of halting the exposure, but fortunately did not. On January 23 and 29, under much better conditions, he took additional plates of the same region.

Tombaugh began to blink the two good plates on the morning of February 18. By mid-afternoon, he had examined a quarter of the plates. Then at 4:00, he detected a faint dot blinking back and forth. "That's it!" he said to himself. Its motion was just right for an object that was farther than Neptune, and with mounting excitement he pulled out the January 21 plate. Although the plate was bad, the faint object was there, too, and in just the right place.

Tombaugh first told Carl Lampland, another Lowell astronomer, and then marched into Slipher's office. "Dr. Slipher, I have found your Planet X." Slipher bolted out of his chair and raced down the hall to see the evidence. He examined the image on the photographic plates. It looked good, very good. The astronomers wanted to observe the planet that night, but the sky was cloudy, and Tombaugh instead spent the evening watching Gary Cooper star in *The Virginian*.

The next night was fairly clear, and Tombaugh took another photograph of Gemini. Using this as a guide, on February 20 the as-

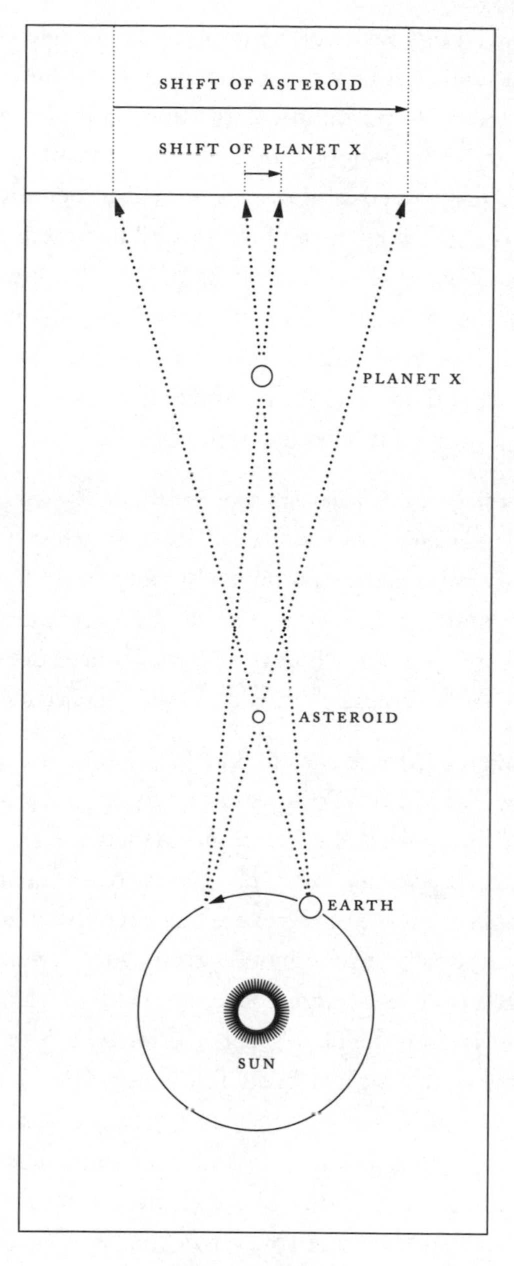

FIGURE 12

Because the Earth moves fast around the Sun, objects on the other side of the Earth from the Sun appear to move from night to night. The farther away the object is, the smaller this apparent shift is—which allowed Clyde Tombaugh to distinguish distant Pluto from a nearby asteroid.

tronomers viewed the new world through a larger telescope. It worried them, though, because the planet was ten times fainter than Lowell had predicted and hundreds of times fainter than Neptune. Furthermore, the planet showed no disk. This meant the new world might be too small to perturb Uranus. Yet it had been found just six degrees from one of two places where Lowell had predicted it.

Slipher was as cautious as Lowell had been flamboyant, and for three weeks the astronomers quietly tracked the distant world as it crept through Gemini. Finally, on March 13—the 75th anniversary of Lowell's birth and the 149th anniversary of the discovery of Uranus—this triumphant telegram appeared:

> Systematic search begun years ago supplementing Lowell's investigations for Trans-Neptunian planet has revealed object which since seven weeks has in rate of motion and path consistently conformed to Trans-Neptunian body at approximate distance he assigned. Fifteenth magnitude. Position March twelve days three hours GMT was seven seconds of time West from Delta Geminorum, agreeing with Lowell's predicted longitude.

The new world garnered front-page headlines, and over a thousand people wrote to Lowell Observatory to suggest names—everything from *Atlas* to *Zymal*, the latter, said the letter writer, being "the last word in the dictionary" for "the last word in planets." But the Lowell astronomers also heard from a less friendly quarter: Lowell's widow Constance, whose lengthy litigation had delayed the Planet X search for over a decade. She concocted a variety of names before deciding that she wanted the planet named after *her*.

"I just despised that woman," said Tombaugh. "All she did was impede the progress of that observatory. She brought on a big lawsuit at the observatory, and they lost thousands and thousands of dollars in court costs and so on, and then she had the gall to want the planet named after her. That was the last straw. That was absolutely the last straw." At one point, she even tried to convince the young Tombaugh that *she* had discovered the planet and Slipher had stolen it from her.

Fortunately, three names other than *Constance* emerged on top: Minerva, Cronus, and Pluto. Minerva, the goddess of wisdom, was the most popular, an appropriate name for a planet that had apparently been predicted prior to its discovery. However, Minerva was already the name of an asteroid. The second suggestion, Cronus, was a son of Uranus and the father of Neptune, but Cronus also had a problem. As Tombaugh later wrote, "Had not Cronus been proposed by a certain detested egocentric astronomer [i.e., Thomas Jefferson Jackson See], that name might have been considered."

The third and winning choice was Pluto, the god of the underworld, a land of perpetual darkness much like the new planet, which was so distant that it received little sunlight. In addition, Pluto's first two letters are the initials of Percival Lowell. Each member of Lowell Observatory had one vote, and each, including Tombaugh, voted to name the new planet Pluto.

After the excitement over Pluto faded, Tombaugh continued to search for more planets. Although his hunt lasted until 1943, it turned up no additional planets. The heavens, which had once smiled on Tombaugh, did not do so again, and Lowell Observatory was becoming equally unfriendly as well. According to Tombaugh, Slipher eventually grew jealous that it was Tombaugh and not Slipher who had discovered Pluto, and in 1946 Slipher gave Tombaugh the shock of his life: he fired him.

The world that Tombaugh discovered is now known to be much smaller than any other planet in the solar system. It has even less mass than the Moon, making Pluto much too small to affect Uranus significantly. Pluto is an innocent bystander, arrested on the strength of circumstantial evidence for a crime it did not commit. Rather than a triumph of Percival Lowell's predictive powers, Pluto is a testimony to the thoroughness of Clyde Tombaugh's search. The true culprit, if there is one, is still on the loose.

4

Planet X

BILLIONS OF miles distant, in the depths of the outermost solar system, a possible unseen planet quietly slithers around the Sun. Far beyond the orbits of Uranus, Neptune, and Pluto, and barely touched by the Sun's light and heat, this purported planet must be dark and cold, a frigid mass clinging to the edge of the solar system. Its gravity, however, may give its presence away, by tugging on the known planets. Both Uranus and Neptune have exhibited apparent irregularities that some astronomers once ascribed to this tenth planet: the elusive Planet X.

During the 1990s, though, this planet's prospects grew as dim as the hypothetical world itself. "The latest on Planet X," said longtime skeptic Brian Marsden of the Harvard-Smithsonian Center for Astrophysics, "is that there is no Planet X." Yet the quest for this world, and the chance that it might even exist, make a fascinating story—one that illuminates the darkest reaches of the solar system.

"We should continue searching," said the late Robert Harrington in 1992, an astronomer at the U. S. Naval Observatory in Washington, D. C., who had led the hunt for Planet X. "I would hate to say, 'Well, it doesn't seem to be there to me; let's just give up on the whole question and forget about it,' and then discover it's been lurking back there all along."

Pluto: Ex-Planet X

The original motivation for the modern Planet X hunt comes from dwindling Pluto, which astronomers now know is much too small to perturb Uranus or Neptune significantly. For decades after Pluto's discovery, many astronomers thought they had caught their culprit. Percival Lowell had used irregularities in the motion of Uranus to pinpoint two possible locations for Planet X and deduce that its mass was about seven times greater than Earth's. When Clyde Tombaugh discovered Pluto in 1930, the planet was just a few degrees from one of the positions Lowell had predicted. Pluto was fainter than Lowell had thought, which suggested it was smaller, but the planet seemed to explain the wobbles in the motion of Uranus and, later, Neptune. Astronomers even used these same irregularities to estimate Pluto's mass. In 1955, for example, these irregularities indicated that Pluto was slightly less massive than Earth. In 1968, a new analysis of problems with Neptune's orbit suggested that Pluto had only 18 percent of the mass of the Earth, and three years later, additional observations reduced that to 11 percent of an Earth mass—about the same as Mars, Lowell's favorite planet.

The first shocker came in 1976, when three astronomers at the University of Hawaii—Dale Cruikshank, Carl Pilcher, and David Morrison—discovered hints of just how minuscule Pluto really is. All along, astronomers had assumed that Pluto was fairly dark, which would explain why an Earth- or Mars-sized planet looks as faint as Pluto. But the Hawaii astronomers found that Pluto's surface bears methane ice, which reflects much of the sunlight hitting it. The presence of this ice implied that faint Pluto might be smaller than the Moon, whose mass is only about 1 percent that of the Earth. If so, Pluto was the smallest planet in the solar system.

The clincher came two years later. In a startling and unforeseen development, James Christy of the U. S. Naval Observatory discovered a bump on photographs of Pluto that proved to be an orbiting satellite. He named the moon Charon, for the ferryman who carried souls across the river Styx. Robert Harrington then calculated the orbit of the moon, which felt Pluto's gravity and therefore revealed the planet's mass. Harrington found that Pluto was a planetary featherweight, being even lighter than the Earth's Moon: a mere 0.2 percent of the mass of the Earth—only 1/3000th the mass of Lowell's Planet X, and much too small to perturb Uranus or Neptune significantly.

Great X-pectations

During the 1980s and early 1990s, however, continuing problems with both Uranus and Neptune convinced some astronomers that another planet, beyond the orbits of both Neptune and Pluto, might exist. It was no coincidence that the astronomer who led the search, and who was taken most seriously even by the skeptics, was the one who had first determined just how puny Pluto was.

"We cannot accurately predict the motions of Uranus and Neptune based on the planets we now know about," said Harrington in 1990. "In the case of both Uranus and Neptune, whenever we try to predict where they're going based on their observed motion, we discover that after a few years the actual observed motion is off the predicted."

During the 1980s, P. Kenneth Seidelmann, another astronomer at the Naval Observatory, also confronted this problem. "I think there is something affecting our observations of the outer solar system," Seidelmann said in 1987. Calculating planetary orbits produces what astronomers call an ephemeris, a list of a planet's past and future positions. Uranus should have had an accurately determined orbit, for it had revolved around the Sun two and a half times since the planet's discovery in 1781 and three and a half times since its first sighting in 1690. But the Uranus observations prior to 1900 had to be thrown out, because they did not agree with the present ephemeris. Astronomers had resorted to the same desperate remedy back in the 1820s, when observations prior to 1781 could not be fit to the ephemeris. The culprit then was the unknown planet Neptune, which was pulling Uranus off course.

There was also trouble with Neptune itself. Although all observations since its discovery in 1846 could be fit to a single ephemeris, the planet eventually departed from its predicted path. Neptune had not completed a full revolution since its discovery, so its orbit was not as well known, which meant that its trouble was less serious than that of Uranus. Even so, Seidelmann considered it surprising that with

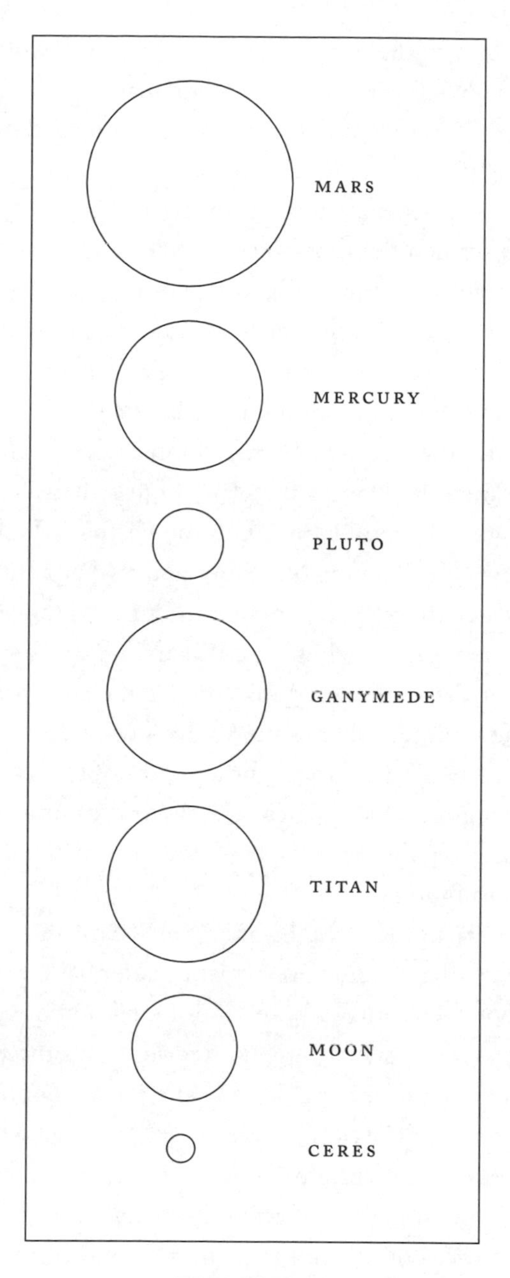

FIGURE 13

Pluto is the smallest planet in the solar system. It is even smaller than seven moons, including Jupiter's Ganymede, Saturn's Titan, and the Earth's Moon. Pluto is, however, larger than Ceres, the largest asteroid.

around 90 percent of the orbit completed, Neptune would still deviate from the predicted path. Thus, in Seidelmann's view, both Uranus and Neptune had problems. (Unfortunately, Pluto cannot lead astronomers to Planet X, because Pluto has covered so little of its path since discovery that its exact orbit is uncertain.)

Harrington studied the residuals in Uranus and Neptune—that is, the differences between their observed and predicted positions—and calculated what Planet X might be like and where it might be hiding. His calculations indicated that the planet's mean distance from the Sun was about a hundred times the Earth's, or a little over three times Neptune's, and it traveled on a highly elliptical orbit. The planet also journeyed far above and below the plane of the solar system. Its mass, said Harrington, was around four times the Earth's, which made it either a super-Earth or a mini-Neptune, and its brightness would be somewhat greater than Pluto's at the time of its discovery. Planet X had eluded previous searchers, said Harrington, because it was around the constellations Scorpius and Centaurus, which are best seen from the southern hemisphere. To search for Planet X, astronomers in New Zealand took photographs of the sky and shipped them to Washington, D. C., where Harrington examined them with a blink comparator, the same type of device that Clyde Tombaugh had used to find Pluto.

Planet X, Harrington thought, might also explain another anomaly in the outer solar system. Every giant planet has a regular satellite system—except Neptune, whose two best-known satellites have strange orbits. Triton, which was spotted less than three weeks after Neptune's discovery, is nearly as large as the Earth's Moon. Yet unlike any other big satellite, Triton circles its planet backward, opposite from the direction of Neptune's spin. In 1949, a smaller but equally odd moon was seen. Named Nereid, it revolves around Neptune in the right direction, but its orbit is far more elliptical than that of any other satellite in the solar system. When closest to Neptune, Nereid is less than 1 million miles from the planet, but at its farthest, the tiny moon scurries 6 million miles from its master.

In the late 1970s, Harrington and Thomas Van Flandern, then also at the Naval Observatory, developed a possible explanation for the odd behavior of Neptune's moons as well as the strange path of Pluto, whose orbit around the Sun is so elliptical that at times the planet dashes inside the orbit of Neptune. Suppose, they conjectured, that Triton, Nereid, and Pluto were once normal moons of Neptune, going around Neptune in the correct direction on nearly circular orbits. Then a large planet—Planet X—brushed by Neptune. The intruder's gravity reversed the motion of Triton, giving it a backward orbit; elongated the orbit of Nereid, giving it an extremely elliptical orbit; and kicked Pluto away from Neptune altogether, sending it on a highly elliptical orbit around the Sun. Starting in 1934, other astronomers had proposed that Pluto might be an escaped moon of Neptune, but Harrington and Van Flandern calculated the properties that their hypothetical planet must have to instigate the proper amount of damage.

"That's one of the things that made all of this so interesting," said Harrington. "We did that article back in 1978, and I thought we were doing it purely as a mathematical exercise—until the discovery of the satellite of Pluto, which revealed Pluto's mass and immediately made it obvious that Pluto wasn't the thing they were looking for back in 1930. And then somebody pointed out to me later that if you do the solution for the planet from the residuals of either Uranus or Neptune, you get something that looks very similar to what we were talking about for this interloper that produced Pluto."

Enter the Skeptics

Harrington and other Planet X proponents faced great opposition from their peers. One staunch critic was outspoken British-born astronomer Brian Marsden. Ironically, as head of the International Astronomical Union's Central Bureau for Astronomical Telegrams in Cambridge, Massachusetts, Marsden might well be the one to report the discovery of such a planet. It was through one of his predecessors that Lowell Observatory had announced the discovery of Pluto back in 1930.

"I suppose potential discoverers know very well that if I'm involved with the announcement, I shall be very skeptical, but I hope that wouldn't deter them from announcing it here," said Marsden. "I may be skeptical, but I am fair, and I am prepared to consider any evidence that is presented in any reasonable way." In contrast, he said, the 1930 announcement was outrageous. "No blame can be attached to Clyde Tombaugh," he said. "Clyde found the images of Pluto, and he measured them. And what did his superiors do? They waited three weeks before saying anything, so that they could make an announcement with no information, no proof, on the 75th anniversary of Percival Lowell's birth. That was a disgrace!" In particular, Lowell Observatory provided only one position of Pluto, preventing other astronomers from trying to calculate its orbit. "This whole thing of calling it a major planet should not have been done," said Marsden, who thinks Pluto is too small to be a true planet. "If I'd been in charge back then, it would *not* have been done."

Marsden's interest in astronomy goes back to childhood. "One of the things I did when I was ten was independently discover Kepler's second law," he said. "I had the equivalent of *The World Almanac*, and by looking in a large number of volumes, year by year, I was able to detect certain patterns that showed me that planets move faster when nearer the Sun. But I soon realized that somebody had already done this!"

Today Marsden spends much of his time overseeing comets and asteroids. As far as he is concerned, there is no unseen planet that is sufficiently massive to make Uranus or Neptune wobble from their proper paths. "Most of it's a bunch of nonsense," said Marsden. "Just to see that there are some little wiggles and say, 'Oh, that's a trans-Neptunian planet that's doing it,' is not playing the game! That's not good science. You find what the real explanation is, the most logical explanation, following William of Occam [famous for Occam's razor—the simplest explanation is usually the best], and that's that. And the most logical explanation is *not* that there's a big planet out there."

Although John Couch Adams and Urbain Leverrier had triumphed in predicting Neptune, said Marsden, that does not mean recent Planet X prognosticators are right. "This kind of thing happens in scientific endeavors from time to time," he said. "Somebody has a wonderful success. Lesser people try to jump on the bandwagon and claim things that have to be withdrawn. The only startling thing is how long this whole myth has gone on." Planet X, said Marsden, is not an astronomical problem; it is a psychological one.

Marsden pointed out that the residuals in Uranus that Adams and Leverrier explained with Neptune were about a hundred times larger than those which troubled astronomers during the 1980s. Simple observational errors, he said then, could fully explain such tiny residuals.

Marsden's view on this matter differed from that of Kenneth Seidelmann, who thought that the residuals were too systematic to result from observational errors. A decade earlier, however, Marsden and Seidelmann had actually joined forces to attack one proposed Planet X. In 1972, Joseph Brady of the Lawrence Livermore Laboratory in California analyzed irregularities in the motion of Halley's Comet, which travels beyond the orbit of Neptune. Brady said that he could explain these irregularities by postulating a Jupiter-sized Planet X that revolved around the Sun backward. The prediction made headlines in newspapers, and a brief search was mounted, but nothing was found. Seidelmann, Marsden, and Lowell Observatory's Henry Giclas wrote a paper castigating Brady's planet: Seidelmann described how it would have perturbed the outer planets in ways not seen; Marsden described how Halley's Comet could alter its own path as the comet's ice sputtered and pushed the comet around haphazardly like the gas from a deflating balloon; and Giclas explained why Clyde Tombaugh would likely have already discovered such a planet. Another team of astronomers pointed out that the gravity of Brady's planet would have sent Neptune flying away from the plane of the solar system.

"It was crazy," said E. Myles Standish, Jr., of the Jet Propulsion Laboratory in Pasadena, California, and perhaps the harshest Planet X critic of all. "You can't put a bull in a china shop without figuring out what it's going to do. He got this monster Goliath out there, and he never even bothered to compute what that planet would have done to Uranus and Neptune. It would have knocked them completely wacko."

Today Standish predicts the future positions of the planets, so that interplanetary spacecraft will reach their proper destinations. He first discovered astronomy while in college. "I majored in math," he said, "and nearly flunked out because I had absolutely no interest in it. I can't stand doing a problem when the answer's in the back of the book. The head of the math department came up to me at the beginning of my senior year and said, 'There's a new astronomy professor up on the hill. Maybe you'd like to work with him and enjoy doing research-type work more than just classroom work.'"

Standish did, and his job today is critical for NASA, because if he predicts a planet's location wrong, a visiting spacecraft may pass too far from the planet or even crash into it. He recalled Voyager 2's 1989 flyby of Neptune with the fresh concern as if it had happened only yesterday. The spacecraft skirted over Neptune's north pole so that the planet's gravity yanked it south, allowing Voyager to study Neptune's big moon, Triton. Shortly before Voyager's encounter, though, three changes conspired to heighten the drama. First, astronomers lowered Neptune's estimated mass, which meant that in order to deflect itself toward Triton, the spacecraft would have to fly closer to Neptune, thereby placing itself in greater danger of burning up in Neptune's atmosphere. Second, estimates of Neptune's diameter increased, again jeopardizing the spacecraft during the encounter. And third, Triton's estimated position was revised southward, forcing the spacecraft to fly still closer to Neptune if it was to meet the satellite. Despite these obstacles, Voyager 2's flyby of Neptune, nearly 3 billion miles from Earth, was a spectacular success, and the planet remains the most distant that NASA has yet reached.

"Sure, people would love to find another planet out there," said Standish. "That's the fastest route to fame that anybody could think of." During the 1980s, Standish maintained that the apparent irregularities of Uranus and Neptune were due to observational errors. Before 1911, astronomers measured the positions of Uranus and Neptune less accurately, so the only trustworthy data came after that year. Those data gave less than one full orbit of Uranus, which showed no trouble; earlier positions did not fit the modern ephemeris only because they were less accurate. In the case of Neptune, the post-1911 data yielded less than half an orbit, making it impossible, said Standish, to determine that planet's precise path. Thus, neither Uranus nor Neptune had any real problems.

Standish did, however, respect Harrington's work, even if he disagreed with its conclusion. "Harrington was a good friend and a nice person," said Standish. "But he was quoted as saying something that did make me very angry": that Standish must tell his boss that the solar system was under control, so anyone who claimed that Standish had overlooked the tenth planet made him and his boss look bad. Said Standish, "He just turned it all around. If you ask me, *I*'m the one that's got to produce. It affects our space missions! If I tell my people I know everything's perfect and it isn't, we're going to get into a lot of trouble. I've got to be as honest as I can. And Harrington, like the other Planet X predictors, can go and claim anything and it doesn't affect anybody. The more sensational it is for some of those guys, the more they justify their work and position, and they get the newspaper there and bring glory to themselves and blah, blah, blah. But I'm the one that has to be realistic about the thing."

In 1988, Harrington published what he considered the most likely location of Planet X, predicting a planet of several Earth masses near the constellation Scorpius. At about the same time, two other investigators postulated that Planet X lay farther north. One hypothetical planet was a bit less massive than Earth and lay in Virgo, and another proposed planet was in Cancer or Gemini, on the opposite side of the sky from Harrington's Planet X.

The three predicted planets did not impress Standish. "Curly says it's here, Larry says it's there, and Moe says it's somewhere else," he said in 1990. "These guys are starting with the same data and ending up with three wildly different Planet X's. That says the data cannot tell you at all where Planet X is." Standish once received a telephone call from someone who wanted to see Planet X. When Standish told him he did not think the planet existed, the caller said, "That's okay; I don't care. I just need some coordinates to point my telescope at."

Harrington acknowledged that simple observational errors, rather than Planet X, might account for the problems with Uranus and Neptune. "I tell everybody I give it fifty-fifty odds," said Harrington in 1990. "There are too many other explanations and possibilities to bet the whole farm on this thing." For that reason, he said, a modest search like his made sense, but an expensive one did not. He also noted that unlike the other predicted planets, his Planet X was plausible, because it was too far south for Clyde Tombaugh to have found.

The Demise of Planet X?

Planet X, however, may have met its end in 1993. Earlier, Harrington had lost funding for his search and, far worse, had been stricken with cancer. "We agreed to disagree on the subject of Planet X," said Marsden, "but I always regarded him as a friend, and I was very sorry when he came down with an illness and died. He did a lot of fine work in stellar dynamics, and we worked together in the campaign to get positional observations of Halley's Comet." Harrington died in January 1993, only fifty years old.

Four months later, Standish published a paper that demolished Planet X, convincing even Seidelmann that the planet probably does not exist. Yet the resolution to the problem was a surprise, even to Standish. When Voyager 2 flew past Neptune in 1989, the planet's gravity deflected the spacecraft in such a way as to reveal the planet's precise mass. That mass turned out to be 0.5 percent less than had been thought, a difference that was a bit less than the total mass of Mars.

"In a way," said Standish, "the guys that were looking for Planet X earlier were right: there *was* a gravitational effect in the orbit of Uranus. The problem is that they assigned it to Planet X, and it was really the missing mass of Neptune. Now, with the correct mass of Neptune, we have the correct forces upon Uranus and can therefore compute a more accurate orbit." That orbit shows no major problems, even in fitting observations before 1900. Standish said the final version of his 1993 paper reporting these results differed somewhat from the original. "The original version showed a little more of my—I don't know—sarcasm toward these guys," he said. "They really, at times, got me angry, and it showed in the original version."

Before writing the paper, Standish had not thought the change in Neptune's mass would prove so critical. "It's not like I said, 'Now that we have a new mass for Neptune, we should go redo the Planet X thing,'" he explained. "For a number of years, I had been kind of irritated by these guys who were screaming about Planet X, and I hadn't even thought that the mass of Neptune might be off. I actually started out writing a paper just to show that the problems with the observations would be enough to give a sizable signature in the orbit of Uranus. And then I said, 'Well, if I'm going to do this right, I'd better make sure that we've got a good orbit,' and I readjusted the orbit of Uranus, and the signature went down a little, and I said, 'Gee, you know, I ought to make sure I've got the right masses in here. I don't know if the Voyager masses are going to make a difference or not,' and I put them in, and honest to God, I looked at those residuals for maybe a couple hours or even a day before it dawned on me that the signature was gone."

Seidelmann initially hesitated to accept Standish's conclusion but has since changed his mind. "I was of the opinion that, yes, there was a problem with Uranus," said Seidelmann. "We had tried changes in the planetary masses to try and resolve it. They weren't successful, but we did not change all the different masses at the same time." He added, "Right now, there are no scientific reasons for searching for another planet." Nevertheless, Seidelmann pointed out that Standish has changed his tune somewhat: "Standish used to argue that pre-1911 observations of Uranus just weren't accurate. Now he's happy to accept them, since they fit the new ephemeris." Standish acknowledged that those old observations, though they have some problems, seem to be better than he had thought. Marsden, too, had suspected those old observations were the culprit and was surprised when Standish discovered the problem with Neptune's mass.

Harrington learned of the new work shortly before his death. "It is certainly true, if you go back and put in the new mass of Neptune as derived from the Voyager flyby and redo the analysis of the Uranus data, some of the key residuals do decrease," said Harrington. "In my opinion, it is not correct to say that they disappear entirely." For that reason, he said, the search for Planet X should continue.

In 1992, astronomers finally did succeed in finding an object beyond the orbit of Neptune—not Planet X, but a chunk of ice one or two hundred miles in diameter. David Jewitt of the University of Hawaii and Jane Luu of the University of California at Berkeley spotted the dim object on August 30. It was about a tenth the size of Pluto and a quarter the size of Ceres, the largest asteroid. Since then, they and other astronomers have discovered many similar objects beyond Neptune's orbit. These objects belong to the Kuiper belt, a zone of cometary bodies named for Dutch-American astronomer Gerard Kuiper, who proposed its existence in 1951. This comet belt supplies short-period comets to the solar system, those which revolve around the Sun in less than two hundred years.

"After all these years, Pluto finally makes sense," said Marsden, who has computed the orbits of the Kuiper belt objects. Some of those orbits are remarkably similar to Pluto's, and in Marsden's view Pluto is simply the largest member of the Kuiper belt. Pluto was discovered first because of its size, which affected its visibility in two ways. First, a larger object reflects more sunlight toward Earth, and second, Pluto is sufficiently large that it has an atmosphere which coats it with bright, shiny ice. Neptune's satellite Triton, the big moon with the backward orbit, resembles Pluto and probably once resided in the Kuiper belt before Neptune captured it. Other Kuiper belt objects are smaller and darker.

But those who seek Planet X are hunting for bigger game—game that Marsden thinks probably does not exist. "I have always had that view, I still have that view, and nobody has convinced me of a reason not to have that view—with the possible exception, but it's only a possible exception, of those Lalande observations of Neptune back in 1795."

The French Connection

Just when it might seem that Planet X could finally be buried, the plot thickens in a way that even hard-core Planet X skeptics like Marsden and Standish admit is at least puzzling. The central character in this plot twist is Joseph-Jérôme Lalande, a French astronomer whose activities extended well beyond astronomy. During the French Revolution, he hid royalist friends in Paris Observatory. Less seriously, to convince the public not to fear spiders, he once swallowed several of the eight-legged creatures. He also christened what turned out to be the fourth nearest star system to the Sun, Lalande 21185. Although he did not discover its proximity, Lalande 21185 was part of an extensive star catalogue that he was compiling.

And it was while making observations for that star catalogue that Lalande's nephew, Michel, unwittingly provided tantalizing evidence in favor of Planet X. On the night of May 8, 1795, he was observing stars for his uncle's catalogue when he recorded what would turn out to be one very special "star." Two nights later, he observed the same "star" again, but either he or his uncle noticed that its position differed slightly from the first night's. The first observation was thrown out, the second one accepted, and thereby did Lalande miss discovering what astronomers now know was the planet Neptune.

Because Neptune has not yet completed a full orbit since its actual 1846 discovery, Lalande's positions for the planet in 1795 are extremely important: they extend the observations back over 200 years, quite a bit longer than Neptune's orbital period of 165 years. Lalande's two positions for Neptune agree with each other, suggesting they are correct, but they differ from where the modern orbit places Neptune in 1795. According to Standish, who has analyzed the observations, Lalande typically erred by 4 arcseconds in recording the positions of stars (1 arcsecond is 1/3600 of a degree). But both of the observed positions of Neptune differ from the predicted ones by 8 arcseconds, twice Lalande's mean error.

"Those two are kind of curious," said Standish. "If they are accurate, they are just a little uncomfortable. They definitely are not consistent with modern observations. If it were only one, you'd say, 'Eh, it doesn't mean anything.' But there are two of them, and they both are off by almost the same amount." That, said Standish, raises an eyebrow.

"The most troublesome fact about this whole Planet X business has always been those 1795 observations," said Marsden. "It may be that one ends up with it really being a problem, and one possible solution to that problem is the introduction of Planet X perturbing Neptune. But I don't think that all avenues have been explored, and we don't know exactly what problems Lalande might have had." Marsden noted that Lalande's observations were visual, not photographic. "What can you do with visual observations?" he asked. "They cannot be as completely reconstructed as photographic observations. We've got Lalande's notes, and we just have to, at some stage, rely on them. There's no independent evidence."

Standish has searched for other prediscovery sightings of Neptune but has failed to find any. "I found a guy who was cheating on his observations," he said. "He was doctoring up the observations, or never took them and fabricated them, or something. It really is bizarre." In times past, explained Standish, observers sometimes got paid according to how many observations they made, a policy that encouraged abuse by the unscrupulous. The case that Standish identified involved a nineteenth-century British astronomer whose positions for stars agreed far too well with one another to be genuine. "This is starting to bother me," said Standish. "It becomes more common than I would have thought."

In 1980, however, prediscovery observations of Neptune did emerge—and this time from one of the greatest astronomers ever, Galileo Galilei. A year earlier, Steven Albers had published an article in

Sky and Telescope listing dates when planets had passed in front of other planets. Such events are rare—the next occurs in 2065—but Albers found that Neptune had passed behind Jupiter on January 4, 1613.

That article caught the eyes of Charles Kowal, an astronomer who had searched for Planet X, and the late Stillman Drake, an expert on Galileo. Both knew that Galileo had discovered the four big moons of Jupiter in 1610, and that these moons had provided him with crucial evidence that, contrary to the Church's belief, not all celestial objects circle the Earth. As Galileo watched the satellites revolve around Jupiter, might he have also recorded the position of the "star" Neptune?

Kowal and Drake searched Galileo's notebooks and found that he had observed Neptune at least three times. The first two occurred December 28, 1612, when Neptune was so far from Jupiter that Galileo's position was probably not reliable. But the third one, in late January 1613, was intriguing. Galileo wrote, "Beyond fixed star 'a', another followed in the same straight line; this is 'b' [now known to be Neptune], which was also observed on the preceding night, but they [then] seemed farther apart from one another." Thus, more than two centuries before the planet's actual discovery, Galileo had not only observed Neptune but also detected its motion.

Galileo's position for Neptune in late January 1613 differs markedly from the modern ephemeris. In 1980, Kowal and Drake suggested that this provided evidence for Planet X. But today, most people, including Kowal himself, regard Galileo's observed position of Neptune as inaccurate. "To his mind, that was just an incidental star," said Marsden, who received his doctorate for studying the orbital motion of Jupiter's moons. "He was interested in the satellites themselves."

In 1990, however, Standish discovered a possible fourth observation of Neptune, on January 6, 1613, which may just be accurate. "I was looking at Kowal and Drake's pictures," said Standish, "and I saw a smudgy-looking thing. I said, 'Gee, that's right where Neptune should be.'" Standish wrote to Anna Nobili at the University of Pisa in Italy. "I wanted to know if it was a real spot or just a blemish on the paper," he said. "So she went and looked at the actual notebook of Galileo under ultraviolet light, and the spot has a special glow because of the ink that Galileo used. Then she looked at it under a microscope, shining a light behind the page, and she said she could see where his pen was actually put down on the paper. So it's an intentional spot made by Galileo." On that date, Neptune was quite close to Jupiter. "We think possibly that he thought it was a satellite, put it down there, and then when he realized the satellites were four other places, he kind of smudged it out," he said. According to Standish, that position, if real, agrees with the modern ephemeris of Neptune, suggesting no Planet X.

Neptune passed behind Jupiter again in 1702. Accurate observations of Neptune from that year could provide evidence for or against Planet X. So far, though, no one has discovered any 1702 sightings of Neptune.

The bottom line, then, is discouraging, for the existence of Planet X is no longer as likely as many had hoped during the 1980s and early 1990s. Except possibly for the Lalande observations of 1795, the evidence suggests that Uranus and Neptune are on course and are not being perturbed by an unseen planet. Furthermore, four spacecraft—Pioneer 10, Pioneer 11, Voyager 1, and Voyager 2—are now beyond Neptune's orbit, and none of the four feels the gravitational force of a tenth planet.

Pluto's discoverer, who died in 1997, had also argued against Planet X. "I think the evidence is very negative," said Clyde Tombaugh, "because my plates recorded stars five times fainter than Pluto and I covered two-thirds of the sky, and nothing more showed up at all. And anything that was fainter than that would hardly be a planet. In my search, I could have picked up a planet like Neptune at seven times Neptune's distance from the Sun."

In short, astronomers seem to have run out of planets—at least in this solar system. If other worlds do exist, they are to be found around the Sun's stellar neighbors, the points of light that decorate the night sky trillions of miles beyond the farthest edge of the Sun's domain.

5

A Shaky Start

TWENTY-FIVE trillion miles of frigid black space separate the nine planets that huddle around the warmth of the Sun from the amber brilliance of Alpha Centauri, the nearest star system beyond. Yet within just a dozen light-years of the Sun shine over two dozen individual stars: some are bright, most faint; some glow white or yellow, and many more orange or red; and some, quite possibly, illuminate planets that harbor water and support life.

But these elusive worlds have repeatedly embarrassed those who sought them. Since the 1940s, astronomers have reported planets circling nearby stars. No such worlds were actually seen, but their stars seemed to wobble as they traveled across our line of sight. A star normally moves through space in a straight line, so the wobble suggested that the star was being tugged by the gravity of an orbiting planet. These purported planets earned headlines in newspapers and magazines, exciting visions of new worlds bearing alien life, but the planets never withstood closer scrutiny.

"It's the Columbus effect," said George Gatewood, director of the University of Pittsburgh's Allegheny Observatory. "Everybody wants to rush out and be the first one to discover a new world. It makes people who are otherwise very academic and calm and reserved go just a little over the edge."

Since 1970, Gatewood's observations have destroyed a number of alleged planets. "One day I got a phone call from a gentleman," said

Gatewood. "He said, 'We've been studying your work, and we must say: you have a reputation rather like that of the Klingons. You wipe out a planetary system every year!'" As the joke goes, you know it's going to be a bad day when the crew from *60 Minutes* shows up on your doorstep. And it's going to be a *really* bad day when Gatewood starts observing your solar system.

The roster of stars whose planets Gatewood and other astronomers have reduced to mere rubble reads like a who's who of the Sun's nearest neighbors: Lalande 21185, the star named for the Frenchman who narrowly missed discovering Neptune; Epsilon Eridani, an orange star that resembles a younger version of the Sun; 61 Cygni, the first star beyond the Sun whose distance astronomers measured; and, most notorious of all, Barnard's Star, a dim red sun that is the nearest star after Alpha Centauri. In each case, the wobble ascribed to an orbiting planet turned out to be due to nothing more than observational error. Nevertheless, this method of planetary detection remains valid, and the specific technique that Gatewood uses—looking for wobbles *across* our line of sight to the star—works best on the nearest stars, where such wobbles stand out the most.

The Sun's Nearest Neighbors

For this and other reasons, the closest stars, said Gatewood, will always be the ones most important to humankind. "To say something as wild as, 'Maybe we'll go there someday,' would be frowned upon by everybody who knows any physics," he said, because even Alpha Centauri is thousands of times farther than Pluto. "But if we were to find an Earth orbiting a nearby star, a little blue ball out there with water and all that; if we were able to tell human beings from the time they were born until the time they grow up, there's another Earth over there: it'd be like a carrot hanging out in front of the most intelligent minds. Who knows what we would do? Humans usually find a way to get to the carrot."

Alpha Centauri, 4.3 light-years from the Sun, is the most famous nearby star. It is actually triple, consisting of two bright Sunlike stars—one yellow, the other orange—that escort a faint red star through space. The red star is slightly closer to Earth and so bears the name Proxima

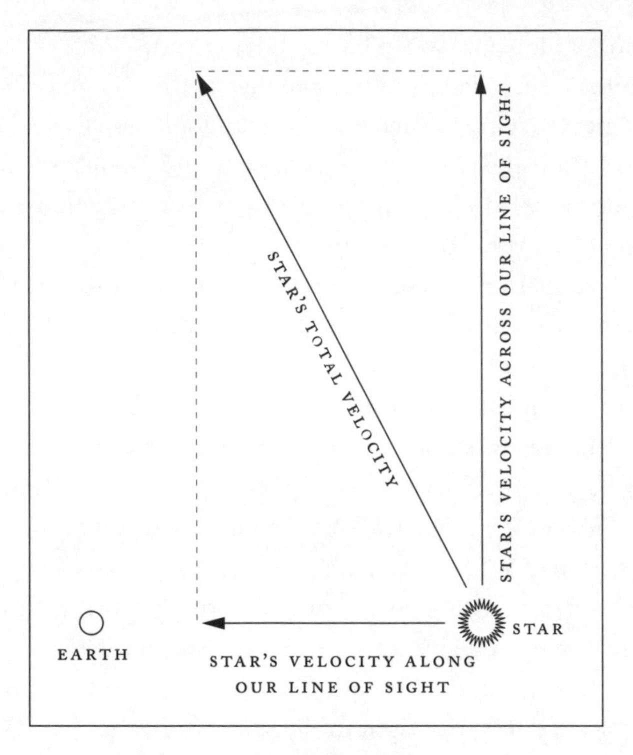

FIGURE 14

A star's velocity through space can be considered to be the sum of two velocities: one across our line of sight to the star and another along our line of sight to the star. A planet's gravity causes its star to wobble in both directions. Astrometrists look only for wobbles in the velocity across our line of sight.

Centauri. At twice the distance of the Alpha Centauri system lies Sirius, the brightest star in the night sky, and beyond that is another stellar luminary, Procyon. Both Sirius and Procyon light up winter skies in the northern hemisphere and summer skies in the southern, Sirius in Canis Major, the Big Dog, and Procyon in Canis Minor, the Little Dog. Two other nearby stars, Tau Ceti and Epsilon Eridani, are sufficiently like the Sun that radio astronomers in 1960 examined them first in the search for extraterrestrial intelligence. Sprinkled among these gems are numerous faint red stars like Proxima Centauri.

The bright star Sirius holds a special place for Gatewood, whose interest in astronomy began in Florida when he was five or six. "The first image that comes to my mind," he said, "is my mother taking me out under a palm tree and saying, 'See those stars? That's Orion. And this one over here is Sirius.'" Sirius is double, with the bright star circled by a faint one, and in the 1970s Gatewood and his wife studied both. He said, "I sent my mother a letter saying, 'Mom, continuing the conversation, we find that Sirius is this far away, it weighs this much, and it has a companion orbiting around it at this distance.'"

Back when he was in college, however, Gatewood told Heinrich Eichhorn, an astronomer then at the University of South Florida, that he wanted to study cosmology, which deals with the entire universe, rather than stars. "His response was extremely negative," recalled Gatewood. "He said there's only one universe, but there's a family of possible equations which might define the universe, so you'll never be able to decide which of this family of equations is correct—because you have only one example. That seemed pretty damned insightful to me! I thought about it for about a week and said, 'What would you do?' And he said, 'I'm an astrometrist. I like to study problems that I can answer during my lifetime.'"

Astrometry is the branch of astronomy in which observers measure the apparent positions of celestial objects. "Some astronomers look down their noses at astrometry," said Eichhorn, for they consider it dull and staid. "They regard it as: if you're not good enough to do anything else, then you become an astrometrist." Like Gatewood, Eichhorn discovered astronomy as a child. One evening in his native Austria, Eichhorn, his father, and his older brothers had missed the last homebound streetcar, which departed at midnight, so they had to walk home in the dark. "I looked up at the sky, and I asked them, 'What are those?'" said Eichhorn. "They said, 'Well, stupid boy, that's a stupid question; those are stars.' Then I asked what a star was, and they said, 'Well, don't you see what a star is? Look up there: that's what a star is.'"

Astrometry provides considerably better answers about the stars, especially those nearby. Without astrometry, astronomers would not know the distances of stars or their motions through space. Moreover, most stars near the Sun were discovered only through astrometry. This is because nearby stars rarely stand out from the crowd. Surprisingly, most stars that are bright are not nearby, and most stars that are nearby are not bright. Of the stars within a dozen light-years of the Sun, only three are real spectacles: Alpha Centauri, Sirius, and Procyon, which are among the ten brightest stars in the sky. Four other nearby stars glow only dimly to the naked eye: Epsilon Eridani, Epsilon Indi, 61 Cygni, and Tau Ceti. All the rest are invisible without optical aid, because they emit little light into space.

Each one of these faint suns had to fight to get its name on the list of nearby stars. For example, 61 Cygni, the first star other than the Sun whose distance astronomers ascertained, is barely visible to the naked eye in the constellation Cygnus the Swan. The star first drew attention in 1804, when Giuseppe Piazzi—the same sharp-eyed astronomer who discovered the first asteroid—noticed that the star moved from year to year with respect to the other stars. Astrometrists call this apparent movement proper motion. It occurs because every star travels through space on a different path from the Sun. Proper motion was discovered in 1718 by Edmond Halley, who found that three bright stars—Sirius, Arcturus, and Aldebaran—had shifted their positions since ancient times. The proper motion of 61 Cygni was far larger than any Halley saw, suggesting that this dim star was near the Sun. In the same way, a motorist sees a nearby road sign whiz by rapidly, whereas distant mountains appear to move lethargically.

Astronomers had long struggled to measure stellar distances, and three more decades elapsed before they succeeded with 61 Cygni. In 1838, Prussian astronomer Friedrich Wilhelm Bessel detected the star's parallax, a minute shift that resulted because he viewed the star from slightly different perspectives as the Earth moved around the Sun. The larger this shift, or parallax, the closer the star, just as walking across a road causes the apparent position of that nearby road sign to shift much more than the distant mountain. Soon after Bessel's success, other astronomers detected the parallax of two much brighter stars, Vega and Alpha Centauri. Whereas those stars had stood out because of their great brightness, most nearby stars resemble 61 Cygni: faint stars that came to notice only through their large proper motions, which spurred astronomers on to measure the stars' parallaxes. Since measuring proper motion and parallax is a job for the astrometrist, astrometry reveals which stars are near the Sun, the very stars most crucial to an astrometric planet hunt. But these same stars are important for other reasons, too.

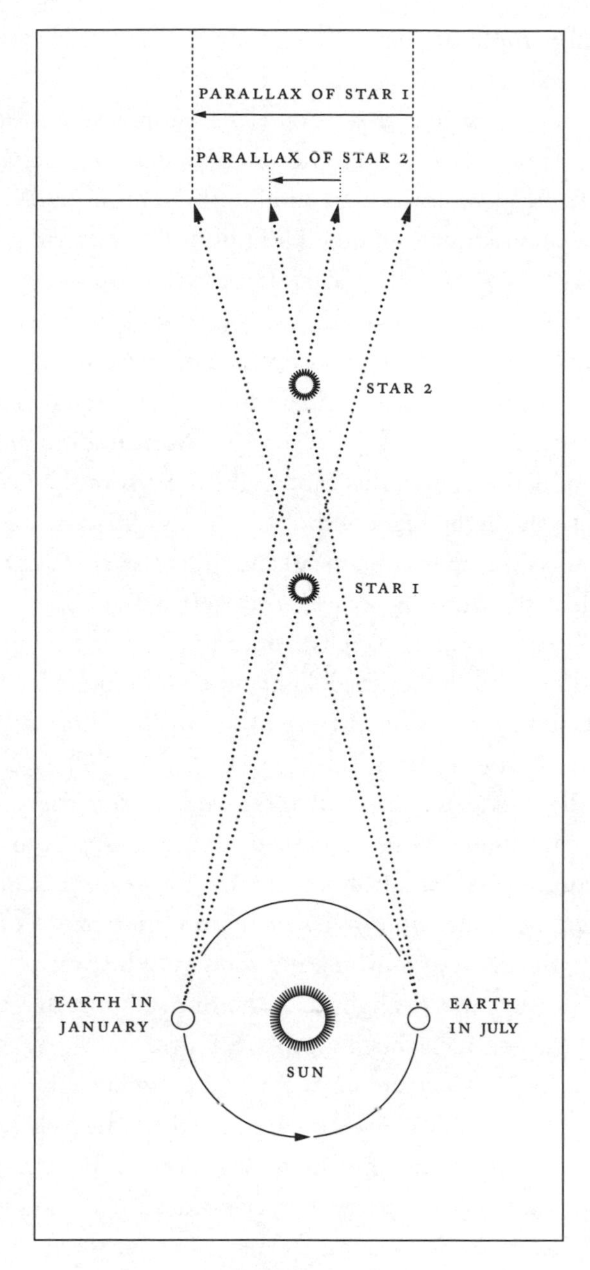

FIGURE 15

Parallax arises as the Earth circles the Sun and causes the apparent positions of stars to shift slightly. The closer the star is to the Sun, the greater this shift, or parallax, so the size of the parallax reveals the star's distance from the Sun.

A Colorful Sample of Stars

Aside from being targets of present planet searches and future starships, the nearest stars let astronomers sample stars in general and thus evaluate the prospects for planets and life throughout the Milky Way. The Galaxy abounds with hundreds of billions of stars and is far larger and brighter than most other galaxies in the universe. Every star visible to the naked eye is part of the Milky Way Galaxy, but for every star the naked eye can see, the Galaxy also contains millions of other stars. The most luminous stars nestle in a blazing spiral-shaped disk measuring some 130,000 light-years across. Surrounding this disk is a huge halo of dark material that outweighs the rest of the Galaxy.

Within twelve light-years of the Sun are 20 different star systems, including the Sun, that contain 30 individual stars. Within twenty light-years of the Sun, there lie about 120 individual stars; within forty light-years, about 1000 stars; and within a hundred light-years, about 15,000 stars. A hundred light-years is minuscule compared with the Galaxy's total size, but even this small volume of space has yet to reveal most of its stars.

This is the most basic reason that the very closest stars are important: they constitute a nearly complete stellar census from bright to faint. In contrast, when astronomers look to greater distances, they see only the more luminous stars. Incredibly, 99 percent of the stars visible to the naked eye outshine the Sun, yet 95 percent of all stars that exist actually emit *less* light than the Sun. It's just that these dim stars can't be seen unless they lie close to Earth.

Most nearby stars, including the Sun, are what astronomers call main-sequence stars. A main-sequence star fuses hydrogen into helium at its core, generating the energy that powers the star. The main sequence usually provides a lengthy period of time during which life might originate on an orbiting planet. The more mass a main-sequence star has, the faster the hydrogen fuses and the brighter and hotter the star burns. The most massive main-sequence stars shine some 10 billion times more intensely than the least massive. Furthermore, massive main-sequence stars are much hotter, which alters their color. To the untrained eye, all stars look white, but to a trained observer, the stars are a rainbow of delight. The hottest stars are blue

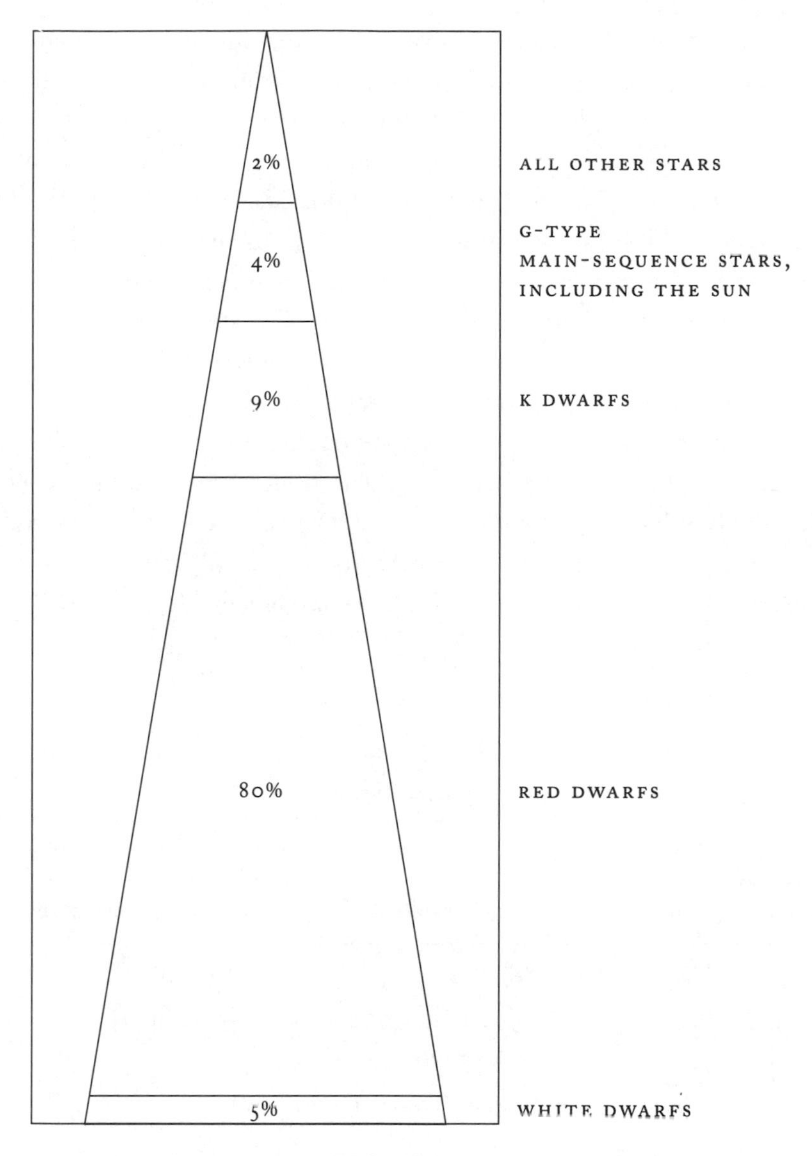

FIGURE 16

The stellar pyramid. Bright stars dominate the night sky, but in reality they make up only a small percentage of the Galaxy's stars. Most stars are red dwarfs, which are too dim for the naked eye to see.

and white; warm stars, like the Sun, are yellow; and the coolest stars are orange and red. For stars on the main sequence, all blue and white stars are very bright, yellow stars are bright, orange stars are modest, and red stars are dim.

A star's temperature affects its spectrum, the detailed wavelength-by-wavelength breakdown of its light, and astronomers classify stars into various spectral types. From hot and blue to cool and red, the seven main spectral types are O, B, A, F, G, K, and M, a line-up memorized via "Oh, Be A Fine Guy/Girl, Kiss Me!" or, especially for students, "Oh Boy, An F Grade Kills Me!" To be more precise, astronomers divide up each spectral type. For example, spectral class G, which includes the Sun, breaks into G0, G1, G2, and so on up to G9. A G0 star is slightly hotter than a G1 star, which is slightly hotter than a G2 star, and so on. The Sun is spectral type G2.

The hottest main-sequence stars are of spectral type O and shine bright blue, but these stars are so rare than none appears within many hundreds of light-years of the Sun. The stars are rare for two reasons: few are born, and they shine so brightly that they burn themselves out in just a few million years. Because intelligent life required 4.6

TABLE 5–1
Spectral Types

Spectral Type	Color	Main-Sequence Stars	Giants and Supergiants
O	Blue	Zeta Ophiuchi	Alnitak A
B	Blue	Regulus; Spica B; Achernar	Spica A; Rigel
A	White	Sirius A; Fomalhaut; Vega	Deneb
F	Yellow-white	Chi Draconis; Gamma Virginis	Canopus; Polaris
G	Yellow	Sun; Alpha Centauri A; Tau Ceti	Capella
K	Orange	Alpha Centauri B; Epsilon Eridani	Arcturus; Aldebaran
M	Red	Proxima Centauri; Barnard's Star	Betelgeuse; Antares; Mira

TABLE 5–2
Main-Sequence Stars

Spectral Type	Mass (Sun = 1)	Output of Visible Light (Sun = 1)	Approximate Lifetime (years)
O	16 to 100	4000 to 15,000	3 to 30 million
B	2.5 to 16	50 to 4000	30 to 400 million
A	1.6 to 2.5	8 to 50	400 million to 2 billion
F	1.1 to 1.6	1.8 to 8	2 to 8 billion
G	0.9 to 1.1	0.4 to 1.8	8 to 16 billion
K	0.6 to 0.9	0.02 to 0.4	16 to 80 billion
M	0.08 to 0.6	0.000001 to 0.02	80 billion to trillions

billion years to arise on Earth, such short-lived stars are hardly good prospects for life. B-type main-sequence stars are also hot and blue, but they are somewhat cooler and less massive than O stars. The nearest is Regulus, 74 light-years from Earth in the constellation Leo. Many other B-type stars decorate the sky, but they owe their prominence not to their numbers but to their intrinsic brilliance, since they emit so much light that they can be seen across vast gulfs of space.

The white stars of spectral type A also contribute greatly to the night sky, and one such star, Sirius, lies within twelve light-years of the Sun. The typical A-type main-sequence star has twice the Sun's mass and lives for around a billion years—probably not enough time for intelligent life to evolve, unless intelligence can arise much faster than it did on Earth. The first plausible life-supporting stars are the yellow-white main-sequence stars of spectral type F. They have somewhat more mass than the Sun and live for a few billion years, which may give enough time for intelligent life to arise on an orbiting planet. The nearest F-type star, Procyon, has already started to evolve off the main sequence slightly.

For seekers of extraterrestrial intelligence, the best objects are the

TABLE 5–3

The Nearest Stars

Star	Constellation	Distance (light-years)	Spectral Type	Magnitudes[a]		Visible Light Output (Sun = 1)
				Apparent Visual	Absolute Visual	
Sun	—	0.00	G2	–26.74	+4.83	1.00
Alpha Centauri A	Centaurus	4.35	G2	0.00	4.38	1.52
Alpha Centauri B		4.35	K1	+1.36	5.73	0.44
Alpha Centauri C		4.24	M5.5	11.09	15.52	0.000053
Barnard's Star	Ophiuchus	6.0	M4	9.55	13.23	0.00043
Wolf 359	Leo	7.8	M6	13.45	16.56	0.000020
Lalande 21185	Ursa Major	8.3	M2	7.47	10.45	0.0056
Sirius A	Canis Major	8.5	A1	–1.46	+1.45	22.5
Sirius B		8.5	WD	+8.4	11.3	0.0026
Luyten 726-8 A	Cetus	8.7	M5.5	12.57	15.43	0.00006
Luyten 726-8 B		8.7	M6	12.7	15.6	0.00005
Ross 154	Sagittarius	9.4	M3.5	10.47	13.16	0.00047
Ross 248	Andromeda	10.3	M5.5	12.29	14.79	0.00010
Epsilon Eridani	Eridanus	10.7	K2	3.73	6.15	0.30

Ross 128	Virgo	10.9	M4	11.12	13.49	0.00034
Luyten 789-6 A	Aquarius	11.3	M5	12.33[b]	14.64[b]	0.00012[b]
Luyten 789-6 B		11.3				
Luyten 789-6 C		11.3				
Epsilon Indi	Indus	11.3	K4.5	4.69	7.00	0.14
61 Cygni A	Cygnus	11.4	K5	5.21	7.50	0.086
61 Cygni B		11.4	K7	6.03	8.32	0.040
Procyon A	Canis Minor	11.4	F5	0.38	2.67	7.3
Procyon B		11.4	WD	10.7	13.0	0.0006
Struve 2398 A	Draco	11.4	M3	8.90	11.18	0.0029
Struve 2398 B		11.4	M3.5	9.68	11.96	0.0014
Groombridge 34 A	Andromeda	11.6	M1.5	8.08	10.33	0.0063
Groombridge 34 B		11.6	M3.5	11.07	13.32	0.00040
Lacaille 9352	Piscis Austrinus	11.7	M1.5	7.34	9.56	0.013
Tau Ceti	Cetus	11.8	G8	3.50	5.71	0.45
Giclas 51-15	Cancer	11.8	M6.5	14.79	16.99	0.000014

[a]Astronomers use magnitudes to express stellar brightness. The brighter the star, the smaller the magnitude. The apparent magnitude is how bright a star looks; in these terms, the Sun is obviously the brightest. The absolute magnitude is how bright a star really is: it is the apparent magnitude a star would have if it were 32.6 light-years away. Of the stars in this table, Sirius is the most luminous.

[b]Combined output of Luyten 789-6 A, B, and C.

yellow main-sequence stars of spectral type G, like the Sun. These are usually stable, and they live for billions of years. The Sun, for example, has a total main-sequence lifetime of 11 billion years. Sunlike stars make up only 4 percent of the stars in the Galaxy, but the Milky Way is so huge that it harbors billions of such stars. Alpha Centauri A, the brightest star in this triple system, is a G-type main-sequence star, as is Tau Ceti, the nearest single G-type star to the Sun and the first target in the search for radio signals from extraterrestrial intelligence.

Somewhat fainter, cooler, smaller, and less massive than the Sun are K-type main-sequence stars, which are sometimes called orange dwarfs. K dwarfs account for 9 percent of stars and are possible abodes of intelligent life, if they have planets. Orange dwarfs include such famous stars as Alpha Centauri B, the second brightest member of the Alpha Centauri system; 61 Cygni, the star whose distance Bessel announced in 1838; and Epsilon Eridani, the second target in the search for extraterrestrial intelligence.

By far the most common stars, though, are the humble red dwarfs, main-sequence stars with spectral type M. Faint, cool, and small, red dwarfs range in mass from 8 to 60 percent of the Sun's. Eight of every ten stars are red dwarfs, so they outnumber all other types put together. But red dwarfs are so dim that not a single one is visible to the naked eye.

Altogether, main-sequence stars make up 95 percent of nearby stars, with orange and, especially, red dwarfs accounting for most. Stars that are not on the main sequence do not much interest life seekers, but they help astronomers understand stellar evolution. When a star is very young, before it ignites its nuclear fuel, it glows by converting gravitational energy into heat. Few pre-main-sequence stars exist, however, because they quickly ignite the hydrogen at their cores and evolve into main-sequence stars.

When a main-sequence star uses up the hydrogen at its core, it expands, brightens, and cools. If the star was born with more than eight times the mass of the Sun, it evolves into a supergiant, such as Rigel and Betelgeuse. Supergiants are extremely rare, and the nearest is probably Antares, five hundred light-years from the Sun. When supergiants run out of fuel, they explode as supernovae. Even if planets and life managed

Only giant galaxies like the Milky Way have enough stars to create and retain high abundances of planet-forming—and life-forming—elements, such as carbon, nitrogen, oxygen, and iron. Our Galaxy is a spiral that resembles this one, named NGC 6946. (National Optical Astronomy Observatories)

New stars—and new planets—arise in clouds of interstellar gas and dust that line the Milky Way's spiral arms. This one is the Lagoon Nebula, which lies in the constellation Sagittarius. Each year, our Galaxy gives birth to about ten new stars and perhaps a hundred new planets. (National Optical Astronomy Observatories)

In 1995, astronomers took this dramatic close-up photograph of the Eagle Nebula, a star-forming region in the constellation Serpens. As ultraviolet light erodes the gas and dust, newborn stars emerge, some still connected to the main cloud by a cylinder of gas and dust, thus making the new stars resemble the flames at the ends of candles. (Jeff Hester and Paul Scowen, Arizona State University; NASA)

A massive star—one born with more than eight times the mass of the Sun—shines brightly but briefly: at the end of its luminous career, it explodes in a titanic supernova and shoots most of its remains into space. This supernova remnant, the Crab Nebula, harbors a pulsar, the dense remnant of the exploded star. It was around another pulsar, PSR B1257+12, that astronomers in 1991 discovered the first planets beyond our solar system. (National Optical Astronomy Observatories)

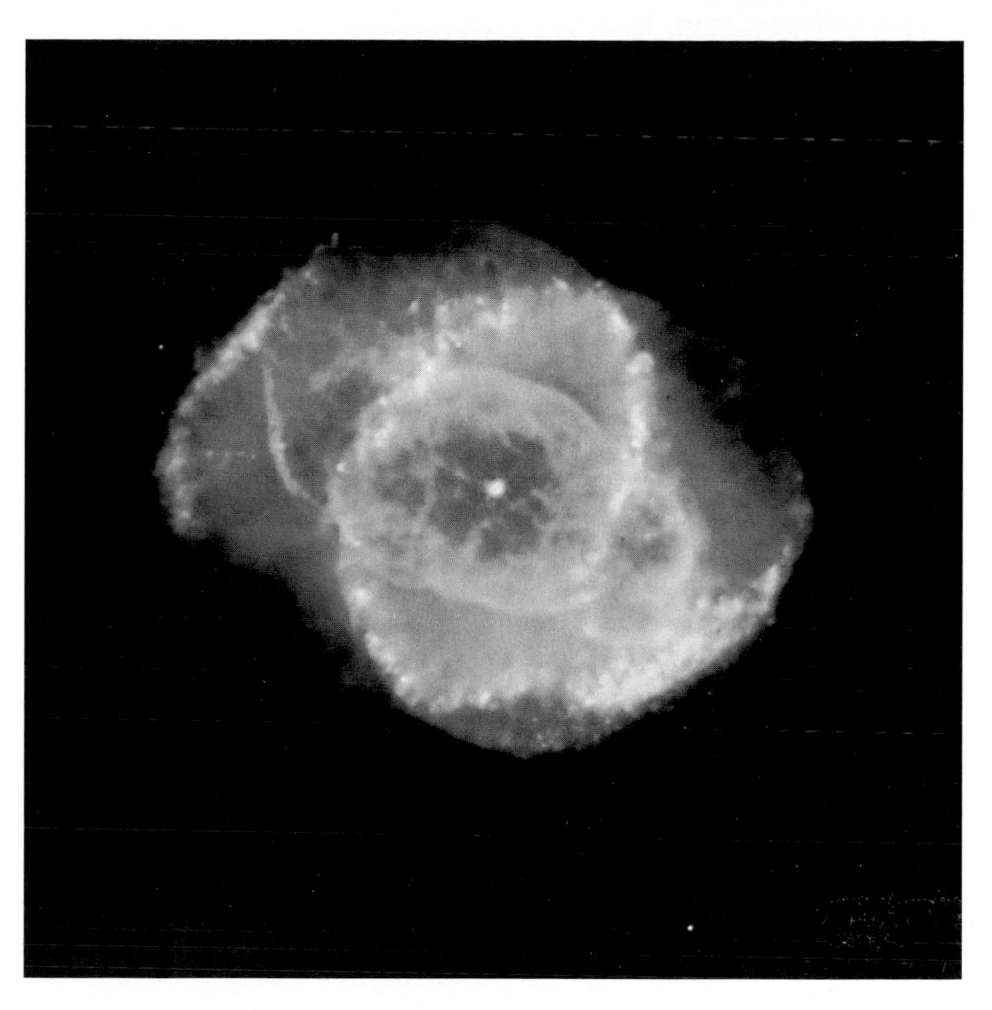

Most stars do not explode. Instead, they end their lives more gently, by casting off their atmospheres and forming what astronomers call planetary nebulae, such as the Cat's Eye Nebula in the constellation Draco. This fate awaits our own Sun, billions of years from now. (J. Patrick Harrington and Kazimierz J. Borkowski, University of Maryland; NASA)

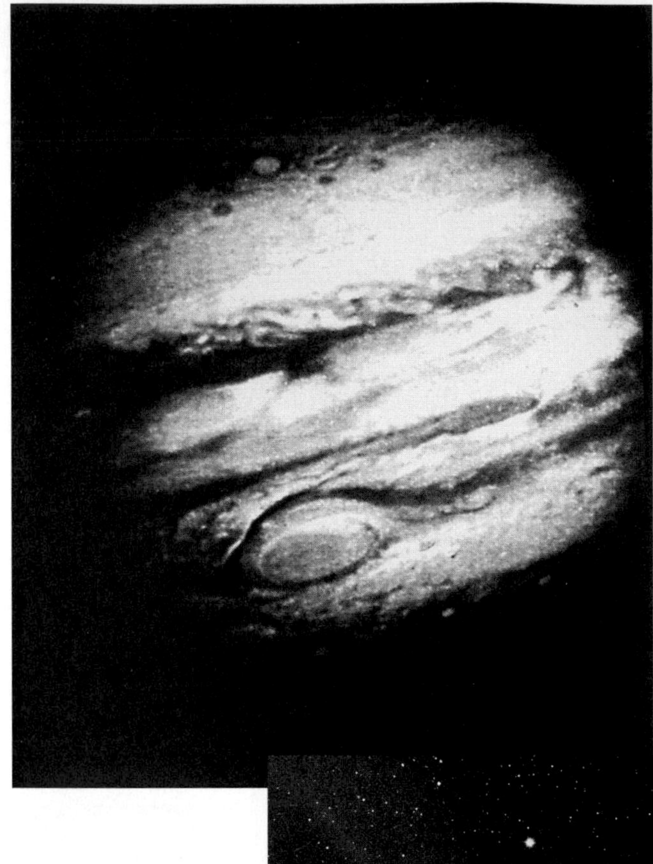

Jupiter (top) is the largest planet in the Sun's solar system, harboring over twice as much mass as all the other planets combined. Long ago, its gravity, and that of its neighbor Saturn, cast trillions of comets out of the solar system. As a consequence, deadly comets rarely struck Earth, and intelligent life could arise. One of the bright comets that did not get ejected was Comet Hyakutake (bottom), which passed Earth in 1996; Comet Hale-Bopp followed in 1997. (Jupiter photo from NASA; Hyakutake photo from Gerald Rhemann and Franz Kersche, Vienna)

In 1995, astronomers finally succeeded in discovering a planet around another yellow star like the Sun, 51 Pegasi; but the planet turned out to be what no one expected: a giant like Jupiter lying right next to its star. (© 1995 Lynette Cook)

The discovery of the first extrasolar planets ranks as one of the greatest scientific achievements of the twentieth century and implies that billions of planets exist throughout the Milky Way. So far, however, astronomers know of only one world that supports intelligent life. (NASA)

to form around such a short-lived star, the supernova explosion would fling the planets into space, where they would promptly freeze.

When stars that are born with less than eight times the mass of the Sun run out of hydrogen fuel at their centers, they evolve into giants, such as Arcturus and Aldebaran, rather than supergiants. Most giants are yellow (spectral type G), orange (K), or red (M). They too are poor prospects for life, for as they expand, they swallow some of their inner planets and can boil away the oceans on others. Though more common than supergiants, giants are still rare, accounting for less than 1 percent of all stars.

Unlike a supergiant, a giant does not explode. Instead, it gently casts off its outer atmosphere, forming a bubble of glowing gas known as a planetary nebula—not because it has anything to do with planets, but because through a small telescope it looks like a planet's disk. The planetary nebula expands into space and leaves the star's core behind. The core is small, hot, and dense—a dead star called a white dwarf. As its heat leaks away, a white dwarf cools and fades, so despite their name, white dwarfs come in various colors, with the youngest white dwarfs the bluest and the oldest the reddest. White dwarfs constitute 5 percent of all stars, and two white dwarfs lie less than a dozen light-years from the Earth, circling Sirius and Procyon. Like red dwarfs, though, white dwarfs are so faint that not a single one is visible to the naked eye.

As Sirius and Procyon illustrate, many stars are double and some, like Alpha Centauri, triple. The Sun is single, and single stars are the best candidates for having planets. In a double or triple system, one star's gravity can kick planets away from the other. Worse, the gravity of one star may prevent planets from forming around the other, just as mighty Jupiter disrupted the formation of a planet between itself and Mars, reducing that region to asteroidal rubble. For this reason, planet seekers prefer single stars, or else double stars in which either the two stars are very close together, so that planets can orbit both stars, or the two stars are very far apart, so that each star can have its own planets with no interference from the other.

Nevertheless, two double stars that lie near the Sun actually inspire the astrometrists who are on the front lines in the battle for finding new planets, because the double nature of both stars first came to light through astrometry.

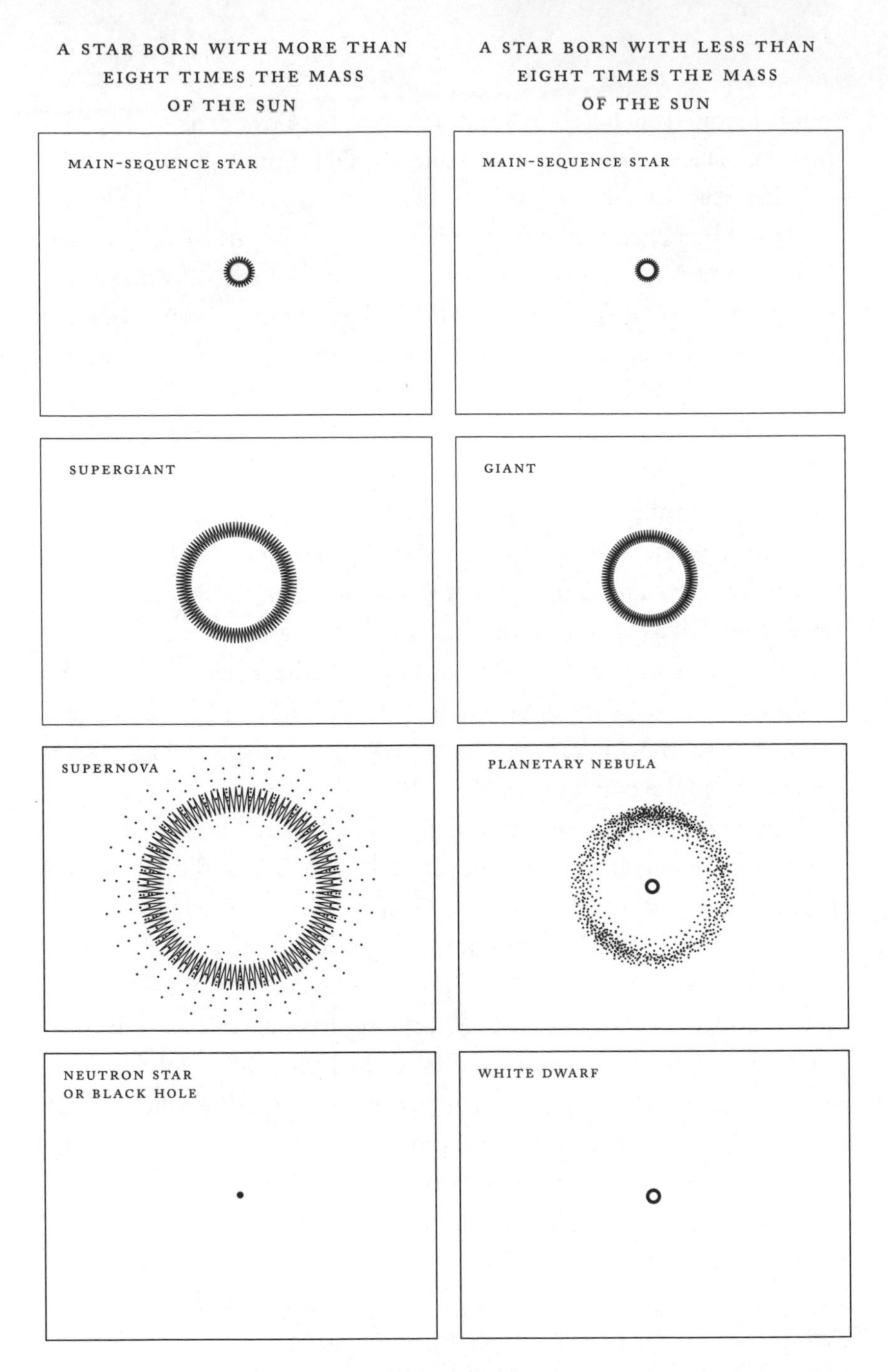

FIGURE 17

Stellar evolution. A star's life depends on the mass it was born with. If a star is born with more than eight times the mass of the Sun (left), it starts out as a bright blue main-sequence star, evolves into a supergiant, and then explodes as a supernova, leaving behind a neutron star or black hole. A star that is born with less than eight times the mass of the Sun (right) leaves the main sequence by evolving into a giant, rather than a supergiant, and casts off its outer atmosphere to form a planetary nebula, which leaves behind a white dwarf.

Planets by Astrometry

Sirius and Procyon were once thought to be single stars, but they earned their place in history when astrometry disclosed their unseen companions. "The subject which I wish to communicate to you, seems to me so important for the whole of practical astronomy, that I think it worthy of having your attention directed to it," wrote Friedrich Wilhelm Bessel, who had earlier measured 61 Cygni's distance, to John Herschel in 1844. "I find, namely, that existing observations entitle us without hesitation to affirm that the proper motions, of *Procyon* in declination, and of *Sirius* in right ascension, are not constant; but, on the contrary, that they have, since the year 1755, been very sensibly altered." Bessel proposed that each star had an orbiting companion whose gravity tugged on the main star but was too faint to be seen. He died two years later, before anyone proved him right, but in 1862 and 1896, respectively, astronomers did detect dim companion stars to Sirius and Procyon. Sirius B, as astronomers called Sirius's mate, was hundreds of times fainter than the Sun, and Procyon B was even dimmer.

If astrometry can reveal unseen stars, it might also reveal unseen planets as they too tug on their suns. The problem, though, is that planets' much smaller masses induce much smaller wobbles. After all, even the solar system's largest planet has only a thousandth the mass of the Sun. If astronomers viewed the Sun from a distance of ten light-years, they would see Jupiter cause the Sun to wobble by only 1.6 milliarcseconds (a milliarcsecond is 1/3,600,000 of a degree—the apparent thickness of a human hair seen from 2 miles). The other planets perturb the Sun less, because they have less mass, and small Earthlike planets would be almost impossible for astronomers using this method to detect. In other solar systems, much depends on the star itself, because the *less* massive it is, the *more* a planet kicks it around. Thus, a red dwarf with a Jupiter would exhibit a much larger wobble than the Sun; however, smaller stars may have smaller planets, which would eliminate the advantage.

More than mass affects the wobble's size, because the farther a planet is from the star, the greater is the astrometric wobble. For example, Uranus and Neptune have similar masses, but Neptune causes nearly twice as large an astrometric wobble. This is because a star and

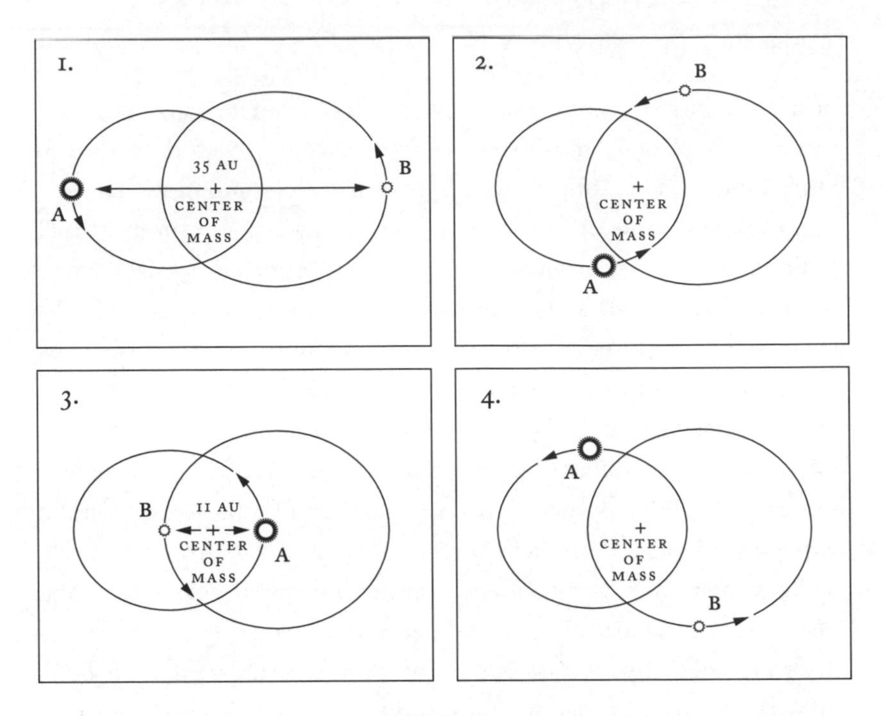

FIGURE 18

Two stars, or a planet and a star, orbit around their center of mass. These diagrams show how Alpha Centauri A and B orbit each other. Alpha Centauri A is more massive, so the center of mass lies closer to it, and A's orbit around the center of mass is smaller. As A and B revolve, the center of mass always lies exactly between them. At maximum separation, A and B lie 35 astronomical units apart (first panel), but at minimum separation, they are 11 astronomical units apart (third panel).

planet revolve around their center of mass, which is the point where one would place a fulcrum to balance a seesaw holding one object on one end and the other object on the opposite end. The center of mass therefore lies between the two objects and is closer to the one more massive. The center of mass in the Sun-Jupiter system is a thousand times closer to the Sun than to Jupiter, because the Sun is a thousand times more massive. It is the center of mass around which the Sun wobbles. If Jupiter were farther from the Sun, the center of mass would also lie farther from the Sun, making the Sun execute a larger and easier-to-detect astrometric wobble. Thus, to an astrometrist, the ideal planet is a massive one far from its star. This is

TABLE 5–4
The Wobbling Sun from Ten Light-Years

Planet	Distance from Sun (astronomical units)	Mass (Earth = 1)	Wobble (milliarcseconds)
Mercury	0.3871	0.055	0.00002
Venus	0.7233	0.815	0.00058
Earth	1.0000	1.000	0.00098
Mars	1.5237	0.107	0.00016
Jupiter	5.203	317.83	1.6
Saturn	9.54	95.16	0.89
Uranus	19.2	14.54	0.27
Neptune	30.1	17.15	0.51
Pluto	39.5	0.002	0.00008

good news, because in our solar system, at least, all the massive planets lie far from the Sun.

Astrometrists who search for extrasolar planets focus on the nearest stars, whose wobbles will appear largest and easiest to see—just as a top's wobble is much easier to see if viewed from a foot than from a mile. For example, an astrometric wobble will look ten times larger if a particular star and planet are ten times closer. In fact, the very first reports of astrometrically detected planets concerned two nearby stars, 70 Ophiuchi and 61 Cygni. In the early 1940s, both stars were reported to be wobbling. Their "planets" were about ten times more massive than Jupiter, so in modern parlance they would probably be considered not planets but instead brown dwarfs—stars much like red dwarfs but too small to ignite their nuclear fuel. A similar wobble led to the report of a similar object around the red dwarf Lalande 21185, the fourth nearest star system to the Sun. Then in the early 1960s

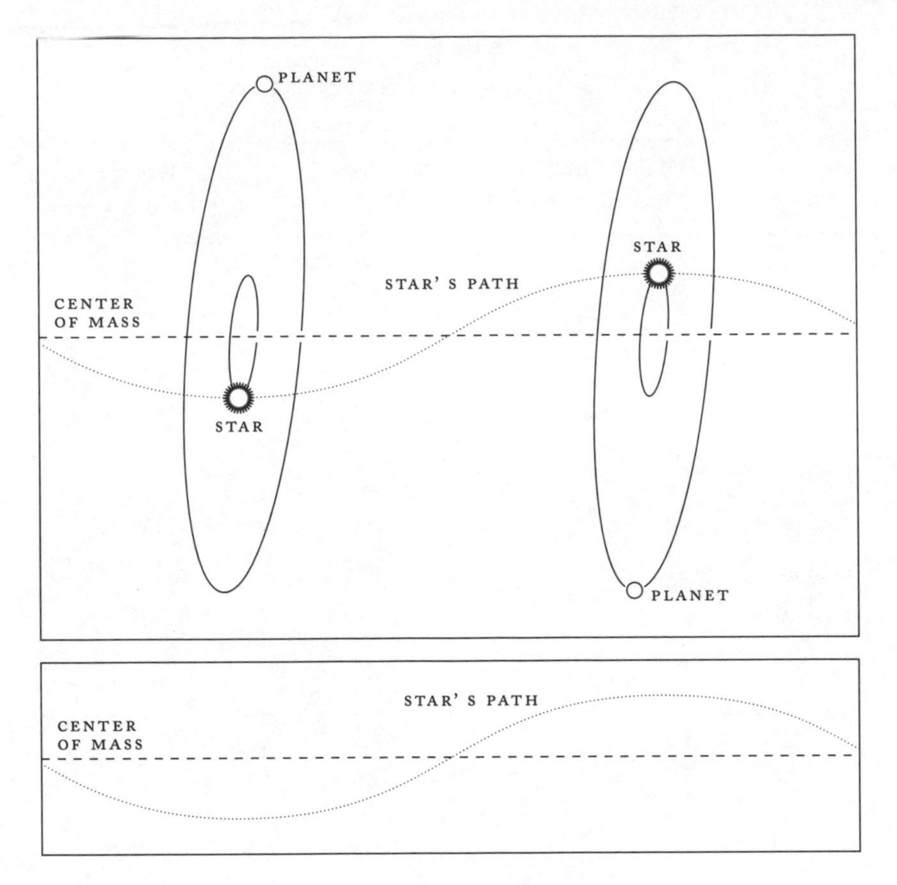

FIGURE 19

The center of mass of a star-planet system does not wobble; however, both the planet and the star revolve around it (top). Thus, astronomers can discover the planet if they observe the star's wobble around the center of mass (bottom).

came the most celebrated case of all, an apparent planet, similar to Jupiter, circling the infamous Barnard's Star.

The Saga of Barnard's Star

No other star is more associated with extrasolar planets than the one named for American astronomer Edward Emerson Barnard, and no other star better illustrates the pitfalls that planet hunters have run into time and time again. ("Barnard" is pronounced opposite from "Bernard"—the stress is on the first syllable.) Barnard himself had noth-

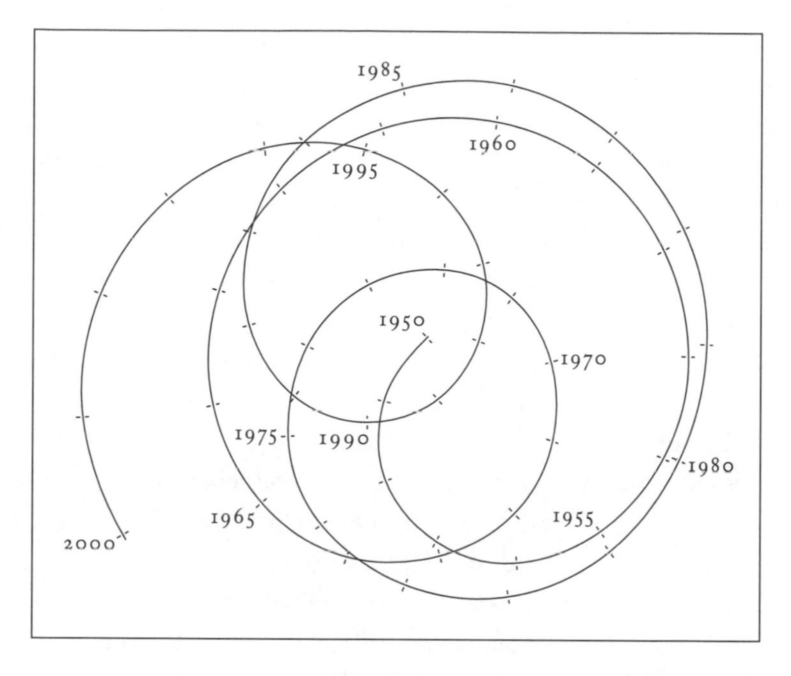

FIGURE 20

If viewed from afar, the Sun itself would wobble—primarily in response to the gravitational tugs of its two largest planets, Jupiter and Saturn.

ing to do with the false claims concerning its planets. Born in Tennessee just before the American Civil War and just after his father's death, Barnard faced a childhood of poverty and received almost no formal education. During the 1880s, he discovered several comets, which meant badly needed money, since at that time a comet earned its discoverer, if an American, two hundred dollars. Barnard's discoveries also caught the eye of professional astronomers, who offered him a job.

In 1916, while at Yerkes Observatory in Wisconsin, Barnard made one of his most important finds—not of a comet but of a peculiar star. "I have found on my photographs a small star of the 11th (photographic) magnitude which has a large proper-motion," he wrote. Barnard had photographed the constellation Ophiuchus and noticed that the star had moved from its position on an 1894 photograph. In fact, it was dashing across the sky faster than any other star ever seen, with a proper motion of 10.3 arcseconds per year—equivalent to a full lunar diameter every 180 years. The enormous proper motion suggested that Barnard's Star was nearby, and astrometrists rushed to measure its parallax. Their work revealed that Barnard's Star was a mere 6.0 light-years from the Sun, closer than any other star but Alpha Centauri.

Invisible to the naked eye, Barnard's Star is a red dwarf, frugally burning its fuel and casting a weak ruddy glow into space. It has only a sixth of the Sun's mass. The Sun emits more light in a month than Barnard's Star does during an entire century. Beyond Barnard's Star, in the same general direction, lie two of the brightest members of the solar neighborhood: the white A-type stars Altair, 16.6 light-years from the Sun, and Vega, 25.1 light-years from the Sun. Someday, starships traveling from the Sun to Altair and Vega may stop to refuel at Barnard's Star. As viewed from Barnard's Star, the spectacular constellation Orion would be even more so, for in addition to the bright red star Betelgeuse, the bright blue star Rigel, and the other dazzling stars that adorn this constellation, there would sparkle a brilliant yellow star: our Sun.

Despite its name, Barnard's Star really belonged to the late astrometrist Peter van de Kamp, who scrutinized it for half a century. "He was a very charming man and also an excellent lecturer," said Heinrich Eichhorn. "He was as good in the classroom as in the laboratory as at the telescope as in the salon. When he gave a talk, you could always look forward to it being not only scientifically solid but also entertaining."

Van de Kamp began observing Barnard's Star in 1938, from Swarthmore College's Sproul Observatory in Pennsylvania. Because Barnard's Star is small, single, and nearby, it is an ideal target for astrometrists who seek planets. By the early 1960s, van de Kamp had accumulated over two thousand photographic plates taken with Sproul's 24-inch telescope, and these indicated the star was not moving straight through space. Instead, its proper motion showed a wobble—a wobble "which admit[s] of no simpler explanation than that of a perturbation caused by an unseen companion of Barnard's star," wrote van de Kamp. From the size and period of the wobble, he concluded that the unseen planet was 60 percent more massive than Jupiter and revolved around the star once every twenty-four years. The orbit was wild, however, for the planet's distance from the star varied from 1.8 to 7.1 astronomical units. If it belonged to our solar system, this planet would have swung from the orbit of Mars to between the orbits of Jupiter and Saturn.

In 1966 van de Kamp visited Florida, where George Gatewood,

just out of college, first met him. "I'd read about the planet," said Gatewood. "I thought that was fabulous: somebody had actually detected a planet. I complimented him on this discovery, and it seemed pretty certain. I didn't see anything wrong with the data; I was just a kid, just a student." The wobble in Barnard's Star was large, 24.5 milliarcseconds—just what a giant planet would produce around a small star six light-years away.

"When I got back to the campus," said Gatewood, "one of the students said, 'Do you know that for all of the perturbations that van de Kamp has detected, you can divide all the periods by 8.3 years?' And that began to worry me a little bit! Why should all the planets in the universe have this pattern? And he was finding them around many stars. It turned out that he would do things to his telescope about once every 8 years, and in the long run we would discover that in fact was the problem."

By 1969, van de Kamp had gathered over three thousand photographic plates of Barnard's Star, and he proposed a new model, which he soon came to favor. Now, he said, there were two planets, similar to Jupiter and Saturn, that revolved around the star every twelve and twenty-six years. Van de Kamp later revised the orbital period of the outer planet to twenty years. These planets, presumably gas giants, had circular rather than elongated orbits and thus seemed more like the worlds of our own solar system. They lay closer to Barnard's Star than Jupiter and Saturn do to the Sun. If they belonged to our solar system, they would have been between the orbits of Mars and Jupiter. They took so long to revolve because Barnard's Star has little mass. Barnard's Star might also possess smaller planets, the size of Earth, that van de Kamp could not detect. His discovery of another planetary system so close to our own hinted that planets abounded throughout the Galaxy.

Gatewood began to investigate Barnard's Star in 1971, as a graduate student at Allegheny Observatory, but it was against his will. "I just didn't think Barnard's Star was all that important," said Gatewood, who wanted to work on another project. "And you know, I liked van de Kamp; he was a nice old guy." Nevertheless, under pressure from the Allegheny astronomers, Gatewood made Barnard's Star the subject of his Ph.D. dissertation, and he asked Eichhorn—who had earlier steered Gatewood into astrometry and supervised his master's thesis at the University of South Florida—to be his advisor.

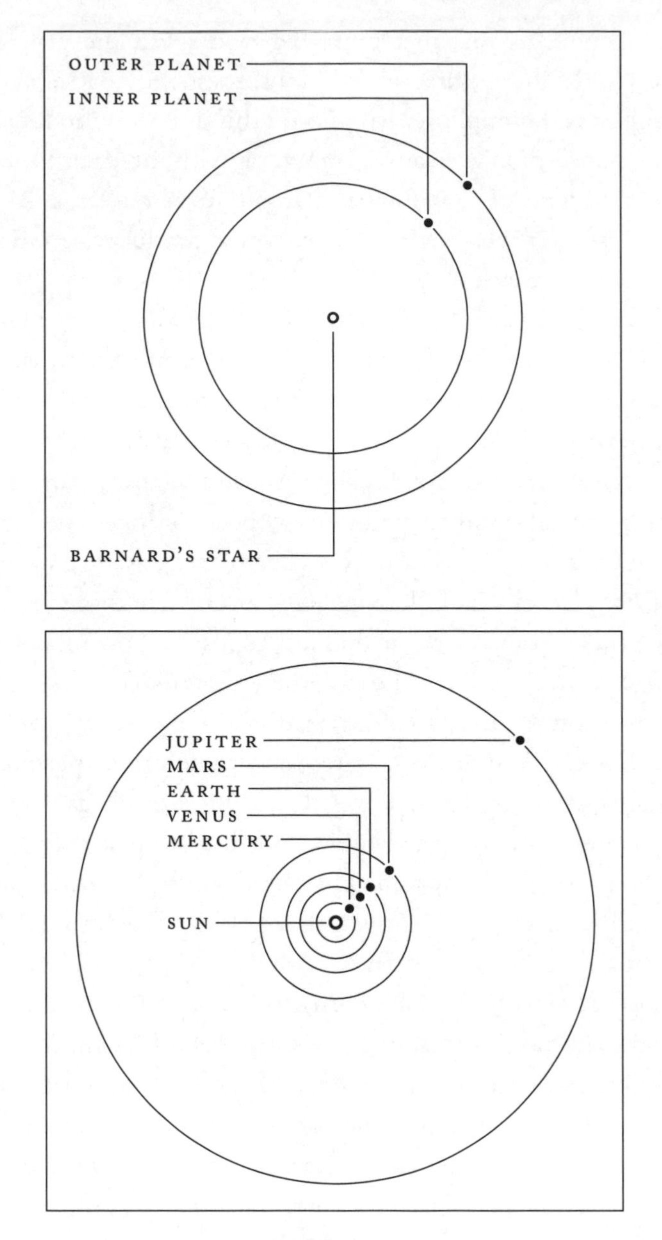

FIGURE 21

The planets envisioned by Peter van de Kamp around Barnard's Star had orbits larger than that of Mars and smaller than that of Jupiter.

Eichhorn thought the project a bad idea. Said Eichhorn, "I asked him, 'Who was the second person who flew solo across the Atlantic?' He said he didn't know. And I told him he'd be as well remembered as that person." That's because Eichhorn was certain that Gatewood would simply repeat van de Kamp's work. "I fully expected the planets to be confirmed," said Eichhorn. "I hadn't the slightest doubt. Look at van de Kamp's original papers: they're perfectly convincing that the effect is *there*."

What Gatewood knew, but did not tell Eichhorn, was that Allegheny Observatory's former director, Nicholas Wagman, had obtained disturbing evidence that van de Kamp's planets did not exist. Wagman measured crude positions of Barnard's Star on plates that had been taken with a different telescope, Allegheny's 30-inch refractor. Those positions showed no perturbation in the star's motion, which suggested that van de Kamp's wobble lay not in the star but in the telescope he used.

Strangely, Wagman never published the result. "I'm conjecturing here," said Eichhorn. "Wagman was a very competent but extremely shy man, who was more comfortable with the janitor than he was with other academics. So he probably thought that he was not going to stick his neck out and contradict one of the luminaries in the field—although Wagman was every bit as good as van de Kamp. But he was not as charming as van de Kamp, he was not as good a speaker, he was not as polished in the parlor; and he was probably afraid that people wouldn't take him seriously. I think that he thought: 'Let Eichhorn get stuck with it.'"

Eichhorn had had past conflicts with van de Kamp. "I'd been bucking van de Kamp on a couple of occasions," said Eichhorn. "For instance, he wanted me to come to Swarthmore and work for him for the typical starvation wages that he paid." Eichhorn had said no. "And then I started doing things differently from the way that he did," said Eichhorn. "From his standpoint, his way was not only the best way, it was the only way. He got a little bit sour on me and tried to sabotage what I was doing, sometimes subtly, sometimes not so subtly. For instance, he once went to the University of Virginia for a talk, and he referred to me as a space cadet, which didn't amuse me very much."

Eichhorn agreed to work with Gatewood on Barnard's Star, not because he expected to contradict van de Kamp, but because the star posed an interesting challenge. "I get my jollies from data reduction," said Eichhorn. "This may sound strange, but for me it is more interesting to squeeze the last drop of information out of a data set than to find out something about nature." In the end, Gatewood and Eichhorn would use only 241 photographic plates of Barnard's Star, whereas van de Kamp had over 3000. To Eichhorn, the challenge was to derive as precise a path for Barnard's Star from their small data set as van de Kamp had from his much larger one. Along the way, Eichhorn expected to confirm van de Kamp's planets, but as he warned Gatewood, that would hardly interest anyone.

The plates that Gatewood and Eichhorn used came from two different observatories, Allegheny in Pittsburgh and Van Vleck in Connecticut. To pinpoint the position of Barnard's Star on the plates, Gatewood used more reference stars than van de Kamp and employed a more accurate machine to measure stellar positions. In addition, Eichhorn had developed a more powerful technique to reduce the data. With these advantages, Gatewood and Eichhorn achieved a precision that actually matched van de Kamp's. Thus, if the planets of Barnard's Star really existed, Gatewood and Eichhorn should have found them.

For Gatewood, the project's culmination was an emotional roller coaster. "It was a strange experience; it all happened so quickly, within minutes," said Gatewood, who had gathered with other students to see the computer print out the paths of Barnard's Star and the stars near it. "We were watching this stuff come out of the printer, and we didn't know which one was Barnard's Star. And there was one star that had a perturbation of about 30 milliarcseconds. I was exhilarated. God, there it is! We found it! This is fantastic! And we're all gathered around looking at it, and we're talking about it, and then I noticed the star number." It was *not* Barnard's Star. Instead, it was just a double star whose companion was tugging it. "So then we dug around in the papers and said, 'Which one's Barnard's Star?'" said Gatewood. "We found it, and its path was straight. It was a real letdown." Gatewood and Eichhorn published the result in 1973: "Thus we conclude, with disappointment, that our observations fail to confirm the existence of a planetary companion to Barnard's star."

Eichhorn viewed the negative result philosophically: "I was born in Vienna, and there you say, 'What it weighs—that's what it weighs.' If van de Kamp had been very kind to me in the past, I probably would have called him first and said, 'Peter, this is what the numbers say, and I'm terribly sorry that's the way it has turned out.' Since it was rather the opposite of that, I didn't feel any need to apologize. But I was disappointed that the discovery of a planet had been set back." Eichhorn said other astronomers, especially van de Kamp's friends, were extremely critical. "Van de Kamp was a very entertaining person, and he was generally liked," said Eichhorn. "There were some people who told me frankly, 'Well, if it's between Peter and you, I choose to believe Peter.'"

Other astronomers took aim at the dim red star. Said Eichhorn, "Everybody and his mother-in-law started observing Van De Kamp's Star—I mean, Barnard's Star." Among them were two of van de Kamp's former students: Naval Observatory astronomer, and Planet X hunter, Robert Harrington, and Laurence Fredrick, of the University of Virginia. Seeing all the activity, Eichhorn suggested that the various astronomers pool their data to achieve a stronger and more precise result.

"They almost tore me to pieces alive," said Eichhorn. "Nobody was going to give *his* plates to somebody else and forego the chance of discovering a planet around another star." None of these studies ever confirmed van de Kamp's planets, and eventually even his former students gave up on them.

Van de Kamp, however, continued to believe in the unseen planets of Barnard's Star. A 1977 paper ended with a quotation from John 20:29: "Blessed are they that have not seen, and yet have believed." In 1983, he spoke about "my dear old friend, Barnard's Star" at a conference in Connecticut. "I know there are people who can't believe all this—and why not?" he asked. "Then why don't you start making observations for forty years because you need about forty years to separate the two orbits [of] 12 and 20 years. If this is real though, we have a situation similar to the Jupiter and Saturn situation, and we find this because Barnard's Star was close enough and because it pleased Barnard's Star to be a single star with planets."

Van de Kamp argued that the wobble he saw did not result from any problem with his telescope. "But of course, some other people felt it was a problem, because there were some other people who, well, were looking for trouble," he said in 1987. "I still hold out for the two-planet interpretation. That was based on over forty years of observations." He called Barnard's Star his favorite star of all. "I've devoted my time mostly to stars nearer than five parsecs [sixteen light-years]," said van de Kamp, "of which Barnard's Star was the prettiest one." But by then, most astronomers were not listening to what he had to say about it.

"I don't think he ever forgave me," said Eichhorn. "He didn't talk to me for about ten years. I think he blamed me for pricking his balloon." Eichhorn added, "I couldn't look into his innermost thoughts, but it's difficult for me to believe that toward the end of his career, he couldn't see the light himself. He saw that nobody confirmed his results. So I think he must have seen himself that the pelts were floating down the river away from him."

Van de Kamp died in 1995, and Gatewood hoped that the champion of the planets of Barnard's Star will be remembered for more. "He did so much else," said Gatewood. "He worked with the proper motions of stars; he worked with the luminosities of stars; he found a number of binary stars; and he wrote what is to this very day the hallmark of astrometric texts. Unfortunately, he put too much of his prestige on that one star." Said Eichhorn, "He really made so many improvements in parallax work that even without the planets of Barnard's Star, his name should be secure in the history of astronomy for being one of the most successful and diligent and careful parallax observers. He deserves that."

Barnard's Star was hardly the only place where purported planets vanished under the assault of better data. In the minds of most, astrometrists also vanquished the planets around Lalande 21185, Epsilon Eridani, 61 Cygni, and other nearby stars—until absolutely no extrasolar planets were left. By the 1980s, the Galaxy began to look considerably less inviting to those who sought planets and life around other stars.

Gatewood admitted that these findings were disappointing. "I think the important thing, however, is to try not to be too emotional about your science," said Gatewood. "For a long time, I managed that by saying that it would be just as profound an answer to find out there were no other planets as to find that there were. And even to this moment, I find it impressive that there are apparently stars that do not have planets—I find that fascinating."

Of course, those who long to explore new worlds may not be so fascinated. Barnard's Star had apparently been the best example of another planetary system; now those planets were gone, a result that knocked the wind out of planet hunters. But then, just as spirits were sagging, came a discovery that startled astronomers and made headlines around the world: the first genuine detection of solid material orbiting another main-sequence star like the Sun.

6

The Vega Phenomenon

Fifth brightest star in the night sky, Vega is a white-hot gem that sparkles brilliantly from a distance of 25.1 light-years and adds luster to hot summer evenings in the northern hemisphere. The Greeks and the Romans associated the star's beauty with music, making Vega part of the small but rich constellation Lyra the Lyre, a seven-stringed harplike instrument whose lovely sounds soothed the gods. In ancient China, Vega starred in a romantic story of love and longing. To the Chinese, Vega was really a heavenly princess who wove exquisite clothes and fell in love with a herdsman represented by another bright star, Altair. Although Vega and Altair appear close together on the sky, running between them is the glowing band of light called the Milky Way, which the Chinese saw as a great river. After meeting each other, Vega and Altair began spending all their time together, which caused them to neglect their heavenly tasks. This angered the Sun god, who separated the two by placing them once again on opposite sides of the Milky Way, where they remain to this day. But on one night of every year—the seventh night of the seventh moon, when Vega and Altair shine high in the sky—a bridge of birds spans the Milky Way's waters, allowing the two young lovers one night together.

Today, millennia later, astronomers know that Vega is indeed a special star, worthy of such lavish tales. Of the four hundred stars lying within thirty light-years of the Sun, Vega is by far the most powerful, casting out over twice as much visible light as the next most luminous star, Sirius, and forty-nine times more light than the Sun. Like the Sun, Vega is a main-sequence star, converting hydrogen nuclei into helium nuclei at its core. The star has two and a half times more mass than the Sun, which explains Vega's greater luminosity, and the star also shines hotter, having a spectral type of A0. A star with Vega's mass remains on the main sequence for only 400 million years, so Vega must be much younger than the 4.6-billion-year-old Sun.

Astronomers who observe optical wavelengths—those visible to the human eye—have long used Vega as a standard reference, comparing other stars with it to ascertain their apparent brightness. In 1850, astronomers at Harvard College Observatory selected Vega to be the first nighttime star ever photographed. Vega's greatest moment, however, came much more recently. In 1983, a satellite observing the sky at a new wavelength stunned the world by revealing, for the first time, solid material in orbit around a main-sequence star other than the Sun.

Vega in a New Light

The invisible matter circling Vega gave its presence away by emitting infrared wavelengths, which are too long for the human eye to see. Infrared radiation was discovered by the same astronomer who discovered the first planet beyond the Earth. In 1800, two decades after spotting Uranus, English astronomer William Herschel used a prism to split sunlight into a rainbow of colors, from the shortest wavelengths visible, purple and blue, to longer wavelengths, green and yellow, up to the longest wavelengths visible, orange and red. Herschel placed thermometers in each color and found that they sensed heat even at the dark wavelengths longer than red.

These infrared wavelengths are everywhere, because all objects that have some warmth, including stars, planets, and people, radiate infrared energy. As Herschel discovered, the Sun does so, but because it is so hot it shines even more intensely at shorter wavelengths, in the visible part of the spectrum, where astronomers usually study it. In contrast, much cooler objects, such as planets and people, radiate no visible light but glow brightly in the infrared. During the day, the Earth receives energy from the Sun, but day and night, the Earth emits this warmth into

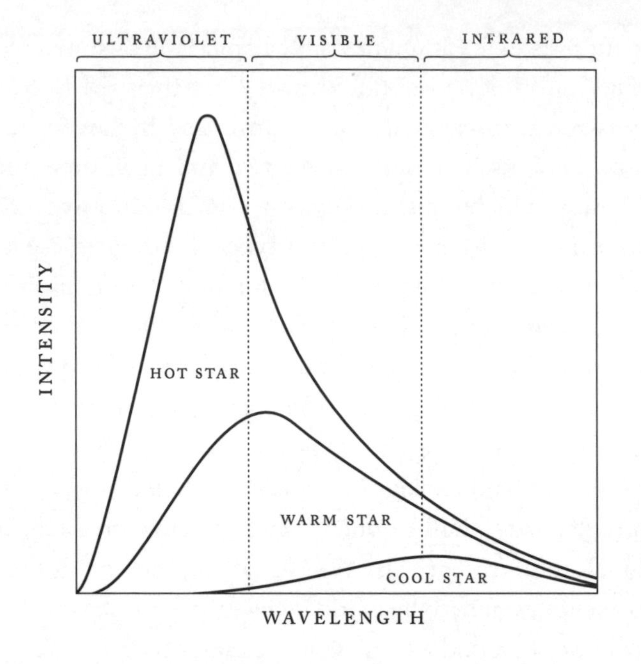

FIGURE 22

A hot star is brightest at ultraviolet wavelengths, a warm star, like the Sun, at visible wavelengths, and a cool star at infrared wavelengths.

space via infrared radiation. The infrared part of the spectrum is therefore the ideal place to investigate cool objects throughout the universe. Unfortunately, infrared astronomers must fight the Earth's atmosphere, whose water vapor, carbon dioxide, and other gases absorb infrared radiation. By no coincidence, these are exactly the same gases that cause the greenhouse effect. As the Earth tries to radiate its heat into space, greenhouse gases absorb this infrared radiation and bounce some of it back to Earth. Thus, the same gases that allow life to exist by warming the Earth also prevent astronomers from viewing the wavelengths that might reveal the possible abodes of alien life.

In January 1983, NASA launched a satellite above the atmosphere to capture infrared radiation from space. The Infrared Astronomical Satellite, or IRAS, was a joint project of the United States, England, and Holland. The satellite carried a container of frigid liquid helium to cool itself to a temperature near absolute zero; otherwise, the satellite's own warmth would have emitted infrared radiation and interfered with the celestial radiation that astronomers wanted to see. The liquid helium ran out ten months later, but by then IRAS had detected over 200,000 infrared sources. One of them was Vega.

At the start of the mission, though, no one much cared about Vega, a hot star whose infrared emission should have been relatively weak. In order to calibrate the telescope, mission scientists pointed IRAS at several cool stars whose output at infrared wavelengths was expected to be strong and well behaved. That way, when IRAS observed other objects, the scientists would know exactly how bright those objects were, by comparing them with these stars. The stars were called radiometric calibration objects.

Because it was so hot, Vega was *not* one of the chosen stars. "We kind of had to sneak in Vega," said Hartmut (George) Aumann of the Jet Propulsion Laboratory in Pasadena, California, "because it was not an approved radiometric calibration object. The radiometric calibration objects were strictly K stars and asteroids." Orange stars of spectral type K are much cooler than Vega and emit profusely in the infrared, so IRAS observed K-type stars such as Arcturus and Aldebaran to calibrate its sensitivity. For the same reason, IRAS took aim at some of the solar system's asteroids, which are also cool and radiate at infrared wavelengths.

Aumann wanted to include Vega, too. "I'm not sure it was a hunch," he said, "but basically I thought, given that other people use it for calibration, why not? Vega was certainly a famous star, and I thought using it for something was going to be useful." In short, Aumann was paying tribute to the star's long history as an optical standard. Vega therefore figured prominently in another calibration test, which aligned the satellite's optical and infrared sensors.

"We certainly didn't expect Vega to be unusual," said Fred Gillett of Kitt Peak National Observatory in Tucson, Arizona. "There was no indication from any previous work on this star that there was anything unusual, and certainly no expectation that there was going to be anything funny at the longer wavelengths."

The satellite observed the sky at four different infrared wavelengths: 12 microns, 25 microns, 60 microns, and 100 microns (a micron is a thousandth of a millimeter and somewhat longer than the longest wavelength of light visible to the human eye). Before IRAS, astronomers already knew that Vega behaved normally at 10 microns, one of the few infrared wavelengths that penetrate the Earth's atmosphere. Vega's peculiar nature came to light at the longer infrared wavelengths, where the IRAS astronomers compared the star with one of the orange standards, Gamma Draconis, the brightest star in the northern constellation Draco the Dragon.

"We expected that Vega wouldn't be as bright as Gamma Draconis at 60 microns," said Aumann. "Fred Gillett was the guy who looked first at the data, and he said, 'Wow! Vega is just as bright as Gamma Draconis.' Then I remember he said, 'Will that cause some headaches.'" Said Gillett, "There was something very unusual going on. You could pick that right off the initial data: Vega was nearly ten times brighter at 60 microns than we expected." The discrepancy was even worse at the longest IRAS wavelength, 100 microns, where the satellite found Vega to be twenty times brighter than anticipated.

Rather than celebrate a major new discovery, however, the scientists worried that IRAS was malfunctioning. The cool K stars had indicated that all was well, because their infrared brightness matched

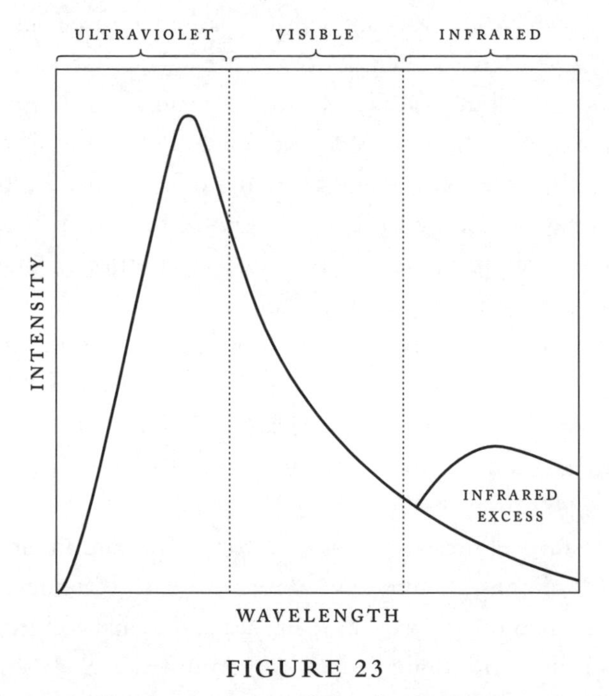

FIGURE 23

An infrared excess can signal dust that is orbiting a star.

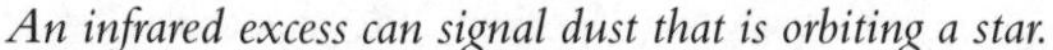

the predictions. But Vega was a hotter star, sending out copious amounts of white and blue light. Perhaps IRAS's detectors were mistaking this visible radiation for infrared. The scientists therefore told IRAS to observe other hot stars. These did not exhibit the same behavior, though, which suggested that Vega's infrared excess was real. Indeed, as the scientists checked possible problems and eliminated them one by one, the scientists grew more and more convinced that the discovery was genuine.

"I woke up very early one morning," recalled Aumann, "and said to myself, 'Why don't I calculate the temperature of whatever that thing is?'" He found that Vega's radiation was coming from two different sources: the star proper, which was sending out all the brilliant white-hot light and a bit of infrared; and something else, which was producing most of the infrared radiation at wavelengths of 60 and 100 microns. Aumann estimated the "something else" to be about -300 degrees Fahrenheit, similar to the temperature of Saturn. "So I thought, 'Well, that's interesting,' and I went to bed again," said Aumann. "Then the next morning, I woke up early again, and I said, 'Maybe it's some kind of an asteroid orbiting Vega.'" That was a natural thought, since IRAS had been observing asteroids. "It was very exciting, because I still remember the details," he said. "I woke up at three o'clock in the morning and I said, 'Hey, why don't I plug those numbers into the asteroid model?'" His computer program already calculated the infrared brightness of asteroids in the solar system, and all he had to do was change the program so that the star illuminating the asteroid was Vega rather than the Sun. Using the program, he could then calculate the properties of the hypothetical asteroid orbiting Vega. He found that the wavelength-by-wavelength fit to the IRAS data was perfect, but there was a major problem: in order to emit all the infrared radiation that IRAS saw, the asteroid had to be far bigger than Vega itself. That was impossible.

Nevertheless, Aumann was on to something, because he could use the object's temperature to compute how far it must be from Vega. That distance worked out to about 80 astronomical units, or twice Pluto's mean distance from the Sun. Vega is so bright that its light can keep such a distant object the same temperature as Saturn. But that hardly clarified the mysterious object's exact nature, so the scientists ordered IRAS to observe Vega more carefully. This time they saw that the infrared emission came not from the star itself but from a region that was centered on the star and extended about 80 astronomical units from it. It was certainly not a single asteroid, or even a belt of asteroids, for they would not emit enough radiation. Instead, the radiation must be coming from a ring of dust orbiting the star. The dust grains get heated by the star's light and then radiate this heat away in the infrared, where IRAS detected it. Dust could explain all of Vega's infrared excess, because a given amount of mass ground up into dust emits far more infrared radiation than one solid object having the same mass.

Dust in space is hardly newsworthy, however. Interstellar dust lines the Galaxy's disk and causes distant stars to look fainter and redder than they really are, by absorbing and scattering away their blue

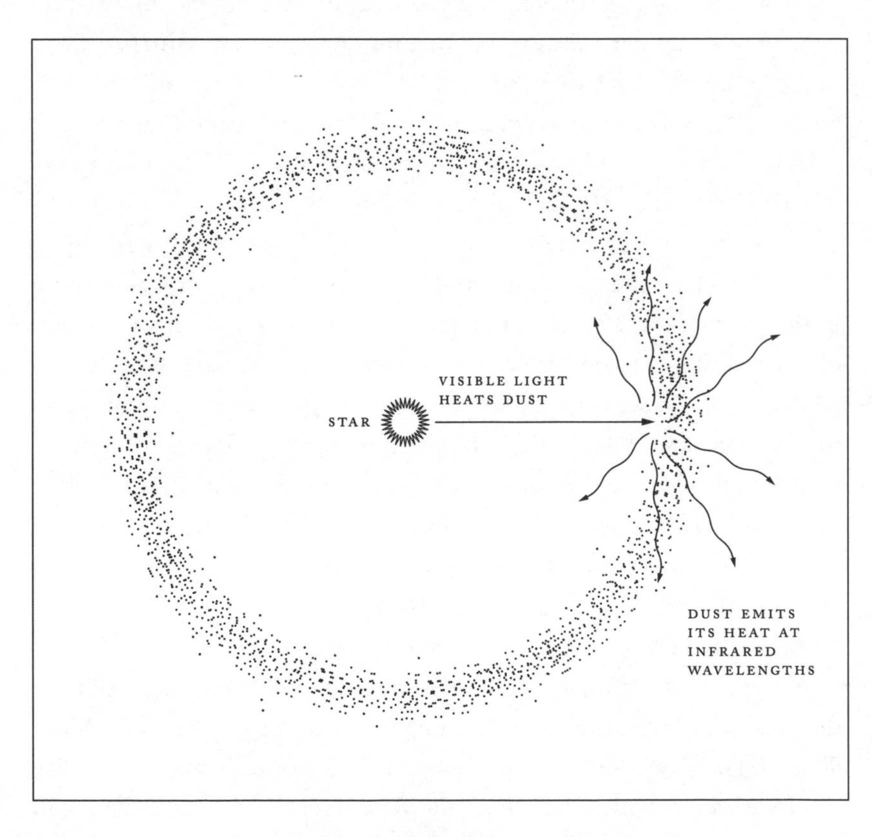

FIGURE 24

Circumstellar dust gets heated by a star's visible light and emits that heat in the infrared.

light. In the same way, the Sun looks redder at sunset, when its light passes through larger amounts of gas and dust in the atmosphere. Some interstellar clouds of gas and dust give birth to new stars and, presumably, new planets.

Still, the dust around Vega *was* special. "That's where we really got excited," said Gillett. "This was a totally different phenomenon. These dust grains had to be much bigger than interstellar dust. Once we understood that perspective, it got to be very, very interesting very quickly." Ordinary interstellar dust is microscopic, for the grains are less than a micron across. In contrast, the dust grains around Vega were over a thousand times larger, big enough for the eye to see. Said Gillett, "The fact that they were large particles suggested some sort of accumulation process, similar to that which we think must have occurred early in the formation of the planets." According to theories of the solar system's origin, dust grains around the newborn Sun stuck together and grew larger, until they resembled those around Vega. These grew larger still, to form asteroids that collided to create the planets. As the astronomers would later write, "These results provide the first direct evidence outside of the solar system for the growth of large particles from the residual of the prenatal cloud of gas and dust."

The astronomers knew that Vega's dust grains were abnormally large by considering how the grains radiate their heat. All objects radiate heat at a variety of wavelengths. For example, the Sun emits a literal rainbow of visible colors as well as large amounts of infrared and ultraviolet radiation. A grain of dust also tries to radiate its heat at different wavelengths, but it cannot easily radiate at wavelengths that are longer than it is. Thus, microscopic grains do not radiate their heat well, so they stay hot. The dust grains around Vega, though, were as cool as Saturn, implying that they radiated their heat efficiently and so were large. The large grains suggested one of two things: they had conglomerated from smaller grains, which meant that still larger objects—like asteroids, comets, and even planets—might have conglomerated from them; or the dust grains were themselves debris created when asteroids and comets smashed into one another around Vega. Either way, the large size of the dust grains hinted that Vega possessed objects even larger—possibly an entire solar system like our own.

"Nobody was supposed to talk about it," said Aumann, "and everybody pressed me: 'Well, what news did you find?' And all I said was that not all calibration stars are as trustworthy as one thinks." The scientists finally announced their discovery in August 1983, when it provoked headlines around the world and excited the public imagination more than any other result from IRAS.

"I was very surprised that there was that much interest in this obscure observation that we'd been carrying out with this funny little satellite," said Gillett. "I didn't think it was all that big a deal. I guess I had not taken that leap to the next level, of planetary systems and other civilizations." He thought the media did a decent job reporting the discovery. "I don't think they were wildly out of line," he said. "They always desire to build on the possibilities—planets like the Earth and other beings out there and things like that, which is certainly going way beyond what these observations were telling us. But you can't blame them for that. I'm happy that there's that much interest in the kind of science we do."

The interest in Vega's dust ring did not surprise Aumann, however. "There's an almost religious fascination by the press and by all people to look for planets," he said, "and if there's any trace of planets, they make a big deal of it—even if we made it crystal-clear in the paper that it definitely wasn't another planet." Ironically, if Vega does have planets, they would never have shown up in the IRAS data, because the infrared emission from the dust was so great that it would have swamped the radiation from all of Vega's planets put together. As the scientists wrote, "The mass of the shell could therefore be much larger than the mass of the asteroid belt and may be comparable to the mass of the solar system excluding the Sun." In other words, precisely because planets emit less infrared radiation than does dust, the IRAS observations could not rule out their existence.

Furthermore, those same observations bore an especially intriguing feature. The satellite detected only relatively cool dust, which was far away from Vega, but no dust near the star, which would have been quite hot. Something therefore removed the dust from the immediate vicinity of the star. Perhaps, early in the star's life, Vega's own radiation pushed the dust away. But a much more exciting possibility also exists: perhaps Vega has planets that swept the inner part of its solar system clean. Because Vega is only a few hundred million years old, though, any such planets probably lack intelligent life, unless intelligence has evolved much faster around Vega than it did around the Sun.

After they discovered Vega's dust ring, the scientists searched for other stars like it. "That was a very hectic time," said Aumann. "We needed to milk everything out of the satellite before the helium ran out, and we tried our darnedest to make enough special experiments on anything that looked interesting." In November, IRAS finally exhausted the liquid helium that had kept it cool. By then, though, the satellite had detected three other stars resembling Vega. Two were white and spectral type A, like Vega. One was famous Fomalhaut, the brightest star in the constellation Piscis Austrinus. Fomalhaut lies twenty-two light-years away, a bit closer than Vega. It is the third most luminous star within thirty light-years of the Sun, after Vega and Sirius. The other A-type star was the much fainter Beta Pictoris, which lies about twice as far as Vega and belongs to a southern constellation. The final star around which IRAS detected dust was the well-known nearby star Epsilon Eridani, the orange dwarf eleven light-years from the Sun that in 1960 was searched for extraterrestrial intelligence.

In the years since, Aumann and others have analyzed the IRAS data further and found that many nearby stars have infrared excesses. The most prominent are the A-type stars, and three of the four original IRAS stars—Vega, Fomalhaut, and Beta Pictoris—are also spectral type A. But neither Aumann nor Gillett thinks A-type stars are particularly unusual. It's just that these stars vastly outshine cooler main-sequence stars, so they heat up their dust rings the most—just as a bright white Christmas light illuminates nearby branches far better than does a dim red bulb. In short, even though IRAS did not detect it, most stars may have as much orbiting dust as Vega.

"We would walk right past our own Sun," said Aumann. "All the material put together in our solar system would give you a signal in the infrared which is just 1 percent bigger than what you expect the G star itself to produce." That, he said, would render the Sun's infrared excess undetectable to any Vegan astronomers who launched their own IRAS-like satellite. "I sat through hundreds of meetings trying to capitalize on this—how to design fancy equipment that would be optimized to search for other planets," said Aumann. "Once you ask yourself if that equipment would detect our own planets, it gets very frustrating. Virtually all of them would discover Jupiter-sized objects, but they've got to be the only object in the system. As soon as you have more of them, then things get really messy. Or if you have an Oort cloud of comets, you can't see it; or if you have an asteroid belt, it messes it up. It's almost as if nature is conspiring to hide the planets in a cocoon of dust and debris to make people work harder." Indeed, if alien astronomers did succeed in detecting excess infrared radiation from the Sun, most of it would come not from its planets but from its dust.

Nevertheless, IRAS did find dust around the Sunlike star Epsilon Eridani. "That's one of the more intriguing stars," said Gillett. "It's much more of a solarlike star than the others. It's very nearby, so it's by far the closest of the strong examples of the Vega phenomenon." In fact, had the star not been so close—if it had been as far as Vega—IRAS might not have detected its excess at all, because the star is much fainter and cooler than Vega and heats its dust less.

In the end, the most crucial of the four IRAS stars turned out to be the most obscure: Beta Pictoris. Whereas Vega and Fomalhaut were bright and famous, and Epsilon Eridani had a colorful history, almost no one had ever heard of Beta Pictoris. Yet this star would soon reveal, through a direct optical photograph, a disk of orbiting dust—and a possible solar system beyond our own.

Beta Pictoris: A Lust for Dust

Beta Pictoris is the most distant of the four IRAS stars, lying about fifty light-years from the Sun. To Bradford Smith, a planetary scientist then at the University of Arizona, the IRAS results themselves seemed light-years away—even though some of the team members belonged to his own department.

"I have worked on space projects going back to early Mariner days," said Smith, "and I never ran up against a group of scientists like them. The only people who in modern days would match them are the Hipparcos people, who just resisted giving out any information at all on their stars." The Hipparcos satellite measured parallaxes and proper motions of stars, data that were released only in 1997, four years after the mission ended. "Most astronomy departments are pretty leaky places," said Smith. "It's pretty hard to keep information like that quiet. Hell, I was in the same department as these guys, and I still couldn't find out anything about what IRAS was seeing—just these tantalizing rumors."

In mid-1983, as word of the discoveries began filtering out, Smith was well poised to take advantage of them. He owed his interest in planets to Pluto discoverer Clyde Tombaugh, whom Smith had worked with in the 1950s. Smith loved astronomy but thought job prospects were poor, so he had majored in chemical engineering instead. It was Tombaugh who converted Smith into an astronomer, and specifically a planetary astronomer.

To observe the planets, Smith used a device called a coronagraph, which blocks the glare of a planet or star so that the viewer can see faint objects nearby. With the coronagraph, Smith searched for uncharted satellites orbiting the outer planets of the solar system. In this way, he and his colleagues had found two new moons around Saturn. By blocking the light of a star like Vega, the instrument might reveal the surrounding ring of dust. Smith therefore took aim at the three IRAS stars easily visible from Arizona—Vega, Fomalhaut, and Epsilon Eridani—but he struck out every time.

That left only the southern star Beta Pictoris, and Smith didn't expect much. "I thought that Vega was probably one of our best shots at it," he said, "and after not finding anything around Vega, I was certainly pessimistic about Beta Pictoris." Beta Pictoris itself hardly inspired much hope. It is a dim star in a dim constellation, Pictor the Painter's Easel, which is ignored even by stargazers in the southern hemisphere.

In April 1984, Smith and Richard Terrile, of the Jet Propulsion Laboratory, were in Chile, observing Uranus and Neptune. Smith wanted to give Beta Pictoris a chance, so the two of them used the coronagraph to study it. Once again he saw nothing—at least, at the telescope. Beta Pictoris seemed to be another loser.

After the astronomers returned home and Terrile began processing the data, what he saw was startling: extending out from the star was an actual disk of orbiting material—not in infrared light, but in good old-fashioned optical light. "I was very, very excited," said Terrile. "You look at the fundamental questions of astronomy—the origin of the universe, the origin of life: those are the things that we really, deep down in our souls, want to know, the things that keep us up at night. Here's a direct link to that.

"Everybody had this feeling that we understood how planets formed, from a flattened disk of material. But there was no real, hard, physical evidence—no picture that you can see. And suddenly, in this blazing, obvious thing, is this picture which is an absolutely classic example of exactly what we thought happened when stars and planets formed. It was a slap in the face that said, 'Wake up! Don't you see what's going on all around you?'" Terrile told Smith, and the two announced the discovery later that year, reporting that Beta Pictoris's dust disk extended to over 400 astronomical units from the star, or over ten times Pluto's mean distance from the Sun. Depending on the star's exact age, planets might be forming, or have already formed, around Beta Pictoris. It was the best evidence yet for a genuine solar system around another star, and it was only fifty light-years away.

Media interest was intense. "It was really amazing," said Terrile. "Limos to your house at four o'clock in the morning to get you on *Good Morning America*. It was an extraordinary amount of coverage—dozens of programs." He said his experience was positive. "Many of my other colleagues have been burned in the past," he said, "but I didn't feel burned at all. A news show tends to use a sound bite. If you can say something in one sentence that's meaningful and conveys the message, then they'll use it. If you say something that's unmeaningful but comes off as a good sound bite, they'll use that, too, so you have to be a little careful. And some reporters, just because they want a more flamboyant story, will tend to do that. There were things that were inaccurate, but less than I expected. I thought that we'd see *National Enquirer* articles—SCIENTISTS DISCOVER ALIENS—and I was almost disappointed that we didn't get that."

After the discovery, Smith and Terrile tried to repeat it, by observing over a hundred other nearby stars. But their search failed; Beta Pictoris remained unique. Part of the reason, said Smith, is that Beta Pictoris's disk happens to be turned edge-on to Earth, making it easier to detect. In contrast, astronomers view Vega nearly pole-on.

More than just orientation is at work, however. "There's something peculiar about Beta Pictoris," said Smith, because it has much more dust than Vega or Fomalhaut. "We still don't fully understand just why there is so much of this remnant material left around an established main-sequence star. It may have been a very dusty object to begin with, but something has to be holding that material in place, or creating new material." Smith believes that Beta Pictoris probably has planets that maintain the disk, in the same way that certain satellites of Saturn, called shepherds, gravitationally sculpt that planet's rings. As with Vega, the disk around Beta Pictoris has a central hole that orbiting planets may have swept clean. Because Vega has much less dust than Beta Pictoris, Smith said that even if Vega's dust ring were viewed edge-on, his coronagraph would not be able to detect it. So Beta Pictoris is doubly lucky for astronomers: it has a lot of dust, and its disk is almost exactly edge-on to Earth.

Beta Pictoris differs from its peers in other ways, too. For one thing, its dust grains are smaller and hotter than those around Vega and its cohorts. For another, French astronomers have detected changes in the spectrum of Beta Pictoris that they interpret as comets which are falling into the star and getting vaporized by its heat. In 1995, other astronomers suggested that Beta Pictoris may be peculiar because, unlike Vega, it is not yet a main-sequence star. Instead, said these astronomers, the star may be so young—perhaps only 12 million years old—that it has not ignited its nuclear fuel. If so, its dust disk would be a remnant of the star's birth. Smith does not favor this idea, though, saying that the star's spectrum looks like that of a typical A-type main-sequence star.

Whatever the star's exact age, Smith's colleague saw the discovery of its disk of dust as part of a larger mission. "We scientists explore planets not just for ourselves but for a much broader community," said Terrile, who grew up in the Apollo era when science was prominent in newspapers and magazines. "The real people who benefit are the people like I was—the kids, the public. The public has to get stirred on by this constant bombardment of new ideas, the fact that there are frontiers out there, the fact that we don't know things."

Exciting though the dusty disks around Vega and Beta Pictoris were, they still left planet seekers short of their ultimate goal: a solar system that definitely harbors genuine planets. At about the same time, however, astronomers were gearing up for a new hunt—a hunt for small, dim stars called brown dwarfs, which would put planet-hunting techniques to the test and open a new investigation into the nature of the nearest stars.

7

Dark Star

NOT FAR from our solar system may lurk a cold, dark "star," completely invisible to human eye. This ghostly object drifts among the Galaxy's luminous stars, dragging with it a small retinue of planets that enjoy no light or warmth from the star they circle. When the star was young, it did glow dim red, but it had too little mass to ignite its main supply of nuclear fuel. Instead, the star steadily cooled and faded, like an ember plucked out of a fire, and the planets watched helplessly as it turned black and their temperatures plunged to near absolute zero. Billions of years later, these planets still faithfully revolve around their failed star, its gravity anchoring them to its lifeless embrace; and any light that does reach their frigid surfaces comes from beyond—from stars like Sirius, Alpha Centauri, and our own Sun.

Although they seem exotic, these dark stars, known today as brown dwarfs, may constitute an important part of the Galaxy, accounting for some of the mysterious dark matter that astronomers have sought for decades. Brown dwarfs are not planets, but like planets they are dim and challenge astronomers simply to detect them. Moreover, brown dwarfs have less mass than main-sequence stars, so their existence suggests just how readily the Galaxy creates small objects.

(Don't) Light My Fire

That failed stars might exist was recognized soon after astronomers discovered what made stars like the Sun such a success: nuclear fusion. In the late 1930s, scientists confirmed earlier speculations that the Sun and other main-sequence stars fuse hydrogen nuclei into helium nuclei at their cores. The most massive main-sequence stars had the hottest centers and so burned their hydrogen the fastest; these were the bright blue O and B main-sequence stars such as Regulus in the constellation Leo. Less massive main-sequence stars had cooler centers and fused their hydrogen more slowly; these were the white and yellow stars of spectral types A, F, and G, like the Sun. The least massive main-sequence stars had such low central temperatures that they hoarded their fuel and cast only a weak orange or red light into space; these were the main-sequence stars of spectral types K and M, the orange and red dwarfs.

Some stars might be born with so little mass that their centers stay too cool to spark a nuclear flame. In the early 1940s, astrometrists discovered what they thought were dark companions to the nearby stars 70 Ophiuchi and 61 Cygni. The dark objects were said to have roughly 1 percent of the mass of the Sun, or 10 times the mass of Jupiter. Although these reports proved false, they motivated Princeton astronomer Henry Norris Russell to investigate such small objects. He correctly recognized that their centers would be too cool to fuse hydrogen, and in 1943 he wrote, "At a temperature of 1,000,000° and with densities like those inside the sun the nuclear reactions which maintain the main-sequence stars would be completely inactive." He then went on to say, incorrectly, that such stars could engage in other nuclear reactions. Astronomers later learned more about the nuclear reactions that power stars, and in 1958 Harvard astronomer Harlow Shapley, who had once been Russell's student, briefly mentioned the dim objects in a popular book, *Of Stars and Men*, calling them Lilliputian stars.

The first to compute the actual evolution of such a star, however, was Indian-born astronomer Shiv Kumar. "As a kid," said Kumar, "my earliest memories are of looking at the sky—the Sun and the Moon and the stars—and wondering: what's going on out there? I had this deep urge to find out." By the age of five or six, he had already decided to go into science. After college, he came to the United States, and in 1958 he began to investigate low-mass stars.

"There was nothing known about how far we could go down in

mass," said Kumar, "so I did the calculation by trial and error. I started at the higher masses, and then I kept going down." The first star whose evolution he computed had 15 percent of the mass of the Sun. His calculations showed that it readily ignited its hydrogen and became an ordinary red dwarf star. Over the next four years Kumar therefore tried model stars with lower and lower masses, until he finally reached a mass—about 8 percent of the Sun's—where the star failed to burn. "I had absolutely no reason to think that the stellar domain stops there," said Kumar. "I'm saying it because even thirty-five years later, people still don't appreciate this. The question was: what happens below that critical mass? Is this the end of the stellar mass range? I had absolutely no reason at that time, and there's absolutely no reason right now, to say that that's the end. It's *not* the end. So I kept going, and I went down as low as 4 percent." Kumar published his results in 1963.

Whatever its mass, a star forms in space from a huge cloud of gas and dust that is many times larger than the Sun. The main ingredient in the cloud is the lightest element, hydrogen. Clouds of different masses speckle star-forming regions like the Orion Nebula and spawn a wide array of different-sized stars. To appreciate the distinct evolution of a failed star, first consider a typical star that successfully reaches the main sequence, a red dwarf with 20 percent of the mass of the Sun. Initially, the cloud that will form the star is colder than Pluto, but under the weight of its own gravity it begins to collapse, which warms the cloud as bits of the cloud fall toward its center. In the same way, gravity causes a cannonball atop a hill to gain energy of movement as it rolls down the hill. The cloud begins to glow red, and a new star takes its place in the heavens. Through this simple mechanism, our Galaxy gives birth to about ten new stars every year.

Though luminous, the collapsing star has yet to ignite its hydrogen-1 fuel, for its core is still too cool. Powered by gravitational energy, the star shines more brightly than the Sun, because the star is much larger. As it collapses and shrinks further, the star fades, but gravitational energy continues to heat up the core. Eventually the core attains a temperature of 2 or 3 million degrees Kelvin, and ordinary hydrogen, or hydrogen-1, begins to fuse. The core becomes a giant nuclear reactor, with nuclear power replacing gravitational power as the star's source of light and heat, and the heat's outward pressure holds the star up against the inward force of gravity. The star therefore stabilizes itself and assumes its place on the main sequence, where it will shine for hundreds of billions of years. For a typical red dwarf, the time from the start of the collapse to the ignition of hydrogen-1 is around 100 million years.

Now imagine the birth of a star having only 4 percent of the mass of the Sun. Like the red-dwarf-to-be, this object begins life as a collapsing cloud of gas and dust that glows red by converting gravitational energy into heat. Within a short time, the star's core does get hot enough to ignite one fuel, a rare isotope of hydrogen called deuterium. This is hydrogen-2, consisting of one proton and one neutron. Unlike hydrogen-1, which has just a proton, deuterium burns at a low temperature, so any star born with more than 1.3 percent of the Sun's mass can burn the isotope. The pressure of the heat generated by burning deuterium temporarily halts the star's collapse.

Because deuterium is rare, however, the star exhausts its supply in only about 10 million years and resumes its collapse. The core grows hotter, hinting that the star might ignite its hydrogen-1 and join its siblings on the main sequence. Instead, the star runs into an insurmountable obstacle: its own electrons. At the beginning of its life, the star was so large and spread out that its electrons barely noticed one another. Now that the star has shrunk and become denser, its electrons have started to repel other electrons. This repulsion has nothing to do with the electrostatic repulsion between like charges but instead arises from the Pauli exclusion principle. This holds that electrons cannot all be in the same energy state, just as all passengers on a bus cannot share the same seat. As the star tries to collapse further, to gain more gravitational energy and heat itself more, the repulsive electron pressure tries to prevent this, and the energy derived from the now-stalling collapse goes into pushing the electrons closer together rather than heating the star. In addition, the star continues to lose heat from its surface. As a consequence, the star's center actually starts to cool, and the star loses any chance to ignite its hydrogen-1 and achieve a long life on the main sequence.

"In the 1950s and 1960s," said Kumar, "this was a totally strange and weird idea. That's why I didn't tell anybody, because I had better back it up with more calculations. And I was also aware that people were not going to understand it. Even when I started telling people, they would look at me, especially the observers: 'What is he talking about?' And I got that kind of response many times, even after my papers were published in 1963. People would look at me, without saying anything, as if I was talking nonsense."

Kumar cited one example as Peter van de Kamp, the astrometrist who that same year reported a planet around Barnard's Star. "He never understood what I was saying," said Kumar. "I would explain to him, 'Look, Peter, this is how it is: a star is formed and it evolves. Some stars will go through hydrogen burning, and some won't.' He said, 'How is it possible? How can you have a star which will not have hydrogen burning?' This was true of other people, too, and I'm afraid that some people still don't understand. If you talk to some of the people who are making a big noise about the brown dwarfs, when they say these are not stars, they don't know what they're talking about."

After failing to ignite its hydrogen-1 fuel, the star shrinks a little more, until it becomes what astronomers call degenerate: its weight is supported solely by pressure among subatomic particles, electrons. The star's diameter is then somewhat smaller than the planet Jupiter, but the star is much denser. Over time, it cools and fades, turning from red to black. The more mass the object has, the longer the heat takes to escape, both because a more massive star has more heat and because, oddly, its greater gravity actually compresses it to a smaller size, so it has a smaller surface from which to radiate its heat. Some of the objects remain luminous for over a billion years, whereas others fade from view much faster. This is considerably shorter than the age of the Galaxy, so Kumar realized that many such objects may have passed through this stage of evolution and vanished from the sky.

In 1963, Kumar called the dim objects "black dwarfs," a term astronomers already used to describe white dwarfs—which are also degenerate—after they had cooled and faded to invisibility. "When I was about to publish the paper," said Kumar, "everybody said, 'Well, why don't you give them a special name?' I said it doesn't seem right to give a special name to objects which are already known. A 'black dwarf' is a completely degenerate object that is approaching the zero-luminosity, zero-temperature state. Because there was a general definition for a group of astronomical objects, I did not feel right at the time to introduce a special name.

"But then in the late 1970s and early 1980s, people started to change the name, as if somebody else had thought of this idea, which is nonsense. Just because you propose a new name doesn't mean you are the originator of the idea. Now the younger people, that's what they are thinking: in 1975 Jill Tarter [then a graduate student at Berkeley] called these objects 'brown dwarfs' as though she was the one. She was *not* the one." Indeed, Tarter herself never made such a claim, for she had cited Kumar's work.

Kumar called the new name "stupid," and other astronomers have described it as "lousy," "ridiculous," and "idiotic." Brown dwarfs are not brown: they start red and turn black. But because of the inaccurate term, magazines sometimes portray the objects as—you guessed it—brown. Better terms might have evoked various shades of red, such as "crimson dwarfs." In 1975, Kris Davidson of the University of Minnesota called them "infrared dwarfs," which correctly indicates that they are so cool they emit nearly all their energy at infrared wavelengths. That name never caught on, though.

"People say I'm making a big thing about semantics," said Kumar. "I'm not. As long as the history is right, as long as the science is right, I don't care what you call them. So if you people want to call them 'brown dwarfs,' that's fine with me—so long as you say that I was the first one to propose them, and that these are stars."

Of Stars and Planets

Actually, however, some scientists do not consider brown dwarfs to be stars. "There's a bifurcation between the two," said Adam Burrows of the University of Arizona. "A star is an object that is creating the same amount of energy that it's losing from the surface. You ask whether those losses can be compensated by the gains from thermonuclear burning. If they can, then it can be a steady object. If it's not steady, then it's going to cool off. So there's every reason to make the distinction between brown dwarfs and stars, and anyone who doesn't make the distinction is rather simple-minded." Burrows said that it's even a stretch to consider *white* dwarfs to be stars, since like brown dwarfs they do not burn nuclear fuel but simply cool and fade over time. Few astronomers, though, would consent to throwing Sirius B, Procyon B, and other white dwarfs out of the pantheon of stars.

David Stevenson, of the California Institute of Technology, has also studied brown dwarfs and like Kumar prefers to consider them stars. "But I wouldn't make a big fuss about it," he said. "I'm a believer in giving things names based on genesis, on how they came to be. If you want to use a terminology that is based on how things formed, then they have more in common with stars."

Historically, of course, *star* involved neither of the sophisticated concepts that Kumar and Stevenson, on the one hand, or Burrows, on the other, cite to support their views, for only in the twentieth century did astronomers learn how stars form and that most derive their energy from nuclear fusion; so neither condition has ever been a requirement for starhood. Indeed, millennia ago, the stars were simply the few thousand points of light visible to the naked eye at night that did not move the way planets did. The Sun was not considered a star, and no one could see the fainter points of light that the telescope would reveal. Since then, the term *star* has continually expanded to embrace a greater number and variety of objects. The Sun was added to the class, and telescopes revealed points of light in such profusion that they outnumbered the previously known stars. These too were considered stars, even those—such as Sirius B and Procyon B—that emitted far less light than the Sun. Today, astronomers know, the fundamental difference between stars and planets lies not in their movement but in their source of light: stars are self-luminous, generating their own light, whereas planets merely reflect the light of the star they orbit. This definition appears in modern dictionaries, and by that definition, brown dwarfs are clearly stars, since a typical brown dwarf glows for hundreds of millions of years, far longer than such indisputable stars as Rigel and Betelgeuse. To argue that brown dwarfs are not stars, one could point out that the definition of *star* has changed before, from a stationary point of celestial light to a self-luminous celestial body, and should change again, to the stricter requirement that the object employ nuclear fusion. Whatever one's opinion, the term *brown dwarf*—or *black dwarf*, *crimson dwarf*, or *infrared dwarf*—allows one to duck the question. To someone who considers brown dwarfs to be stars, the term is simply shorthand for "brown dwarf star," and to someone who considers them to be nonstars, the term puts the objects into their own category.

Whether one views brown dwarfs as stars, failed stars, or nonstars, there does seem to be a clear distinction between stars, including brown dwarfs, and planets. "This is a prejudice," said Stevenson, "but a planet forms in a different manner from a star. This is based on our experience in our solar system, and is why it's obviously prejudiced, because maybe our solar system is not typical. But in our solar system, at least, the evidence points towards a planet starting as an embryo that is composed of heavy stuff—ice and rock—and then putting the gas on top. There's no doubt that Saturn, Uranus, and Neptune are of this character; it is likely, but not absolutely certain, that Jupiter is, too. The difference between the terrestrial planets and the giant planets lies in the ability of the more distant embryos to gather gas, whereas objects like the Earth, fortunately for us, did not gather much gas."

In short, a planet forms from the bottom up, whereas a star forms from the top down. At its birth, even mighty Jupiter was a mere speck of dust orbiting the Sun. That speck of dust stuck to other specks and formed a clump; that clump grew still further, first into a small asteroid and then, as other asteroids hit it, a larger one. Jupiter kept growing until it was an object of ice and rock perhaps a dozen times more massive than the Earth. This object's gravity then attracted large amounts of the hydrogen and helium gas that pervaded the disk around the newborn Sun. Today Jupiter boasts a mass over three hundred times that of the Earth.

Brown dwarfs form differently from planets. "Everybody is looking for planets," said Kumar. "Fine, I want to find planets, too. I'll be as excited as anybody else, but these are stars. All kinds of nonsense have been published that these are planets or planetlike objects. These are *stars*. That's what I said in '63, and that's what I'm saying now. People don't like it, people don't understand it, but that's the way it is."

Unlike a planet, which starts life small and gets bigger, even a little star originates as a huge cloud dozens of times larger than the Sun. It collapses, begins to glow, and becomes a star. In the case of a main-sequence star, hydrogen-1 ignites. In the case of a brown dwarf (star), the hydrogen-1 fails to ignite.

Because stars form from the top down and planets from the bottom up, brown dwarfs should be more massive than planets. Some astronomers have estimated that the smallest possible objects that can form in a stellar fashion have about 1 percent of the mass of the Sun, or 10 times the mass of Jupiter. If so, brown dwarfs would range from 1 to

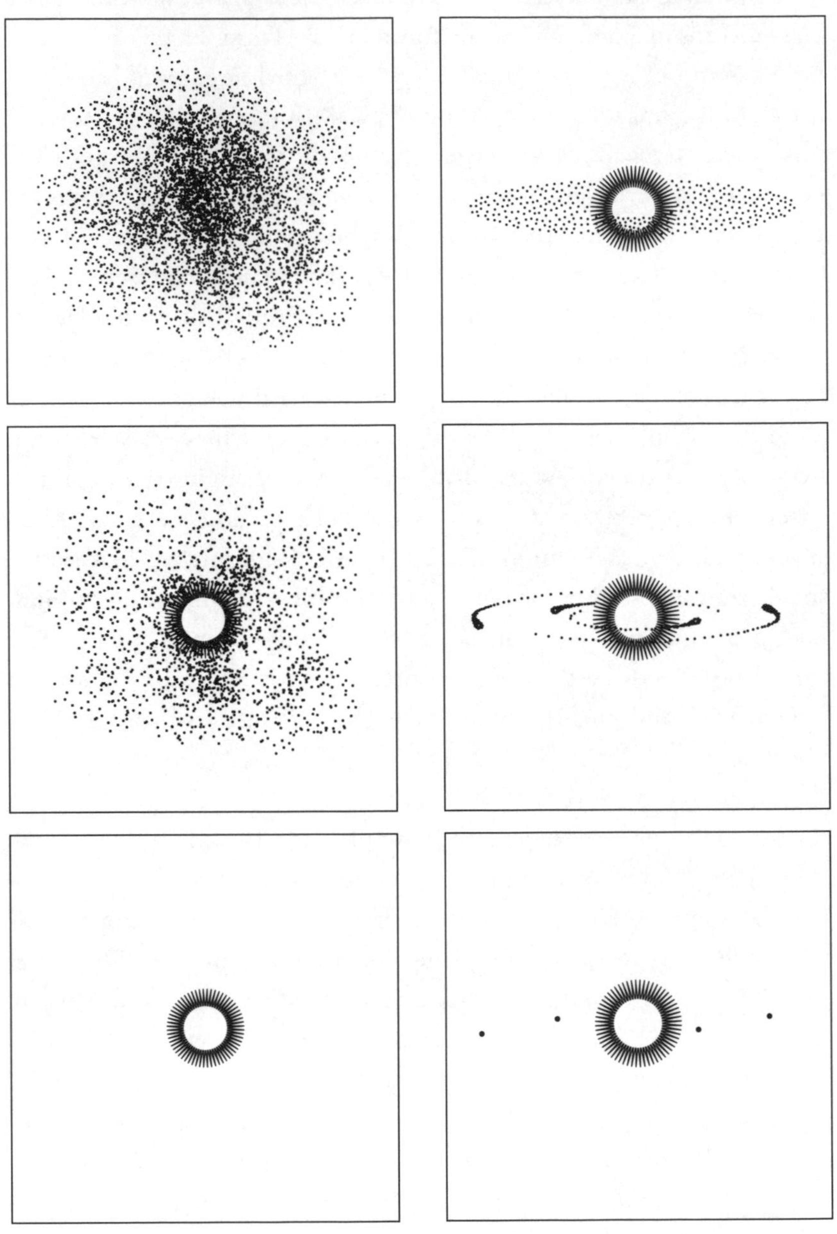

FIGURE 25

Stars form from the top down, from a huge cloud of gas and dust that collapses (left). Planets form from the bottom up, as particles collide inside a disk orbiting a star (right).

8 percent of the Sun's mass, or from 10 to 80 Jupiters. And how big can a planet be? Jupiter proves that planets can have as much as 0.1 percent of the Sun's mass, and some theories suggest that this may be near the upper limit of planetary masses. As a giant planet accretes gas, the gas in the planet's orbit gets used up, and the planet can grow no further. If this theory is right, then there may be a mass gap between planets and brown dwarfs, and astronomers can use an object's mass to distinguish brown dwarfs from planets: if the object is comparable in mass to Jupiter, it's a planet; if its mass is 10 Jupiters or more, it's a brown dwarf.

The real universe, though, may not be so simple, and Burrows said that no one really knows how large a planet, or how small a brown dwarf, can be. "There are people who will say they know, but they don't know," said Burrows. "Since the theory on this is rather immature, I think that we have to keep an open mind." Both Kumar and Stevenson agree that there could be an overlap in mass between planets and brown dwarfs. In particular, the region between 1 and 10 Jupiters—that is, between 0.1 and 1 percent of the mass of the Sun—is a blur. It could abound with planets, or brown dwarfs, or both, or neither.

If mass alone cannot separate planets from brown dwarfs, then perhaps other properties can. "I would say that the distinction between a planet and a brown dwarf lies in its composition," explained Stevenson, "and in particular I would expect a planet to be nonsolar in composition. Jupiter is *not* a piece of the Sun." That's because the planet has an ice-rock core and so contains a greater proportion of heavy elements than the Sun. "Whereas a brown dwarf," he said, "if it forms in a fashion similar to the formation of stars, will be a representative piece of cosmic stuff. It will be like the Sun." The Sun formed from a cloud of gas and dust that contained the elements in about the same proportion as the overall Galaxy, and the same should hold true for a brown dwarf, since it formed in the same way.

As Stevenson readily admitted, however, observers will find this a hard test to apply, because determining an object's composition usually requires seeing the object and obtaining its spectrum. If an observer does not see the object—if its existence is inferred only from the wobble in the star it circles—then astronomers will know the object's mass and orbit, but not its composition, so the composition will not reveal whether the orbiting object is a brown dwarf or a planet.

Nevertheless, the shape of the orbit itself may help distinguish between the two. "That is a very important point," said Kumar, "because it has to do with the star formation process. When we look at double stars, eccentric orbits are the common denominator." The average double star has an orbital eccentricity around 0.30, which means that the separation between the two stars varies greatly. For example, the distance between Alpha Centauri A and B changes from 11 to 35 astronomical units and back every eighty years. Brown dwarfs should therefore have eccentric orbits. In contrast, the giant planets—in our solar system, at least—have nearly circular orbits. The giant planet on the most elongated orbit, Saturn, has an orbital eccentricity of only 0.05, and the planet's distance from the Sun varies from 9 to 10 astronomical units.

Burrows, though, urged caution in using orbital eccentricity to distinguish planets from brown dwarfs: "I think this is being pushed a little prematurely. It's a possibility, but we don't know, and it shouldn't be the basis of our categorizations. I think people are too quick to type and to define and to pigeonhole. We have too many terms that have proliferated about objects we know too little about. We should step back and let the observations speak for themselves."

The Search for Brown Dwarfs

Unfortunately, brown dwarfs have not been exactly eager to reveal themselves. Yet they may be quite common, because small stars outnumber large ones: 80 percent of the Galaxy's stars are red dwarfs. Therefore, brown dwarfs, which are even less massive than red dwarfs, could be even more abundant. If so, they may account for some—perhaps much—of the Galaxy's dark matter.

The first person to postulate the existence of dark matter was Dutch astronomer Jan Oort, who had earlier proved that the Galaxy rotates swiftly around a point far from the Sun. In 1932, Oort analyzed the motions of stars near the Sun. As stars revolve around the Galaxy, they oscillate up and down through the plane of the Galaxy's disk, like a bobbing horse on a merry-go-round. The greater a star's vertical speed, the farther the star can climb away from the plane of the Galaxy; but this excursion also depends on how much mass the disk has, since this mass has gravity that tries to hold the stars tightly to the Galactic plane.

By comparing the stars' speeds with their heights, Oort concluded that the disk had much more mass than met the eye and so contained a large amount of dark matter. In the 1960s, he repeated the work and reached the same conclusion, as did John Bahcall, of the Institute for Advanced Study in Princeton, during the early 1980s. In the late 1980s, however, Konrad Kuijken and Gerard Gilmore in England questioned this conclusion, and today astronomers do not know whether the Galaxy's disk harbors any dark matter.

Nearly everyone, though, believes that huge amounts of dark matter engulf the disk itself, forming an enormous dark halo that outweighs the rest of the Galaxy. The dark halo betrays its presence because its gravitational pull affects stars and gas in the outer part of the disk, which revolve around the Galaxy far faster than they would if the dark halo did not exist. In addition, the Milky Way has at least ten satellite galaxies that orbit it the way the Moon circles the Earth, and these satellite galaxies move so quickly that they would have escaped the Milky Way's grasp if not for the gravitational attraction of the dark halo's mass. Brown dwarfs could be one contributor to the Galaxy's dark halo.

By its nature, astronomy has a hard time studying objects like brown dwarfs that emit little or no light. Astronomers usually express stellar brightnesses in so-called magnitudes. The *apparent* magnitude is how bright the star looks, and like a book's position on the best-seller list, the *lower* the number, the better. The Sun and the next four brightest stars are so bright that their apparent magnitudes are actually negative, while the remaining very bright stars have apparent magnitudes around 0 and +1. The stars of the Big Dipper are somewhat fainter, with apparent magnitudes of +2 and +3, and the dimmest stars the naked eye can see are around apparent magnitude +6. Each successive magnitude corresponds to a factor of 2.5 in brightness, so a magnitude +2 star is 2.5 times brighter than a magnitude +3 star, which is 2.5 times brighter than a magnitude +4 star, and so on.

Apparent magnitude alone does not reveal how much light a star really emits, because even a luminous star looks faint if it is far away. Astronomers therefore compute a more meaningful number—*absolute* magnitude—to correct for the effect of distance. Astronomers imagine how bright the stars would look if they were all exactly the same distance from Earth, 32.6 light-years. For example, the blindingly bright Sun, if viewed from this distance, would have an apparent magnitude of +4.83—visible to the naked eye, but hardly a spectacle. This, then, is the Sun's absolute magnitude.

Red stars that are far less luminous than the Sun might be brown dwarfs. In 1915, Robert Innes in South Africa discovered what was then the least luminous star known, the famous Proxima Centauri, whose absolute magnitude is +15.52. Proxima Centauri is a red dwarf, not a brown dwarf; nevertheless, it has so little mass—probably around 10 percent of the Sun's—that it burns its fuel frugally and glows dimly. Proxima's reign as the least luminous star known ended three years later, when German astronomer Max Wolf discovered an even fainter red dwarf, Wolf 359. It is the third nearest star system to the Sun, after Alpha Centauri and Barnard's Star, and has an absolute magnitude of +16.56. Like Proxima Centauri, Wolf 359 is almost certainly a main-sequence star.

In 1940, Georges Van Biesbroeck at Yerkes Observatory in Wisconsin mounted a search for dim stars by examining faint nearby stars for even fainter companions. Because the brighter star's distance was already known, Van Biesbroeck could immediately infer the companion's absolute magnitude. In 1943, Van Biesbroeck spotted a star in orbit around a red dwarf in the constellation Aquila, nineteen light-years from Earth. Van Biesbroeck's Star, which also goes by the name VB 10, has an absolute magnitude of +18.67, over two magnitudes fainter than Wolf 359. As Van Biesbroeck noted, the star is so dim that if put in place of the Sun, it would look as faint as the Moon. The star retained its dubious honor as the dimmest known star until the early 1980s, when even fainter stars were sighted.

Unfortunately, neither redness nor dimness guarantees a star's brown dwarf status. When a brown dwarf is young—before it has faded—it actually outshines many red dwarfs, which makes distinguishing brown dwarfs from red dwarfs difficult. To do so, astronomers could measure a red star's mass and see whether it is above or below 8 percent of the Sun's. But astronomers can weigh a star only if it orbits and tugs on another star. Many stars are single, and Van Biesbroeck's Star has such a long orbital period around its mate that astronomers do not know its mass, either. These stars could be red dwarfs rather than brown dwarfs. As they age, brown dwarfs cool and fade, becoming more distinct from red dwarfs, but also more difficult for astronomers to detect.

Early searches for brown dwarfs turned up nothing, and brown dwarfs, which should have been common, began to seem rare. Then, in the 1980s, came an apparent discovery that gave brown dwarf hunters just what they were looking for: a star so red and faint it seemed to be a genuine brown dwarf.

The VB 8 B Fiasco

VB 8 is a faint red dwarf star twenty-one light-years away, another of the dim suns that Georges Van Biesbroeck discovered. Appropriately, it lies in Ophiuchus, the same constellation that housed the phantom planets of Barnard's Star, and it starred in a major astronomical embarrassment.

In 1983, Robert Harrington of the U. S. Naval Observatory and his colleagues reported that VB 8 exhibited a wobble in its proper motion, suggesting that the star had a companion with a few times the mass of Jupiter. However, Harrington deliberately expressed the mass in "milli-Suns" rather than Jupiter masses, to avoid the connotation of planets, and the report attracted almost no attention.

The following year, a team led by Donald McCarthy, Jr., of the University of Arizona, claimed to see the companion to VB 8. The astronomers achieved this apparent feat by using a technique known as speckle interferometry, which compensates for the blurring effects of the Earth's atmosphere by taking split-second images of a star. Applying this method at infrared wavelengths revealed the "companion" to VB 8. The object was too cool—1360 degrees Kelvin—to be a main-sequence star, and the astronomers estimated its mass at roughly 5 percent of the mass of the Sun, clearly a brown dwarf. Although the paper contains those words, it goes on to state, "These observations may constitute the first direct detection of an extrasolar planet." A press release was issued, a news conference held, and the media swallowed it up, with some magazines even going so far as to run cover stories about planets around other stars.

Shiv Kumar was outraged that a brown dwarf was being called a planet. "Astronomy is already very exciting," he said. "Why do people have to go and lie to the public to make it more exciting? I find it very disturbing. Perhaps it's tied up to getting more money, but I don't understand it. This, to me, is not science." Kumar put out his own press release denouncing the "planet" and wrote a letter of protest to *Physics Today*, which had run a story on VB 8 B. "I know some people were unhappy that I wrote that letter," he said. "I won't mention any names, but I know at least one person who was consulted about that letter and told the editor not to publish it. I met him at a meeting once, and he gave me a dirty look." Kumar laughed. "That's why I don't like to go to meetings any more, because some of these people are wandering around, and they don't know any history, they don't know any science, and they have some weird notions about how the universe should be. They think there's something wrong with me. But my view is, there's something wrong with these people."

Kumar wasn't the only angry astronomer. Those at the Naval Observatory who had first called attention to VB 8 B thought that they received inadequate credit for the "discovery," although the paper by McCarthy and his colleagues did cite their work. The Naval Observatory therefore held its own press conference to showcase the astrometric work.

To David Stevenson, however, the astrometrists had no cause for complaint. "In the broader astronomical community," he said, "the astrometric work has never been widely embraced. It goes back, of course, to Barnard's Star: people looked askance at this sort of thing. They listened to it politely, but they never really fully believed it, because the history was tainted. If you have twenty years of dubious results, then people shrug their shoulders when another result comes along. But if you have a relatively new technique, like infrared speckle interferometry, then people may pay more attention to it. They don't have cause at that point to be as suspicious."

Although astronomers disputed VB 8 B's alleged planetary nature, most believed that the object did indeed exist. After all, the astrometrists had seen VB 8 wobble, and more convincingly, the infrared speckle team had seen VB 8 B itself. But in 1986, French astronomers used infrared speckle interferometry to look for VB 8 B—and could not find it. A team of American astronomers also failed to locate the much-ballyhooed object. As the French politely put it in their paper, "Constraints on the dynamics of the system are deduced; the nature of VB 8B is discussed; and its existence is questioned." Worse, new astrometric data showed that VB 8 was not really wobbling. VB 8 B was gone.

Ironically, the nonexistent object stimulated much interest in brown dwarfs. "Prior to that so-called discovery," said Stevenson, "there were only a tiny number of people who paid brown dwarfs any attention. VB 8 B created an enormous amount of excitement." Before VB 8 B vanished, astronomers even held an entire conference on the subject, at which VB 8 B was proclaimed to be a definite brown dwarf.

The Elusive Dark Star

With the disappearance of VB 8 B, the number of "definite" brown dwarfs plunged from one to zero. In subsequent years, other apparent brown dwarfs emerged, only to suffer a similar fate. For example, in the late 1980s, astronomers reported red stars that seemed to be young brown dwarfs in the constellation Taurus, but they turned out to be more distant yellow G and orange K stars that looked red only because intervening dust was absorbing their blue and yellow light. In another case, a white dwarf named Giclas 29-38 was found to emit excessive infrared radiation that seemed to signal an orbiting brown dwarf, but later work suggested the infrared radiation arose instead from a cloud of dust heated by the white dwarf's light. Brown dwarfs came and went even faster than planets once had; if you didn't like a particular brown dwarf, you could wait a few months and it would probably be gone. Meanwhile, several searches for brown dwarfs turned up nothing. In 1990, an article in *Sky and Telescope* bore the headline BAD NEWS FOR BROWN DWARFS. A headline in *New Scientist* on Giclas 29-38's brown-dwarf-turned-to-dust-cloud read: ANOTHER BROWN DWARF BITES THE DUST.

A few possibly credible examples also appeared, however. In 1988, Eric Becklin of the University of Hawaii and Ben Zuckerman of the University of California at Los Angeles took an infrared image and discovered a very red star around a white dwarf in the constellation Boötes named GD 165. Unfortunately, though the star is cool and dim, GD 165 B could still be an ordinary main-sequence star—a red dwarf rather than a brown dwarf.

That same year, David Latham of the Harvard-Smithsonian Center for Astrophysics and his colleagues discovered a low-mass object orbiting a yellowish main-sequence star of spectral type F9. The star, HD 114762, lay in the constellation Coma Berenices, and the companion caused the star to wobble ever so slightly. This wobble was not an astrometric one, *across* the line of sight, but a spectroscopic one, *along* the line of sight. As the companion caused the star to move toward and away from Earth, the star's spectrum exhibited a minute Doppler shift: when HD 114762 moved toward us, its light waves were compressed to slightly shorter wavelengths, creating a tiny blueshift, and when it moved away, its light waves were stretched to slightly longer wavelengths, creating a tiny redshift. HD 114762's companion revolved around the star once every eighty-four days in an elongated orbit. First reports put the orbital eccentricity at 0.25, but later work raised this to 0.38.

"I thought that was a solid discovery," said Stevenson, "and I was always puzzled that people paid rather little attention to it. When I published a review paper about brown dwarfs in 1991, I included a figure of that object." Still, there were reasons to be skeptical. Although HD 114762 B undeniably exists, a spectroscopic wobble differs from an astrometric one because astronomers cannot ascertain the exact mass of the orbiting body. All they can determine is a lower limit, which the Doppler shift's small size indicated was a mere 1 percent of the mass of the Sun, well below the threshold of a main-sequence star. This value, however, applies only if we view the star-companion system exactly edge-on. We probably do not, so the companion's mass is greater than this. To see why, imagine viewing such a star system exactly pole-on. Then, even if the companion is quite massive, the main star never moves toward or away from us and shows no Doppler shift at all. If HD 114762's orientation is closer to the pole-on case, then its companion could have more than 8 percent of the Sun's mass, in which case it would be just another red dwarf.

Kumar, who had published the first papers describing the evolution of what are now called brown dwarfs, watched these developments with mixed feelings. "In one sense," he said, "I'm very glad and very happy that so many people have worked in this area, and more or less things have been confirmed according to what I said so many years ago. My only problem is that some of these people are not giving me credit for what I did." For example, the paper on HD 114762 did not cite his work but rather that of Tarter, the one who coined "brown dwarf," and other papers have done the same. "In science, people generally don't behave like this," said Kumar. "If you're the first person to say something and it turns out you're right, people give you full credit for it and your name is associated with it. I don't understand it. I explain things to some of these people, that I did it back in 1963, and they look at me as if either I didn't do it or it wasn't right. The conversation becomes sort of uncomfortable for me. Some of these people were kids in the sixties, and now they're acting as if they know everything about the history and the science of the subject, which they don't." Kumar said neither Stevenson nor Burrows has given him the credit he deserves. He was especially angry at Stevenson's 1991 article on brown dwarfs, because just one sentence mentioned Kumar's work. That gave him credit only for calculating the minimum main-sequence mass, not for computing the structure and evolution of stars below that mass.

"I'm aware of his annoyance," said Stevenson, "and in fact I knew that after I wrote my review paper—which *does* reference him—that nonetheless he felt slighted and he wrote a letter expressing annoyance to the editor. You know, what can I say? I think you or anyone else writing a history of the subject should acknowledge his contribution, and he must be at least partly right that people have given him somewhat less acknowledgment than he deserves. But I also think he's overstating the case. It's the kind of thing that was regarded by some people as a sort of self-evident truth: that of course there was this lower mass for main-sequence stars, and you could estimate it on the back of an envelope. The distinctive thing that Kumar did was that he actually wrote it down carefully and published it, and you must not take that away from him."

"He gets credit, he gets credit," said Burrows, noting he has also cited Kumar's work. "He has this bee in his bonnet, and I don't think it's going to be easy to liberate. So why don't we just leave it at that?"

When Stevenson published that 1991 article, no firm brown dwarf candidate had yet appeared, and four more years would elapse before astronomers finally spotted an ironclad brown dwarf, one that Kumar, Stevenson, and Burrows all considered to be unassailable. Before then, however, astronomers were shocked and delighted by a revolutionary discovery, one that would herald the century's greatest decade of planetary discovery: actual *planets*, similar in size to the *Earth*, were found in a most unexpected location, circling a pulsar.

8

Pulsar Planets

SOMEDAY, astronomers knew, their great quest to find alien worlds would finally be fulfilled, and the first planets beyond those of the solar system would at last emerge from hiding, hailing a new era of astronomical discovery. Exactly when this momentous day would arrive, and around which star the new planets would appear, no one could say; but a safe bet was a star within a dozen or so light-years of the Sun—if not Barnard's Star, then one of the Sun's other neighbors, such as Tau Ceti or Epsilon Eridani. Imagine, then, the enormous shock that jolted astronomy in 1991, when radio astronomers detected the first genuine extrasolar planets over a thousand light-years away, circling a star few thought could even have planets: a fast-spinning neutron star known as a pulsar.

Celestial Lighthouses: The Discovery of Pulsars

A pulsar is a collapsed star, born when a massive star like Rigel or Betelgeuse dies. A massive star, formed with more than eight times the mass of the Sun, shines brightly but briefly, for it soon uses up its fuel and explodes in a titanic supernova that can outshine an entire galaxy. During the supernova explosion, the star's outer layers shoot off into space at millions of miles an hour, enriching the Galaxy with elements such as oxygen and neon. At the same time, the star's core suddenly collapses, because it no longer generates the energy that holds it up against the inexorable inward pull of gravity. The stellar core shrinks and shrinks, until its positively charged protons ram into its negatively charged electrons, to create a dense sphere of neutral neutrons only ten miles across—smaller than the main islands of Hawaii. In this catastrophic event, one of the Galaxy's largest stars becomes one of its tiniest. The huge weight of the neutron star is supported by pressure among the neutrons. In the same way, pressure among electrons supports white dwarf and brown dwarf stars.

White dwarfs themselves are extraordinary stars, little larger than Earth but usually bearing about 60 percent of the mass of the Sun. They form when a star born with less than eight times the mass of the Sun dies. A typical white dwarf is half a million times denser than water, so a mere spoonful of white dwarf matter weighs a ton. Only the star's great gravity keeps the hot material compressed; if a chunk of white dwarf matter were instantly transported to Earth, it would explode in a huge fireball.

Neutron stars, however, make white dwarfs seem tame. A neutron star is over twice as massive, with 1.4 solar masses, but much smaller, so its density is far greater: hundreds of *trillions* of times that of water, or a billion times the density of a white dwarf. One spoonful of neutron star matter weighs a billion tons. The star's gravity is so enormous that if you dropped a pebble from a height of four feet, it would smash into the star's surface at 5 million miles per hour.

These exotic objects were predicted three decades before they were actually seen. In 1932, shortly after the discovery of the neutron, Russian physicist Lev Landau speculated that neutrons could constitute a star's core. In 1934, American astronomers Walter Baade and Fritz Zwicky wrote, "With all reserve we advance the view that a super-nova represents the transition of an ordinary star into a *neutron star*, consisting mainly of neutrons. Such a star may possess a very small radius and an extremely high density."

Right they were. In 1967, Jocelyn Bell, a graduate student at the University of Cambridge, accidentally found the first neutron star. As a child, Bell had liked astronomy, but she almost gave it up when she was told she would have to work nights. Then she discovered radio astronomy. Of all forms of electromagnetic radiation, radio has the longest wavelength, even longer than infrared. Radio waves can be observed

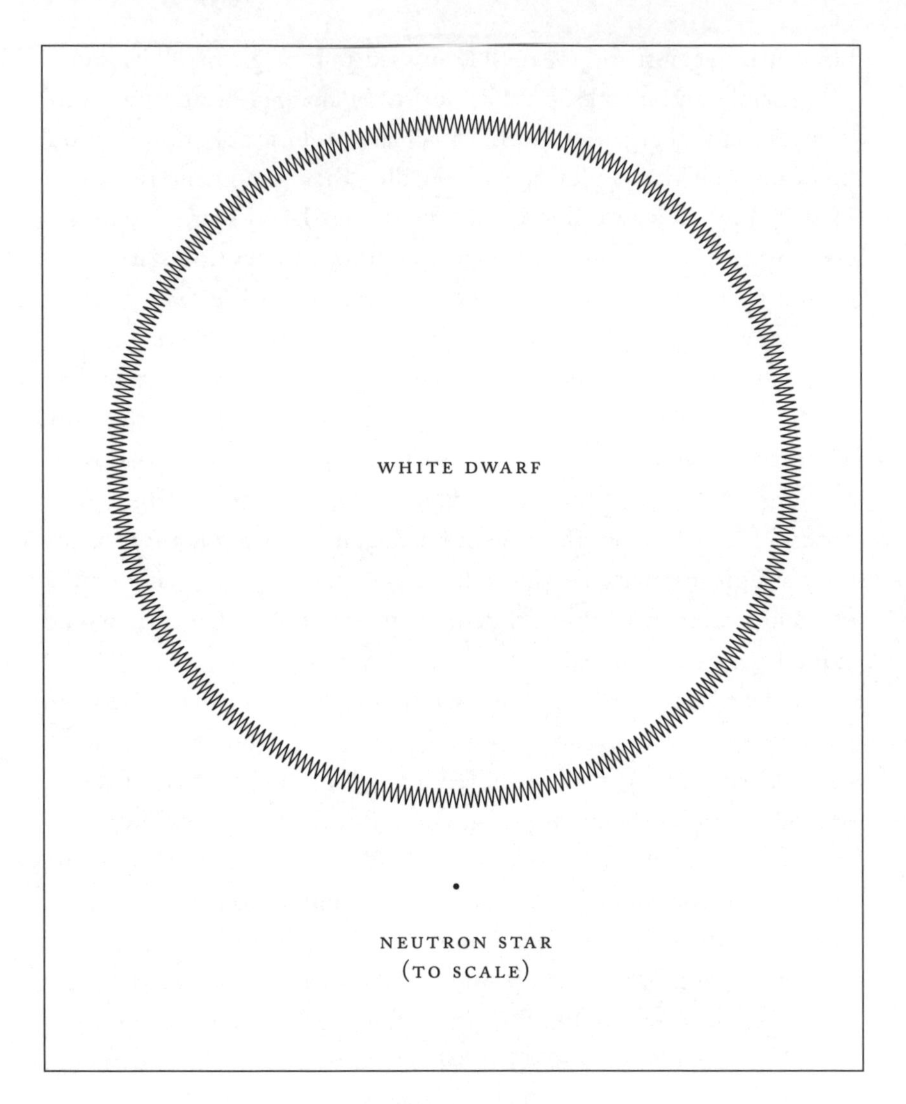

FIGURE 26

A neutron star has over twice as much mass as a typical white dwarf, but it is much smaller in size.

when the Sun is up, so unlike optical astronomers, radio astronomers have the luxury of working during the day and sleeping at night.

Bell noticed a peculiar source of radio waves in the constellation Vulpecula the Fox, and further investigation revealed that the source was beeping every 1.3 seconds. Such regular beeping seemed artificial. Might it be a message from another civilization? If the object were stationed on an orbiting planet, the signal should show a Doppler shift as

FIGURE 27

Pulsars are like celestial lighthouses, emitting a beam of radiation toward us every time they spin.

the planet circled its star, but it did not. Furthermore, Bell soon discovered a second pulsar far from the first, and it seemed unlikely that two different civilizations would erect such similar beacons.

In fact, a pulsar is a rapidly rotating neutron star. It emits radiation in a beam, like a lighthouse. When this beam sweeps past Earth, astronomers see a pulse of radio waves. The period between two successive pulses reveals how fast the pulsar spins. In the case of the first pulsar, this amounted to a mere 1.3 seconds, which means the star spins over a million times faster than the Sun. The beams of many pulsars, however, miss the Earth, so radio astronomers do not see them.

Since Bell's discovery, astronomers have found over eight hundred pulsars. Unlike bright stars, pulsars do not bear romantic names like "Antares" or "Aldebaran." Instead, a pulsar carries the designation "PSR"—for pulsating source of radio—followed by its coordinates. Preceding these coordinates is the letter "B" or "J," indicating whether the coordinates are for the year 1950 or 2000. For example, the pulsar that caught Bell's eye is now named PSR B1919+21. More than one astronomer has wished for something more poetic.

Some pulsars still reside in the remains of the exploded star that gave them birth. The youngest pulsar of all, and the fastest spinning of the "normal" pulsars, lies in the Crab Nebula, a cloud of expanding gas in the constellation Taurus where our ancestors saw a star explode in 1054 A.D. The Crab pulsar spins once every 0.03 second. Another pulsar is in the remnant of a star that exploded many thousands of years ago in the constellation Vela. As they expand into space, these supernova remnants vanish from view in just a few tens of thousands of years. Pulsars therefore outlive their supernova remnants, and most pulsars do not reside in one. But pulsars themselves are also mortal. Over millions of years, a pulsar gradually spins less and less rapidly, until it turns so slowly—about once every 5 seconds—that it ceases to radiate, disappearing from the radio sky.

Pulsar Planets, Part One

The 1967 discovery of the first pulsar came at just the right time for Andrew Lyne, a radio astronomer at the University of Manchester's Jodrell Bank in Cheshire, England, and now one of the world's experts on pulsars. "It was pure luck," said Lyne. "I was working towards my Ph.D. thesis using the 76-meter telescope here at Jodrell Bank. I was actually observing on the telescope when the discovery was announced, and I was persuaded to give up some of my valuable telescope time to look at this thing." His observations helped confirm the pulsar's existence. "That's where it all started," said Lyne, "and I've been with them ever since, apart from a year that I had to take out to finish writing my thesis."

Lyne found pulsars attractive for good reason: "They are wonderful things for physicists to play with. You can measure so many things and learn all sorts of things about so many different areas of physics that you can't do elsewhere." Not only do pulsars reveal how bright stars end their lives, but their enormous gravity allows physicists to test Einstein's general theory of relativity, which describes how objects behave in strong gravitational fields. In addition, as radio pulses speed through space, interstellar gas disturbs them in ways that reveal its characteristics. Several pulsar scientists have received Nobel prizes for their work, with the prominent exception of the scientist who started the whole field, Jocelyn Bell.

Not long after Bell's discovery, astronomers thought they made another one, equally exciting: the Crab pulsar seemed to have a planet that circled it every eleven weeks. The evidence lay in the exact times when pulses from the pulsar reached the Earth. The gravitational pull of a planet would cause a pulsar to move periodically toward and away from Earth, just as a planet around a normal star does. When the pulsar moves toward Earth, one pulse takes slightly less time to reach Earth than the one before, because the second pulse travels a shorter distance; consequently, the apparent period between two successive pulses decreases. Likewise, when a pulsar moves away from Earth, the period between two successive pulses increases. Thus, a planet causes the pulsar's period to oscillate. In 1969, the Crab pulsar's period was oscillating in just this way, suggesting the presence of a planet. Unfortunately, the planet proved to be a mirage, caused by what astronomers call timing noise. Internal disturbances in the fast-spinning pulsar, which is a mere thousand years old and still settling into its final state, affected the rotation rate and mimicked the periodic behavior induced by a planet. Ten years later, in 1979, astronomers reported a possible planet around a pulsar in the constellation Camelopardalis named PSR B0329+54. This planet was said to revolve every three years, but later work questioned its existence.

Pulsar planets then languished for over a decade. After all, how could a pulsar—the remnant of an exploded star—have planets? The explosion would jettison any planets into space. In 1991, however, pulsar planets returned with a vengeance, as first one pulsar and then a second were reported to have planets.

The first pulsar was PSR B1829-10, which lies in the constellation Scutum the Shield, some 35,000 light-years from Earth. "It was one of a few dozen pulsars we had discovered in about 1985," said Lyne, "and it was quite weak." In part because of its weakness, Lyne and his colleagues, Matthew Bailes and graduate student Setnam Shemar, had trouble modeling the pulsar's pulse behavior. In May 1991, Shemar told Lyne he suspected that PSR B1829-10's pulse period was varying periodically due to an orbiting body, but Lyne dismissed the variation as mere timing noise. Shemar continued to work on the pulsar, and Lyne

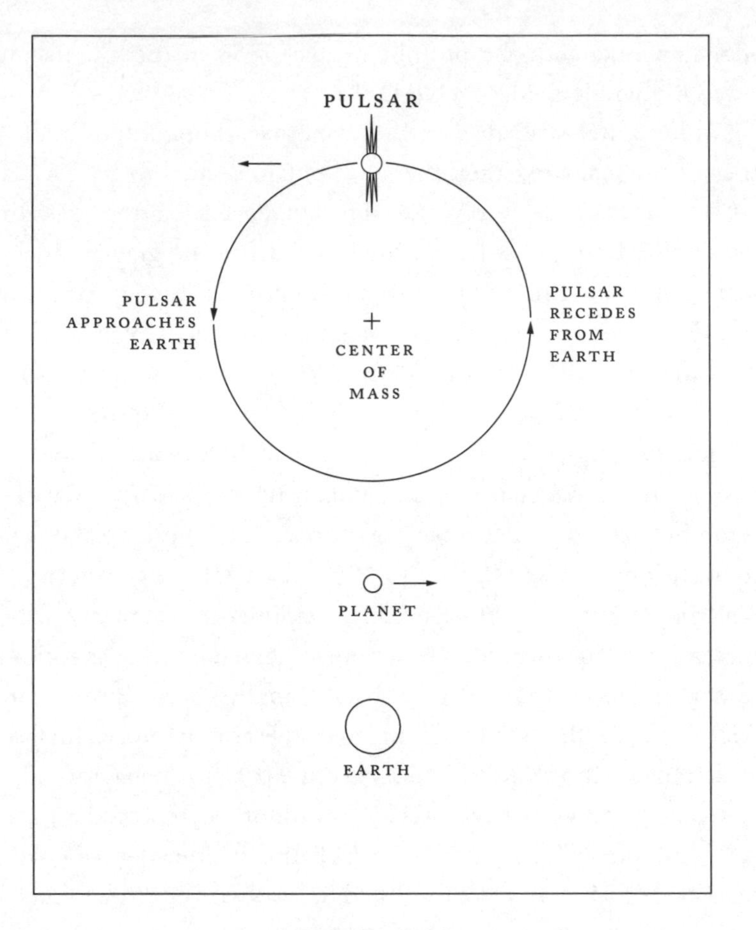

FIGURE 28

As a pulsar with a planet circles the center of mass, it can move toward us, which bunches up its pulses, so that astronomers receive more pulses per second than normal. When the pulsar moves away from us, the pulses spread out, and astronomers receive fewer pulses per second than usual.

pointed out that the oscillation affected the pulsar's period so slightly that if it was really caused by an orbiting object, that object would have the mass of a planet. Lyne showed Shemar how to fit a planet model to the pulse data, and Shemar found that the model matched the data perfectly; all the problems that had dogged this pulsar suddenly vanished. It seemed to be the first extrasolar planet ever found. On the evening of May 25, Shemar calculated the properties of this planet: a world with the mass of Uranus but on an orbit like that of Venus, circling the pulsar every six months. Shemar did not sleep very well that night.

"We searched what we thought was very thoroughly for all possible causes of this oscillation," said Lyne. "We hadn't seen anything like that in any of our other pulsars, and eventually we concluded that, provided we hadn't made a mistake, the most likely explanation for it was that the pulsar was moving due to the pull of an orbiting planetary body." The announcement of the discovery, which stunned the world, appeared in the July 25 issue of *Nature*, whose cover proclaimed FIRST PLANET OUTSIDE OUR SOLAR SYSTEM.

"I think the reaction of other astronomers was much the same as ours," said Lyne: "one of surprise. We really didn't know what to make of this. It was so unexpected that one of these neutron stars, particularly a young one, should have any bodies revolving about it." As the paper stated, "the existence of the planet challenges conventional theories of the formation of neutron stars from supernovae and has important implications for the existence of planetary systems around other stars."

The announcement excited astronomers not only because it seemed to be the first extrasolar planet but also because a pulsar was the last place anyone expected to find such a planet. There are two reasons for this. First, a massive star that will eventually become a pulsar begins life as a bright blue main-sequence star of spectral type O or B and then normally expands into a red M-type supergiant like Antares or Betelgeuse. Red supergiants are the largest stars in the universe. If put in place of the Sun, a typical red supergiant would engulf Mercury, Venus, Earth, Mars, and the asteroid belt. Thus, during its life, a massive star swallows its inner planets, such as the one that Lyne's team had reported close to the pulsar. Second, all the planets face trouble when their star explodes. Contrary to media reports in the summer of 1991, the supernova explosion does *not* destroy the planets. They survive, but are flung into space, because if a star suddenly loses over half its mass, its gravity becomes too weak to retain any orbiting bodies. Were the Sun to suffer such a sudden loss of mass, Earth and the other planets would sail out into the void. For this reason, pulsars rarely have even companion stars, since the supernova explosion usually frees any partner—unless that partner has so much mass that *its* gravity can hold the system together.

Because pulsars should not have planets, astronomers concocted a flurry of theories to explain how the apparent pulsar planet had originated. Most of these ideas avoided the problems resulting from the supernova by arguing that the planets arose after the star exploded. Perhaps some of the debris from the exploded star fell back toward the newborn pulsar and formed a rotating disk around it in which planets were born. Or perhaps the neutron star had arisen not in the conventional way, from the collapsing core of an exploding star, but from the merger of two white dwarfs, which created a disk that formed planets. Or perhaps, most improbable of all, a neutron star had crashed into another star, stolen its planets, and destroyed the star.

All along, however, the planet's six-month period troubled astronomers. "That was the hint that was sitting in front of us all the time," said Lyne, "but we didn't know what to do about it. We just couldn't see why it should come about." Six months—half an Earth year—suggested that the astronomers somehow had made an error that simply reflected the Earth's motion around the Sun. "It was a sort of nagging worry," said Lyne, "but we were all aware of coincidences which occur in nature, and eventually we became convinced enough we thought it was just one of nature's coincidences." Their paper discussed this concern but noted that no other pulsar showed the same behavior, not even one that lay just two degrees away.

Meanwhile, in America, work on another pulsar was about to divulge not one but two pulsar planets.

Pulsar Planets, Part Two

When Andrew Lyne's team announced the discovery in July 1991, another astronomer was hot on the trail of pulsar planets—worlds that would prove to be the first genuine planets found around another star.

Alex Wolszczan became interested in astronomy while a child in his native Poland. "My parents started to show me the sky and the constellations," he said, and he began to view the moons of Jupiter through a small telescope, and found Orion an especially impressive constellation. "I felt it was a pity that it disappeared from the sky at the end of winter," he said, "and it was fun to see it come back in the fall." After a brief stint as an optical astronomer, Wolszczan switched to radio astronomy when his university tried to build an array of radio telescopes.

"I was part of the design team," said Wolszczan. "The vision was great, but it was completely unrealistic. The technology, money, and experience weren't there." He added, "Doing astronomy in Poland, especially in the old communist times, was not that easy. Funding was poor, and your ability to travel was limited, not only for financial reasons but also for political ones." However, Poland had a long tradition of astronomy, with the most famous Polish astronomer being Copernicus.

In late 1981, under threat of a Soviet invasion, Poland's communist dictator decreed martial law. "After the imposition of martial law, a lot of people lost the incentive to live and to work under such a harsh regime," said Wolszczan, who in 1983 moved to Puerto Rico, home of the great Arecibo radio telescope. "I wasn't afraid of living in the middle of nowhere. In fact, both my family and I were so tired of all of the political turmoil in Europe that going to a tropical paradise and staying away from the world for a while looked like the perfect thing to do."

Built by Cornell University, Arecibo is the world's largest single-dish radio telescope, an antenna a thousand feet across, nestled in a natural bowl of the landscape. In early 1990, however, a routine inspection had revealed cracks in the telescope's structure. The operators immobilized it, and the telescope could not point to specific objects in the sky. Instead, it could detect only objects that happened to drift over it, so astronomers who planned to target specific radio sources were out of luck.

Arecibo's problem actually gave Wolszczan the opportunity to search for a peculiar, extremely fast-spinning breed of pulsar. The first of these so-called millisecond pulsars had been found in 1982, spinning 642 times a second. That gave it a period of just 1.6 milliseconds and made it over twenty times faster than the fastest "normal" pulsar, the Crab pulsar. By 1990, only a few more millisecond pulsars had been found, in part because they were much fainter than normal pulsars. Pulsars congregate along the plane of the Galaxy, so pulsar seekers usually looked there.

But Wolszczan wanted to search away from the Galactic plane. "In normal conditions, it would be very difficult to get such a proposal approved," he said, because the chances of success seemed slim. "The only reason I did that search was precisely because the telescope was broken. Under conditions of almost unlimited access to the telescope, I had a very good chance to do what I thought was the thing to do."

Wolszczan's search paid off handsomely with the discovery of two millisecond pulsars. One, in the constellation Serpens, was immediately interesting, because it had a companion that was another neutron star. The two whirled around each other in such a way that they would allow scientists to test Einstein's general theory of relativity better than any other pulsar system then known.

The other millisecond pulsar, PSR B1257+12, was more subtle, but it held a far greater treasure. The pulsar lay in the constellation Virgo, 1300 light-years from Earth, and spun once every 6.2 milliseconds.

The pulsar was cantankerous. "It was very difficult to match any timing model that you would normally fit to data for a millisecond pulsar," said Wolszczan. "It just didn't work." A timing model is supposed to predict exactly when a pulsar's pulses reach Earth, but Wolszczan's timing model failed to do so. Part of the problem, he thought, was the pulsar's measured position on the sky. That position might be slightly wrong, by just enough to ruin the timing model.

Wolszczan therefore asked Dale Frail, a young astronomer at the National Radio Astronomy Observatory in Socorro, New Mexico, to measure a better position for the pulsar. "I'm a horrible night person," said Frail. "I'm up in the morning at six. That's one of the reasons I went into radio astronomy." Jocelyn Bell had found radio astronomy attractive for the same reason. "I sort of wake up every morning thinking, 'I get paid to do what I do, and it's absolutely *fun*,'" said Frail. "And yet, to get paid to do this, you have to be good at it, because there are so many other people who would like to get paid to do this sort of fun; so I feel enormous pressure to work hard and do well. A lot of my colleagues who are brighter than me have had to leave astronomy. I'm just one of the lucky ones."

Frail likes pulsars because of their connection to stellar evolution. "I'm interested in the whole cycle of how stars are born, how they live, and ultimately how they die," said Frail. "The thing that interests me the most is the end point of the evolutionary cycle. Pulsars give me a tool that I can use to study the evolutionary history of stars." It did not take much to convince Frail to measure the position of the new millisecond pulsar. To do so, he used the Very Large Array, a network of twenty-seven radio dishes that work together to yield a much more precise position than the single Arecibo dish can.

"I tried on several different occasions to look for the pulsar," said

Frail, "but I couldn't find it." Because interstellar gas affects radio waves, radio sources twinkle, and PSR B1257+12 was twinkling below the threshold of detectability; so in January 1991, Frail and Wolszczan submitted a formal request for enough observing time on the VLA to track down the recalcitrant pulsar. They did not suspect that the problem with the pulsar's pulses was caused by planets. Instead, they thought it had a companion star whose gravity perturbed the pulsar and affected the arrival of its pulses.

Meanwhile, as he waited for the VLA proposal to be approved, Wolszczan gathered more data with the Arecibo telescope, to understand better the problems with the pulsar. Despite their rapid spins, millisecond pulsars usually behave themselves. That is because they are old and stable, so old that they do not reside in supernova remnants. They are ancient neutron stars that were once pulsars and have died and then been revived by material spilled onto them from a companion star. This material makes them spin extremely fast, so that they begin radiating again, but because of their advanced age, their spins are stable. In contrast, normal pulsars are young and still suffer disturbances that rock their interiors and affect their spins.

By June 1991, Wolszczan began to see what looked like two distinct periods in the data, possibly caused by planets. "The only real difficulty that I had was the lack of an accurate position for the pulsar," he said. "If the position is not quite correct, then you get extra residuals in the pulse arrival times, and you may actually be fooled into believing that the timing model that you have is correct."

Late July saw the sensational report from England of what then seemed to be the first pulsar planet. Wolszczan received the news with mixed feelings. "It was very exciting," he said, "but at the same time, I had something interesting in my hands, so I certainly felt a little bit disappointed. At that time, however, I was not confident enough that I had anything planetlike in the data. I had all kinds of suspicions, and I was actually quite desperate about coming up with ideas of how to explain what I saw in terms of problems with the model, problems with the equipment, problems with the analysis code. You normally think about things like that first, and only after you have eliminated all those potential problems can you start doing physics. I was actually getting close to that moment, when Andrew Lyne's paper came out."

Meanwhile, Dale Frail had succeeded in locating the pulsar with the VLA, and in August he analyzed the observations. "I had just read the paper on the pulsar planet that weekend," said Frail. "That was very much on my mind when I reduced these data. I faxed Alex the pulsar's position, and I remember making a joke to him: 'Don't go finding any planets.' And he actually sent me email back, saying there were two planets. I thought he was joking."

Once Wolszczan had the improved position, it took him only a day to identify the planets. Problems still persisted, though, for even with two planets his model failed to predict the exact times of the pulsar's pulses.

In the model, Wolszczan had assumed that the planets' orbits were circular: "So I thought that maybe there's nothing wrong with my model—it's still planets—but the orbits are not exactly circular." In September, Wolszczan traveled to Cornell and modified the computer code, adding small eccentricities to the planets' orbits. "That was the magic moment when it would either work and basically prove that it was planets, or it would show that it was something else," he said. "I ran the code again and looked at the results." The new model fit perfectly. Said Wolszczan, "I knew then that it could hardly be anything other than terrestrial-mass planets around that pulsar." They were the first planets ever discovered beyond those of the solar system.

Wolszczan began to give talks about the planets, and the first detailed report of the momentous discovery appeared in the December 14 issue of *New Scientist.* "Radio astronomers in the US have discovered two and possibly three planets around a nearby pulsar," said the article. "The discovery is the second reported case of planets orbiting a pulsar and compounds the mystery of how planets can exist around these objects." The two planets that Wolszczan and Frail felt confident about had just a few times more mass than the Earth, with the inner planet slightly more massive than the outer. The inner planet circled the pulsar once every 67 days, and the outer one revolved every 98 days. The inner planet was slightly closer to the pulsar than Mercury is to the Sun, and the outer planet was slightly farther. (The possible third planet, which had a one-year period, was later disproved.)

The *New Scientist* report noted that the new planet pulsar, PSR B1257+12, differed from the first by being a millisecond pulsar. This could explain the planets' existence, because a millisecond pulsar acquires its rapid spin when a companion star dumps material onto it. Unlike many millisecond pulsars, PSR B1257+12 was single, so its companion had been completely destroyed, and Wolszczan and Frail suggested that the planets they had found might be "second-generation" planets formed from blobs of gas left over from the companion's death.

Astronomers quickly hailed the new discovery. The following week, in the Christmas issue of *New Scientist*, a lengthier article featured views of independent astronomers who were asked to assess the reality of the discovery reported the week before. The headline said it all: PULSAR'S PLANETARY ENTOURAGE IS REAL, SAY ASTRONOMERS. The periods of the two planets were especially encouraging; neither was a simple fraction of an Earth year, suggesting that neither could arise from a simple error in the model. Furthermore, the two planets were in a 3:2 resonance: the inner planet completed three orbits in about the same time that the outer planet made two. A similar 3:2 resonance holds in our own solar system, between Neptune and Pluto. This resonance would cause the pulsar planets to interact gravitationally, because they would repeatedly be in the same place with respect to each other and would therefore perturb each other. If astronomers could detect this perturbation, they would prove once and for all that the planets really existed.

The new pulsar planets made astronomers feel better about the *first* pulsar planet, the one that Andrew Lyne's team had reported around PSR B1829-10; but the *New Scientist* article also contained these ominous words: "Many astronomers were sceptical of PSR 1829-10's planet, because its orbital period is almost precisely six months. This suggests that the perturbation detected in the pulsar might be produced by the Earth's motion around the Sun."

Within days, those words would prove devastatingly true.

One of Our Planets Is Missing

In England, Andrew Lyne heard about the new pulsar planets with great interest. The American find had come so soon after the British one that even more pulsar planets seemed to be in the offing. But Lyne was about to make an awful discovery concerning his own planet, the one with the six-month period that he and his colleagues had reported back in July.

Around January 2, 1992, Lyne was examining the pulsar. "I was doing some work on the data, trying to study the orbit in more detail," he said. "I looked at a couple of files and noticed that the two positions in the two files were somewhat different. It surprised me a bit." The old and new positions for the pulsar differed by a ninth of a degree. "Normally our procedure was that once we had found a new position for a pulsar," said Lyne, "we would update the old position to the new one and repeat all our processing. But for this pulsar, that had not been done." An incorrect position for a pulsar could affect the timing model, since the exact time when a pulsar's pulse reaches Earth depends critically on both the Earth's exact position in its orbit and the pulsar's exact position on the sky.

Lyne therefore entered the pulsar's new position. "Only a matter of three minutes later," he said, "the computer had reprocessed it, and the six-month periodicity went away. Within half a minute of seeing it disappear, I knew what had happened: there was no planet around that pulsar at all.

"For the next half hour or so, I just sat staring at my work station. I was completely numb and inert, figuring out all the consequences of all the rubbish that we had written about and spoken on in the previous few months. It was a terrible feeling."

Lyne told his colleagues of the result. "This was in the period around Christmas and the New Year," said Lyne, "and not wanting to ruin Setnam Shemar's holiday too much—I knew he would take it very badly—I spoke of it to Matthew Bailes the next day, and we discussed its implications." The next two weeks were difficult. "A lot of people had spent a lot of time thinking about the possibilities of this object and how it might have formed," said Lyne, "and we didn't want people spending further time working on something that didn't exist."

Lyne's team therefore searched for an opportunity to retract the planet. Lyne was already scheduled to give a talk on pulsar planets at the next meeting of the American Astronomical Society, which would begin just a little over a week later in Atlanta. These meetings occur twice a year and typically attract over a thousand astronomers, who report the latest discoveries in their field. At the time, the president of the American Astronomical Society was John Bahcall, of the Institute for Advanced Study in Princeton.

Said Lyne, "I phoned him up and spoke to him, and told him I had this problem. He was very sympathetic, and offered me the opportunity to still give a talk at the AAS, but to put the record straight." Lyne would speak on Wednesday, January 15, the third day of the week-long conference.

Neither Alex Wolszczan nor Dale Frail, whose own pulsar planets had just been reported, knew anything about the first planet's disappearance—until they reached Atlanta. "I was at the AAS meeting, the famous AAS meeting," said Frail. "Rumors among the small group of pulsar people that were there started circulating. We knew Andrew Lyne was coming there for this talk on pulsar planets: Alex was going to describe his discovery, after Andrew Lyne's talk. And word circulated that Andrew was actually in town, but staying at a different hotel, because he had some very important news, potentially that the pulsar planet had gone away." Wolszczan heard the rumors, too, and the day before the talks Bahcall told him that Lyne's planet did not exist.

Nearly everyone else at the meeting knew nothing. The talks took place in a huge, luxurious hotel conference room, bedecked in maroon and lit by gigantic chandeliers. The crowd was equally huge, for most of the thousand-plus astronomers at the meeting were pouring into the room to hear the latest developments in what had been the greatest astronomical discovery of 1991: the first planets found beyond the solar system. Andrew Lyne, who had reported his planet first, would face the crowd first, shortly before noon, and Alex Wolszczan would follow half an hour later.

"I was sitting in the front of the audience," said Frail, "next to Alex and a few other pulsar people. I don't remember Alex's talk at all, but I certainly remember Andrew's talk. I remember him delivering the punch line, and feeling the intake of breath propagating back through this large audience room. Just people going—" Frail drew in his breath "—that propagating. It was a very big shock to most people, and what a horrible thing it must have been for Andrew to stand up in front of those people and give this talk."

At the start of Lyne's talk, all had appeared well; there was no word of any retraction. Instead, Lyne recounted how his team had discovered the six-month oscillation in the pulsar's period, and he then listed five possible causes. The first possibility was that it arose from the Earth's orbital motion, but Lyne said the oscillation differed by two days from half a year, and he noted that another pulsar near this one showed no six-month oscillation. The second possibility was the exciting one, that a planet circled the pulsar every six months, and Lyne sketched out various theories for its origin, saying that the planet probably formed after the supernova rather than before. Lyne then ran through the third, fourth, and fifth possible explanations of the six-month periodicity—that it was due to wobbling of the pulsar, disturbances inside the pulsar, or an interstellar cloud interfering with the pulsar's radiation.

Lyne said his team had continued to monitor the pulsar, and he assured the audience that the six-month oscillation had continued to appear like clockwork. But in looking over the data more closely, he had just discovered that the correct explanation for the six-month oscillation lay in the *first* of the five possibilities listed above.

A gasp ran through the audience. "The planet just evaporated," Lyne said. He went on to explain how two small errors had conspired to produce an apparent six-month oscillation in the pulsar's period. The first error was the pulsar's slightly incorrect position, which the computer tried to correct. The Earth's exact position around the Sun was also crucial, because the Earth moves in an orbit whose radius is eight light-minutes, whereas pulses from the pulsar arrive three times a second. Unfortunately, in correcting the timing model for the pulsar's incorrect position, the computer program simplified the calculation by assuming the Earth's orbit to be a perfect circle. Lay a perfect circle on the Earth's slightly elliptical orbit and the two cross each other four times, so that the difference between the Earth's assumed distance from the Sun and its true distance is first positive, then negative, then positive, and finally negative again—a six-month oscillation that mimicked an orbiting planet that did not actually exist. At the end of his talk, Lyne appeared somewhat shaken. "Our embarrassment is unbounded," he told the audience, "and we are very sorry."

Sitting in the crowd, Frail was impressed. "What a noble thing to do," said Frail. "Most scientists would have issued a correction and would never have shown their face. But Andrew recognized his error, kept quiet about it, and then flew into the United States to take the full brunt of it. It was stunning." The audience felt the same way and met Lyne's talk with thunderous applause.

Lyne had expected a much more subdued, perhaps negative, response. "I was most amazed at the level of sympathy that people had with me," said Lyne. "It was very surprising. I didn't feel I'd deserved it. It was a mistake that I shouldn't have made, if my methodology had all been correct."

The next day, a brief letter from Lyne and Bailes appeared in *Nature*. Bearing the simple title "No planet orbiting PSR1829-10," it concluded: "Our failure to recognize the result of a position difference, and to perform the usual procedure, has resulted in the apparent planet, and we must accept full responsibility for this error. Although there is no planet orbiting PSR1829-10, pulsar timing remains the most sensitive technique for detecting planets outside the Solar System, as exhibited in last week's paper by Wolszczan and Frail."

During his talk, Lyne had said that his error in no way affected the planets around the second, and now only, planet pulsar, PSR B1257+12. Indeed, the precise position that Frail had provided for that pulsar prevented Wolszczan from making the same mistake.

Nevertheless, Wolszczan worried how the audience would receive his talk, which immediately followed Lyne's. "Had I been an outside observer, not knowing much about pulsar timing, I would have probably been quite skeptical," said Wolszczan. "When you speak about something very unusual in such an unusual atmosphere, as it was on that day, you get extra mobilized. I was really concentrating on what I had to say and how to sound as convincing as possible after Andrew's report. Andrew's part was quite emotional, but what I really remember quite distinctly was that when I finished, there were a lot of questions, but there was no question or comment of the kind, 'Well, after what Andrew Lyne said, what you are reporting to us must be nonsense.'"

"To this day," said Lyne, "I'm not quite sure how I could have avoided the error. We were suspicious, you see, right from the start, but we still could not put our finger on what the problem was with this six-month periodicity. So as far as lessons are concerned, I'm afraid that accidents like this will happen." The one thing that he could have done, he said, was to have processed the pulsar data with a different software package, but at the time that had not seemed necessary.

The Planets of PSR B1257+12

Three months after the drama in Atlanta, the scene shifted to California, where Wolszczan, Frail, and several dozen other astronomers gathered for a two-day conference on pulsar planets. The conference was held at the California Institute of Technology in Pasadena, a suburb of Los Angeles. Unfortunately for the astronomers, a day before the conference, the police officers accused of beating Rodney King were acquitted on all but one count, and Los Angeles erupted into riots. "From the airplane, we saw all these fires," said one astronomer, "and we thought, 'What are you guys doing down there?'" Because of the problems, another astronomer required thirty hours to journey from neighboring Arizona. "Now I know what hell is," he said: "eternal transportation."

Nevertheless, the conference was a success, bringing together people who rarely talked with one another: pulsar scientists and planetary scientists. Wolszczan, who spoke first, opened his talk by joking that he had just discovered an error that undermined his planets. "My hope," Wolszczan told the crowd, "is that after this meeting, there will still be planets around pulsars." The most important news came next, when Don Backer of the University of California at Berkeley reported that his team had observed the pulsar with a different radio telescope and had found exactly the same oscillations as those reported by Wolszczan and Frail. Pulsar planets were more real than ever.

Still, some scientists tried to explain the oscillations as arising from wobbles in the pulsar's rotation axis or disturbances in the pulsar's interior. "I thought those alternative models were necessary," said Frail. "I didn't have any problems with them at all." Said Wolszczan, "It's science, isn't it? It's not a healthy situation if you have just one model and nobody questions it. Of course, to me it was kind of comforting that I didn't have to take those models too seriously." Indeed, most scientists thought the alternative models were contrived, and in late 1993 Wolszczan laid any doubts to rest by detecting the gravitational interactions between the two planets that had been predicted. The pulsar planets were thus confirmed as the first extrasolar planets ever found.

Wolszczan also discovered another planet in the system, inside the orbits of the other two. It circled the pulsar once every twenty-five days, less than a third of the orbital period of Mercury. Its mass was little greater than the Moon's, yet it still perturbed the pulsar enough that Wolszczan could detect it. This planet demonstrates how incredibly sensitive pulsar timing is to orbiting bodies, even tiny ones, because a pulsar, especially a millisecond pulsar, is an extremely precise clock in which astronomers can look for telltale deviations. Such a small planet could never have been seen perturbing a normal star like the Sun.

So far, the three planets have no special names, and one astronomer who occasionally deals with celestial nomenclature, Brian Marsden of the Harvard-Smithsonian Center for Astrophysics, is not sure that extrasolar planets even need names. "The popular press would like to have names for them," said Marsden. "In fact, *All Things Considered*, that program on National Public Radio, had a competition for naming the three planets around the pulsar, and just to show you the high level of intellectual capacity of the NPR listeners, the leading candidates were Larry, Curly, and Moe."

Wolszczan has simply designated the planets from innermost to outermost A, B, and C. In science fiction, planets often bear roman numerals instead. For example, in our solar system, Mercury would be Sun I (or Sol I), Venus would be Sun II, Earth would be Sun III, and so on. Although this procedure is logical, it assumes more than astronomers are likely to know about other planetary systems. In general, astronomers will detect the larger planets first, which may not be those closest to their star. If our solar system were observed from afar, the first planet to be discovered would likely be giant Jupiter, not minuscule Mercury; so if extraterrestrial astronomers designated the Sun's planets in the order of their discovery, then Mercury through Pluto would not form the logical sequence I, II, III, . . . , IX but instead a seemingly haphazard one, VIII, V, VI, VII, I, II, III, IV, and IX.

Despite the bizarre nature of their star, the three planets of PSR B1257+12 mirror the three innermost worlds of our solar system. The chief difference is the pulsar system's greater compactness. For this reason, and because their star is more massive, the pulsar planets revolve faster. But if the orbital distances of the three pulsar planets were doubled, they would nearly line up with the positions of Mercury, Venus, and Earth around the Sun. This suggests that pulsar planets, and perhaps solar systems in general, obey the Titius-Bode

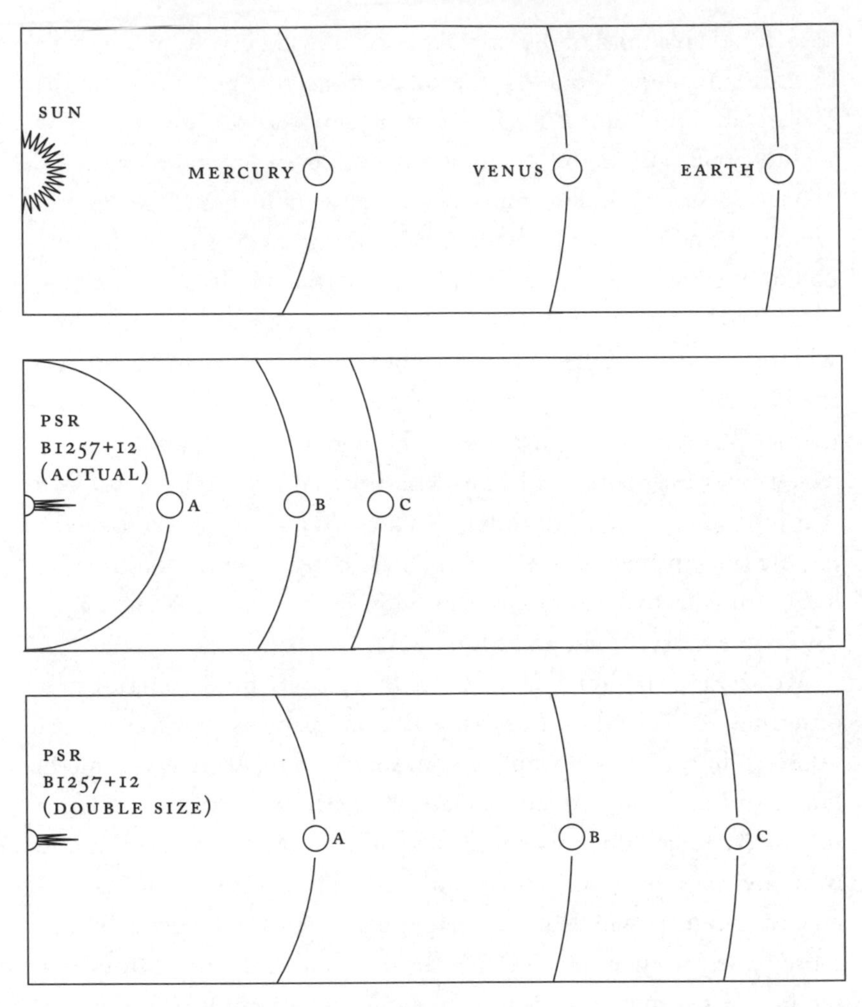

FIGURE 29

The distances of the Sun's three innermost planets (top) resemble those of the pulsar planets (middle); if the pulsar planets' distances are doubled (bottom), they line up almost perfectly with Mercury, Venus, and Earth.

law of planetary distances. The pulsar planets' masses also resemble those of Mercury, Venus, and Earth. In both systems, the innermost world is a featherweight: Mercury has 0.055 Earth mass, and the first pulsar planet has about 0.02. The next two pulsar planets are more massive, like Venus and Earth. In our solar system, the third planet, Earth, is slightly more massive than the second, Venus; in the pulsar system, the second planet is the one slightly more massive. Both of

TABLE 8–1

PSR B1257+12's Planets

Planet	Distance from Pulsar (astronomical units)	Period (days)	Mass[a] (Earth = 1)	Orbital Eccentricity
a	0.19	25.34	0.019	0.0
b	0.36	66.54	4.3	0.018
c	0.47	98.22	3.6	0.026

[a]These are the likely masses if the pulsar planet system is tilted at a random angle to our line of sight. The minimum masses of the planets are 0.015, 3.4, and 2.8 Earth masses, which hold if we view the system exactly edge-on.

TABLE 8–2

The Sun's Innermost Planets

Planet	Distance from Sun (astronomical units)	Period (days)	Mass (Earth = 1)	Orbital Eccentricity
Mercury	0.39	88	0.055	0.206
Venus	0.72	225	0.815	0.007
Earth	1.00	365.256	1.000	0.017

these pulsar planets are more massive than Venus and Earth, though, having around four Earth masses each. All three pulsar planets follow nearly circular orbits, as do Venus and Earth. Here the oddball is our own Mercury, whose orbit around the Sun is rather elliptical.

Although the pulsar planets' orbits are known, their physical characteristics are uncertain. Take, for example, their temperature. "It depends on just one thing," said Wolszczan. "Does the pulsar beam hit the planets or not? If it doesn't, then there's really no substantial radiation of any kind that reaches them, and people who live there may be completely unaware that they are circling something." The planets would then be quite cold. On the other hand, if the pulsar's beam does strike the planets, they could be warm or hot, depending on how much energy they absorb. One of them could even be the same temperature as the Earth.

Still, you probably wouldn't want to set up residence there. "It would be like living inside Chernobyl," said Frail. "The spectrum of energy is mostly gamma rays, high-energy electrons, positrons, ions. It's a nasty, nasty place—not a place to go for a vacation." Of course, in the unlikely event that pulsar planets have intelligent beings, they have probably reached an equally negative conclusion about worlds circling stars like the Sun.

No one knows how dense the planets are or what they are made of, but the worlds are probably rocky and barren. Their composition must depend on how they formed, and astronomers have invented a whole suite of different formation scenarios. In all likelihood, though, the planets were born after the pulsar; they are not survivors of the original star that exploded.

"This pulsar has a very high velocity through space," said Wolszczan, "and a velocity that high must have been generated by something violent that happened in the system. So if there were any preexisting planets around the original star that became the pulsar, they would have been shaken off. For that reason alone, it makes sense to think that these planets must have formed after the formation of the pulsar."

One idea, which Frail favors, works like this. An old neutron star—a pulsar that has died and stopped radiating—is revived when a companion star dumps material onto it. This causes the neutron star to spin faster and faster, until it switches on again and becomes a millisecond pulsar. Such a scenario is plausible, for unlike most normal pulsars, many millisecond pulsars do have companion stars. But the companion may pay a price for its charitable deed, because the reborn pulsar's beam can rip into the companion and destroy it. In 1988, Andrew Fruchter of Princeton University and his colleagues discovered that a resuscitated pulsar in the constellation Sagitta had reduced its helpless companion to only a few percent of the Sun's mass. In 1990, Andrew Lyne's team reported a second example, in an old star cluster named Terzan 5. These are called black widow pulsars, after the spiders that often eat their mates.

"That's the evolutionary link that we need in order to understand millisecond pulsars," said Frail. "The pulsar is destroying its compan-

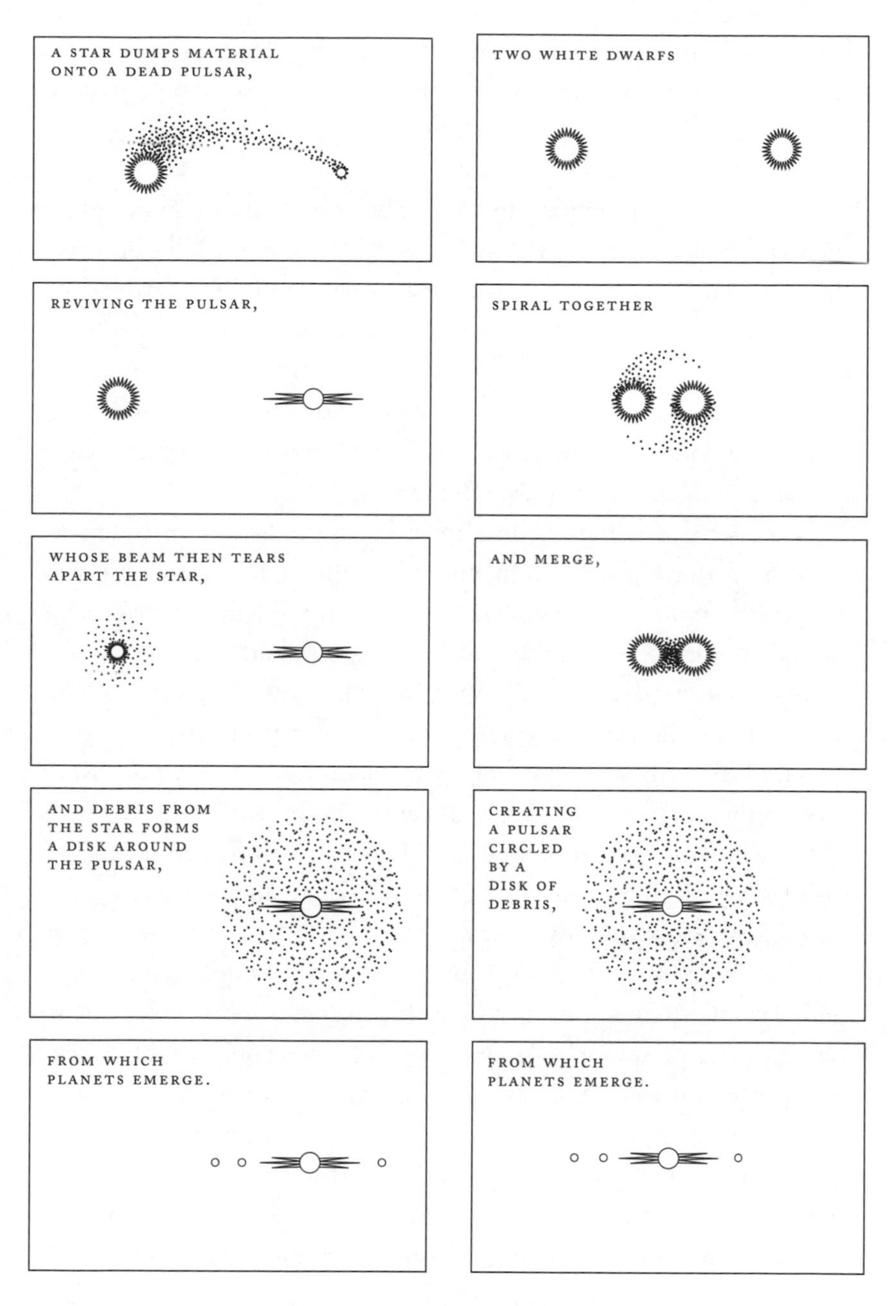

FIGURE 30

Pulsar planet production. Pulsar planets may arise when a reborn pulsar tears apart its companion star (left) or when two white dwarfs merge (right).

ion star, and in so doing a disk of material is created. That's the material that you need to form the planets." Just as our solar system arose in a disk of material around the Sun, so the pulsar planets arose in a disk of material around the pulsar.

Wolszczan prefers another model. "The constraints are so poor that there's a whole big family of ideas," he said. "But the one I like the most is that it could have been a tight binary system consisting of two white dwarfs." Over billions of years, two white dwarfs orbiting close to each other spiral together and merge. If their total mass is great enough—over 1.4 solar masses—they might create a rapidly spinning neutron star, a millisecond pulsar. In the process, a disk of leftover material might appear around the newborn pulsar and give birth to planets.

Both theories have problems, however. In the first theory, the material from the disrupted companion star might not actually gather into a disk. In the second model, the two merging white dwarfs might explode and leave nothing behind, not even a pulsar, let alone planets.

Wolszczan hopes to test the merging white dwarf scenario. "This hypothesis can be verified in a way that I find very attractive," he said. "If you can form a neutron star with planets out of a white dwarf merger, then you can also have a similar system in which the total mass of the white dwarfs is not enough to make a neutron star. Then you may end up making a massive white dwarf with planets around it." For example, if two white dwarfs, each with 0.6 solar mass—the average value for white dwarfs—merge, they may form a single white dwarf with 1.2 solar masses, plus perhaps a few planets. Some white dwarfs actually vibrate, and variations in these vibrations could reveal planets, just as variations in a pulsar's pulses led to the discovery of the first planets outside the solar system. These white dwarf planets would be interesting in their own right, and their existence would bolster the idea that pulsar planets can form when white dwarfs merge.

Because pulsars have planets, so might even more exotic objects. A typical neutron star has 1.4 solar masses, held up by the pressure among its neutrons. If a star's collapsing core exceeds 2 or 3 solar masses, not even this pressure can counteract the pull of gravity, and a monstrous black hole arises—an object with such an intense gravitational pull that nothing, not even light, can escape. Several black holes, most notably the famous Cygnus X-1, have companion stars whose material falls into the black hole. Before it takes the final plunge, however, this material gathers in a disk around the black hole, and that disk could conceivably form planets.

In 1991, after the excitement of the pulsar planet discovery, many astronomers thought that more pulsar planets would soon be detected. "It's been a personal disappointment of mine that there have been no more of these pulsar planets found," said Frail. "We've not found some new branch on the tree of knowledge, if you want to think of it that way. What you're learning is a particular path by which planets form, which is obviously not as common as one would like, so we have this peculiar chain in the evolutionary picture. If I were an anthropologist, it would be like studying some very obscure tribe and trying to apply it to humankind in general." Because of their extremely short and stable periods, millisecond pulsars readily reveal their planets, even those as small as the Moon; but for some reason, most millisecond pulsars do not have planets.

So far, though, astronomers have found only a few dozen millisecond pulsars. "They are quite a bit harder to find than normal pulsars," said Wolszczan, "because they are generally fainter, and they have shorter periods—very sharp pulses—so requirements on your telescope's sensitivity and hardware characteristics are much tighter." Only a few percent of known pulsars are of the millisecond variety, but in reality millisecond pulsars are probably much more common than they seem. Given the small number of millisecond pulsars known, Wolszczan does not find it alarming that only one is known to have planets.

Astronomers occasionally talk about planets around other pulsars. "As you can understand," said Andrew Lyne, "I'm a little bit nervous about publishing anything with the word *planets* in it, but it is certainly a possibility that there are planets around this pulsar." The pulsar Lyne mentioned was PSR B1828-11, which lies close on the sky to the ill-fated PSR B1829-10. At the 1992 Caltech conference, Lyne's colleague, Matthew Bailes, reported that the pulsar displayed unusual behavior. Lyne said this has continued, and three oscillations seem to appear in the data, none of them corresponding to an Earth year. However, the pulsar is young, so Lyne warned that the oscillations may arise from disturbances in its interior rather than from orbiting planets.

In 1995, Russian astronomer Tatiana Shabanova published data on another young pulsar, PSR B0329+54, for which claims had been made back in 1979. According to her, this pulsar has at least one planet, with a period of seventeen years, and possibly another, with a three-year period. However, these claims are in dispute; Wolszczan, Frail, and Lyne all want more data before they declare that the planet, or planets, really exist.

Meanwhile, Wolszczan continues to monitor the one pulsar that definitely has planets, PSR B1257+12. "There is a new, unaccounted-for effect," said Wolszczan, "residuals that stay after you have subtracted out all the known planets." One possibility, he said, is that the pulsar has a much more distant and massive planet, similar to Saturn, that takes over a hundred years to circle the pulsar. More data are needed to confirm this possible planet's presence.

"Real" Planets

Strangely, the first extrasolar planets discovered have elicited only a lukewarm response from those who look for such worlds. "There's been very little windfall that's come to the pulsar community as a result of this discovery," said Frail. "There have been no huge increases in funding to help go looking for planets, and the rest of the planetary community has kind of ignored this result. I don't know why that is."

Some scientists grumble that pulsar planets aren't "real" planets, because they don't circle a "real" star like the Sun. A major article on extrasolar planets that appeared a full four years after Wolszczan and Frail's 1991 discovery flatly stated, "the results to date are that *no other planetary systems have been detected*," and then went on to cite the Wolszczan and Frail work, only to minimize its importance.

"I don't get terribly upset about it," said Wolszczan, "but I think it is important to correct statements like that. Pulsar planets are the only evidence that terrestrial-mass planets exist outside the solar system. They will remain so for quite a while." In stark contrast, any detectable planet-induced wobble in a Sunlike star must be caused by a much more massive planet, like Jupiter.

"The pulsar planets are like a carbon copy of the inner solar system," said Wolszczan. "The only thing that is odd and unusual is that the central object is a pulsar. If we replaced the pulsar with a normal star, people would be hopping around like mad saying there must be life there and water and blue lakes and waterfalls and whatever. If you ignore the fact that pulsar planets exist, not even saying that they were the first to be found, it's quite strange.

"And I do get a lot of queries about it. It's kind of embarrassing, because I don't feel like calling up someone who did this to say, 'Hey, you have forgotten something.' I don't think that's the way to do it."

Wolszczan said, "Look at it psychologically. Some of these people have spent half their careers searching for planets around Sunlike stars, and then all of a sudden someone who wasn't even looking for planets runs into the first extrasolar planetary system." He laughed. "I can understand that there may be some bitterness, but there's really nothing we can do. Pulsar planets are there and will be there. You just have to deal with the fact."

Part of the problem, said Wolszczan, stems from one element that drives planet seekers: the search for extraterrestrial intelligence. "There, pulsar planets are not directly helpful, especially if you look for life as we know it, which is the very strong undertone in those planetary searches," he said. "They are all very anthropocentric, which is okay—it's understandable—but maybe it's a little narrow-minded." He said it would be wiser to search for planets around as many different types of stars as possible, because such worlds hold the potential to teach astronomers the most about planets in general. In any case, benign conditions, such as those on Earth, have never been a prerequisite for planethood; if they were, the Sun would have only one "real" planet.

Frail also resents the implication that pulsar planets aren't "real" planets. "But history is generally pretty kind to things like this," he said. "It's just the noise that's going on right now." To Frail, the most important contribution made by the planets of PSR B1257+12 was philosophical. "The search for planets around other stars is going to be one of the big things in the next century," he said. "It's going to have an enormous impact on astronomy. Now if you had been asked to predict the properties of the planets that people were going to find, you would never have come close to pulsar planets. So if you can find planets around such unlikely objects as pulsars, you can find them anywhere, because planet formation is going to occur wherever it wants to. If I were a planet searcher, the message I would take from this is: 'I'm on the right track, and this work I'm going to do for the rest of my life *is* going to pay off.' "

In the years before and immediately after the discovery of the pulsar planets, however, all searches for planets around normal stars came up dry, and astronomers started to worry. A 1992 article in *Astronomy* bore the title DESPERATELY SEEKING JUPITERS. At the very least, the largest and easiest-to-detect planets began to look rare. Planet seekers could console themselves by noting that the failure to find Jupiterlike planets said nothing about the prevalence or absence of Earthlike planets, the worlds that intelligent life presumably requires. After all, who needs a gas giant like Jupiter?

Then came a remarkable idea that suddenly upped the ante: lifeless Jupiter might be the reason that intelligent life exists on Earth.

9

Thanks, Jupiter

JUPITER AND Saturn, the farthest planets the ancients knew, patrol the vast domain of the outer solar system. Jupiter sports a vibrant atmosphere whose great red spot dwarfs the Earth, while Saturn's brilliant rings encircle a golden globe that commands a retinue of at least eighteen moons. Because of Jupiter's brightness and its slow, majestic movement—it revolves around the Sun only once every twelve years—the ancients named it for the king of the gods. Saturn paces the sky even more ponderously, taking nearly thirty years to complete an orbit, and its name salutes the god of time.

As astronomers know today, these weighty names fit both worlds perfectly, for their most impressive property is not color or beauty but sheer mass. Together, the two gas giants harbor twelve times more mass than all the other planets combined. Indeed, only the enormous sizes of Jupiter and Saturn make these distant worlds stand out in the sky. If Jupiter and Saturn were instead as small as Earth, Jupiter would lose most of its luster, and Saturn would be so faint that it would likely have eluded the ancients.

Although astronomers have assumed that other solar systems resemble ours and thus have worlds like Jupiter and Saturn, this need not be the case. "There are all sorts of ways you could mess up the formation of a Jupiter," said George Wetherill, a planetary scientist at the Carnegie Institution of Washington, "and maybe nature just doesn't care whether it makes a Jupiter or not." For this reason, most solar systems could lack such large planets. "If this is true," said Wetherill, "it might be surprising that we happen to live in a planetary system which has a Jupiter and a Saturn. But maybe it's not so surprising, because perhaps if it weren't for Jupiter and Saturn, we wouldn't be here."

When Wetherill originated this idea in 1992, the notion that Jupiter and Saturn are vital to intelligent life was startling, even radical. After all, both planets are lifeless gas giants hundreds of millions of miles away, locked in the frozen reaches of the outer solar system; they seemed irrelevant to the inhabitants of a small, wet world bathing in the warm glow of sunshine. Wetherill's idea actually followed from theories of how the solar system formed. When the planets were developing, 4.6 billion years ago, countless comets roamed the solar system. Many of these collided with one another to form the cores of the four giant planets—Jupiter, Saturn, Uranus, and Neptune—but the mighty gravitational force of Jupiter and Saturn tossed trillions of other comets away. As a consequence, few comets remain to hit the solar system's planets today, sparing Earth in particular from the devastating impacts that could have thwarted the development of intelligent life.

"I admire George Wetherill's work greatly," said David Stevenson of the California Institute of Technology. "I certainly think that one part of his idea is very sensible. There's no doubt that the presence of Jupiter and Saturn has a major consequence for the flux of bodies hitting the Earth. Those planets are indeed protectors of life.

"The issue of whether Jupiter and Saturn are hard to make is, in my opinion, an open question. In order to understand how Jupiter and Saturn formed, you need to know more than just how to aggregate solid bodies; you have to understand in detail how the gas aggregates." That is because Jupiter and Saturn, unlike the other planets, consist mostly of hydrogen and helium.

Alan Boss, one of Wetherill's colleagues at the Carnegie Institution, called the idea very attractive. "It's something that was very novel," said Boss. "In the depths of our depression not too long ago, when no one was finding any Jupiters, it really struck a chord. If Jupiters don't exist around most stars, yet we have one here, maybe that's the explanation: somehow it takes a Jupiter to allow a sentient being to evolve."

But Wetherill himself urged caution. "The first thing I would warn you about is not to take this idea real, real seriously," he said. "The reason I'm saying this is because I was in a conversation with some NASA officials, and I was embarrassed to learn that it was being described as a new paradigm. But I just thought it was a cute idea."

Jupiter and Saturn: Guardians of Life?

If that idea is correct, however, it raises the stakes in the search for extrasolar giant planets. If astronomers examine other stars and fail to find Jupiters—the planets that are easiest to detect—those solar systems may possess no life at all, even if they have warm, wet planets like Earth. Conversely, if planets like Jupiter abound, they will boost the hope that life exists elsewhere.

George Wetherill followed a circuitous route to the study of planetary systems, a route that went through geology rather than astronomy. "I really wanted to be an astronomer," said Wetherill, who as a kid in the 1930s had watched meteors streak across the sky, "but I was told that there was practically no chance—I was sort of a B student—because there were only a thousand astronomers in the country, and they're all geniuses; so why should I expect to be an astronomer?" Instead, Wetherill decided to become an engineer. However, after he served in World War II, the GI bill enabled him to attend the University of Chicago, which did not have an engineering department, but where he found another subject, nuclear physics, more exciting.

After he earned his doctorate, though, further job opportunities in that field didn't appeal to him. "I didn't want to make bombs," he said. "I recognized the seriousness of the nuclear threat of the Soviet Union. But it would be a very depressing way to spend your life, figuring out better ways to kill hundreds of millions of people. It's like: I'm against capital punishment, but even if I were in favor of it, I still wouldn't want to be a hangman."

Wetherill instead applied his knowledge of nuclear physics to rocks, some of which have minute quantities of radioactive substances that decay in ways which betray the rocks' ages. These ages help geologists reconstruct the Earth's history, and Wetherill later became a professor of geology, though he had never taken a geology course.

"In geology, you learn about the Earth by looking at the record in the rocks," said Wetherill. "As you go back in time, this record gets more and more obscure, and almost vanishes by the time you get back to 4 billion years or so. But there's another way to try and get at the history of the Earth, and that's to go from the other end—understanding how the Earth formed and what its initial conditions might have been, which then in turn control its further evolution."

Today Wetherill uses a powerful desktop computer to simulate the formation of the Earth and the other planets. "What I think is of some relevance to the search for extrasolar planets," he said, "is to try to develop a general theory for the formation of planetary systems, of which our solar system would be but one example and might be similar to others in some ways and different in other ways. There'd be no specific need for all planetary systems to develop in exactly the same way. So I've been looking into how variations of the initial conditions for the formation of a planetary system might lead to variations from one system to another."

Wetherill starts his model with small bodies orbiting a star. These bodies collide and grow into larger ones, which in turn become the planets of a solar system. Terrestrial planets—rocky worlds similar in mass to Earth—form near their star, where the disk of gas and dust orbiting the star is hot and only substances with high melting points, rock and iron, condense into solids. In 1991, Wetherill's work indicated that such planets are fairly easy for nature to produce, if nature operates the same way that his computer simulations do. Proponents of extraterrestrial life greeted Wetherill's result, since it meant that many if not most stars should have small planets like Earth. Wetherill even found that a typical simulation produced four terrestrial planets that match the pattern in our solar system: the first planet (e.g., Mercury) was small, the next two (Venus and Earth) were larger, and the final one (Mars) was again small. Also in 1991 came the stunning discovery of the first two pulsar planets, whose masses resemble Earth's and demonstrate that nature can indeed manufacture terrestrial-mass extrasolar planets—even in exotic locales.

The story changed, though, when Wetherill turned to the giant planets. According to an idea that Japanese astronomer Hiroshi Mizuno and his colleagues published in the late 1970s, these planets formed in a more complicated way than did the terrestrial planets. Far from the star, the disk was cool, so ices of water (H_2O), methane

(CH_4), and ammonia (NH_3) condensed. These far outweighed the rock and iron, because the ices contained three of the most common elements in the universe—oxygen, carbon, and nitrogen—joined with hydrogen, the most abundant element of all. Due to all this material and the large volume of the outer solar system, enormous objects of ice and rock formed that had roughly 10 times more mass than the Earth. In the case of Jupiter and Saturn, these objects formed quickly, and Mizuno said their gravitational pull grabbed huge quantities of the hydrogen and helium gas that pervaded the disk. Today, Jupiter has 318 times the mass of the Earth and Saturn 95 times, most of it hydrogen and helium. Uranus and Neptune, which today have only 15 and 17 Earth masses, grabbed little if any gas, presumably because their ice-rock cores formed later, after the Sun had blown away the hydrogen and helium gas in the disk.

Mizuno's model for the formation of the giant planets explains why all four have similar cores. It also agrees with the planets' observed atmospheric abundances. For example, the theory correctly predicts that all four planets should have more carbon relative to hydrogen than the Sun. This is because their cores had methane, which contains carbon, and some of that leaked into the planets' atmospheres. Furthermore, the two planets that captured the least hydrogen and helium—Uranus and Neptune—have the greatest carbon-to-hydrogen ratios, just as the theory predicts, because their methane was least diluted by the infalling hydrogen and helium.

In 1992, Wetherill ran his model on the computer—and it failed. "Instead of forming a system that looked like Jupiter and Saturn," he said, "I usually got a large number of objects moving in highly eccentric orbits, and it's rather difficult for this to develop into the story we like to tell where a ten-Earth-mass core develops and starts to capture gas. So the possibility occurred to me that maybe it doesn't arise very often. Maybe Jupiter is a fluke, which actually did occur in some of my calculations, but only as a fluke rather than as a rule of thumb.

"Now this is very egotistical: to think that just because *I* don't know how to make Jupiter implies that *nature* doesn't know how to make it. But this was at the same time that people were failing to find any Jupiterlike planets around other stars. It began to look—and I think it still does look—as if Jupiter is an unusual object.

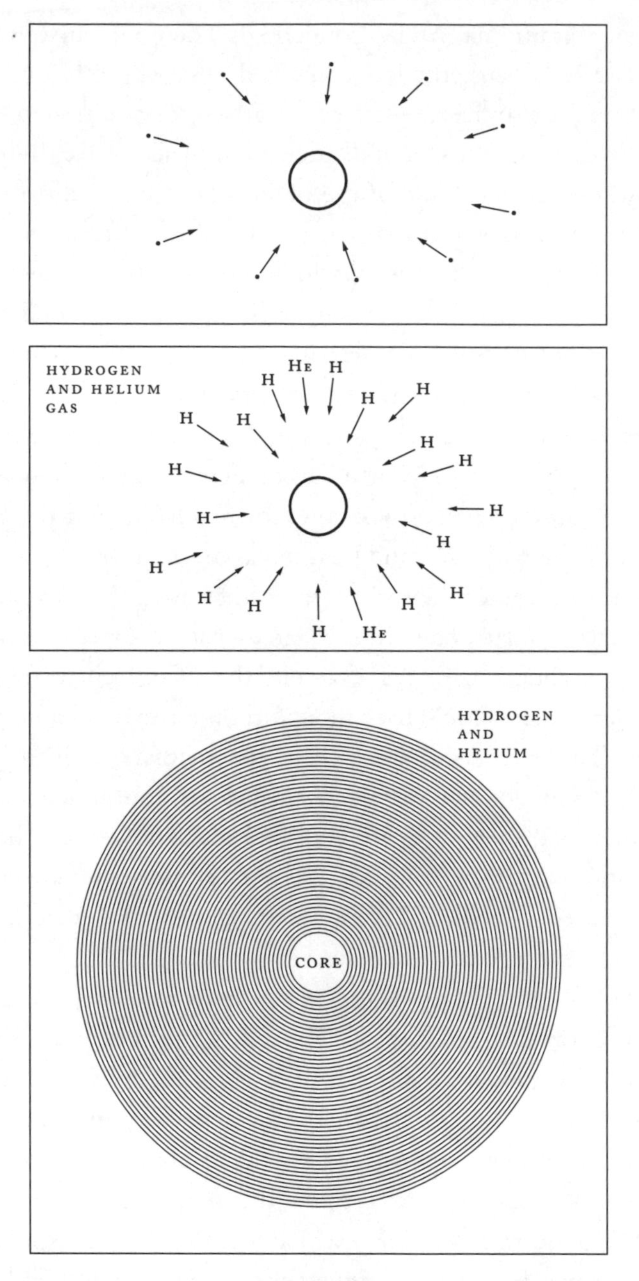

FIGURE 31

Gas giant creation. In the first stage of the formation of a gas giant planet, like Jupiter or Saturn, particles collide to create a large core (top); the gravity of this large core then attracts large amounts of hydrogen and helium gas (middle), which dwarf the core (bottom).

"So if that's the case, then that raised the question: why do *we* have a Jupiter? The possible answer was that our solar system may be a highly biased sample, biased by the fact that we're here to see it." In contrast, a solar system that lacks a Jupiter and a Saturn is not self-observable, if it never gives birth to intelligent life.

Long ago, Jupiter and Saturn cleaned the solar system of most cometary debris, including comets that may have been quite large, possibly as large as planets. Consequently, catastrophic impacts—such as the one that killed the dinosaurs and most other species living 65 million years ago—now occur only rarely. In the early solar system, said Wetherill, Jupiter and Saturn worked as a team, playing ball with comets and passing them back and forth from one planet to the other, boosting the comets' velocities. Most of the comets got cast clear out into interstellar space, while some were deposited in the Oort cloud, the vast comet reservoir that surrounds the solar system. Uranus and Neptune, which are smaller than Jupiter and Saturn, treated comets more gently. They transported comets toward Jupiter and Saturn, which then got rid of them. In 1994, the world witnessed a dramatic event that symbolized Jupiter's role as protector of terrestrial life: the planet took a direct hit from Comet Shoemaker-Levy 9, an impact that left Jupiter's atmosphere scarred for months. Had the planet not intervened, the comet might someday have collided with Earth.

To see what would happen if Jupiter and Saturn did not exist, Wetherill tried simulations in which the two planets failed to accrete much hydrogen and helium. This could occur if a solar system lost its hydrogen and helium disk before the cores of Jupiter and Saturn captured much gas. Many solar systems may therefore have "failed" Jupiters and Saturns—planets with the masses of Uranus and Neptune, but located at the orbital positions of Jupiter and Saturn.

Wetherill found that such planets are dangerous. "Failed Jupiters and Saturns are not really effective in removing material from the solar system," said Wetherill, "but they're effective enough that they perturb this material over the lifetime of the solar system into orbits which come into the inner solar system, into the region of the Earth." Comets would therefore bombard this hypothetical Earth as much as a thousand times more often than they do the real one. On the real Earth, devastating impacts occur roughly once every 100 million years, leaving long intervals of relative calm, during which life can evolve. On an Earth without the protective shield of a full-fledged Jupiter and Saturn, these catastrophes could strike every 100,000 years—a time shorter than *Homo sapiens* is old.

Life might still originate on such a world, Wetherill said, but it might not develop into complex life. "Even on the Earth, it wasn't until around 600 million years ago that multicellular organisms became common," he said. "So we went for 4 billion years without doing much; it might not take much more to prevent intelligent life from arising at all. Apparently, it's not really easy, even on a very nice planet like the Earth."

Without Jupiter and Saturn, the Earth might also be buried under water. Much of the water now on Earth came from comets, and even with Jupiter and Saturn protecting the planet, seawater covers 71 percent of the Earth's surface. If Jupiter and Saturn did not exist, and comets rained down on Earth more often, the Earth might be a completely water-covered world, and any life would forever remain in the sea—another situation that might have prevented the emergence of intelligent life. Although dolphins have a fair degree of intelligence, they have not developed a written language, preventing future generations of dolphins from studying and building on the knowledge of their ancestors.

Interstellar Comets

Astronomers can try to estimate how many extrasolar Jupiters and Saturns exist by studying comets themselves, specifically those from other solar systems. "If other solar systems have Jupiters and Saturns," said Paul Weissman of the Jet Propulsion Laboratory, "then they should also eject comets from the system, which would contribute to the flux of interstellar comets." Therefore, as our solar system moves through space, it should occasionally encounter an interstellar comet.

Unfortunately, even if Jupiters and Saturns are common, no one can yet say how many comets should roam interstellar space. "The biggest uncertainty is simply the fraction of comets that get captured in a star's Oort cloud versus the fraction ejected to interstellar space," said Weissman. "The more that get captured in the Oort cloud, the fewer interstellar comets any solar system will create." The Sun's

Jupiter and Saturn may have ejected some fifty times more comets into interstellar space than into the Oort cloud; in that case, astronomers should have seen about half a dozen interstellar comets entering our solar system over the last two centuries, if planets like Jupiter and Saturn are common. On the other hand, Jupiter and Saturn may instead have tossed only three times as many comets into interstellar space as into the Oort cloud. In that case, astronomers should have observed no more than one interstellar comet.

What astronomers do know well is the number of interstellar comets they have actually seen: exactly zero. Because a comet from beyond the

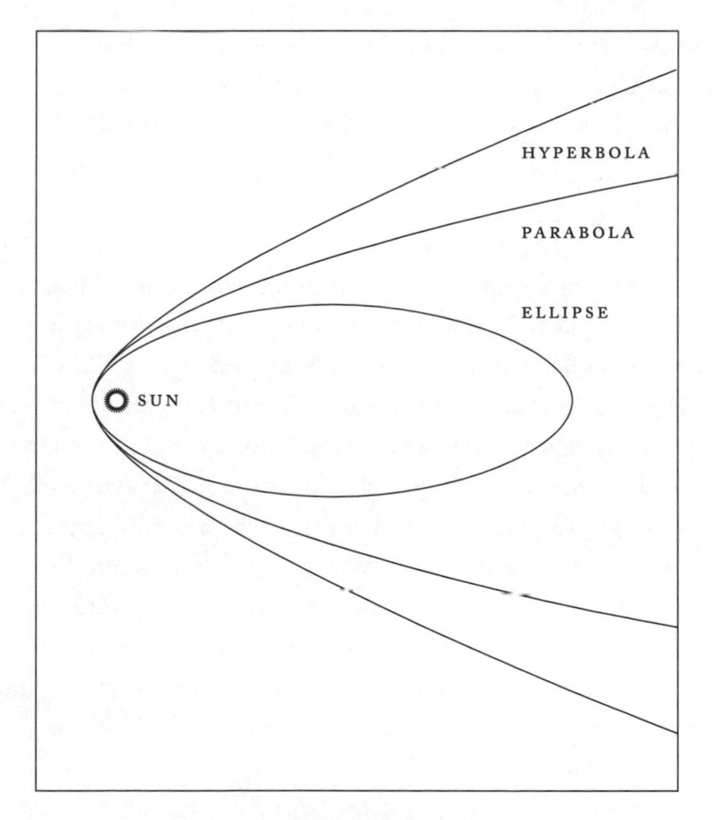

FIGURE 32

Objects that are gravitationally bound to the Sun follow elliptical orbits, with eccentricities of less than 1; a circle is simply an ellipse with an eccentricity of 0. Objects that are not gravitationally bound will escape the Sun's grasp; they follow a parabolic orbit, with an eccentricity of 1, or a hyperbolic orbit, with an eccentricity of greater than 1.

solar system would not be gravitationally bound to the Sun, it would give away its nature by traveling fast, on a so-called hyperbolic orbit.

"The eccentricity of the orbit would be very high," said Weissman. "An elliptical orbit has an eccentricity of between 0.00 and 1.00; a parabolic orbit has an eccentricity of exactly 1.00; and an interstellar comet would have an orbital eccentricity that was probably approaching 2.00, so it would be very obvious that it was coming from interstellar space." Yet no comet has ever been found to have a hyperbolic orbit.

"It's really on the borderline," said George Wetherill. "The problem is rather uncertain, but the fact that we haven't seen any interstellar comets tends to point in the direction that there's not an awful lot of interstellar comets around." If so, Jupiter and Saturn may indeed be rare, and most Earthlike planets are now getting bombarded by the same comets that full-fledged Jupiters and Saturns would have cast into interstellar space. But Weissman warned that this conclusion is premature because of the difficulties in estimating the expected number of such comets.

Lunar Legacies

Prior to Wetherill's speculation about Jupiter and Saturn, those planets were never considered essential to life. They were just there—nice sights in the sky and good worlds for astronomers to study, but little more.

What else is special about our solar system? One spectacular object lies much closer to home than Jupiter and Saturn: the Earth's very large satellite, the Moon. Although three of Jupiter's satellites and one of Saturn's are larger, the Moon circles only a small planet, and its presence has stirred more than just the hearts of romantic lovers gazing up at it. Its gravity raises tides that slosh water onto land, and these may have pushed life from the oceans, where it probably originated, to the continents. Without the large Moon, Earth would experience only the weak tides from the Sun, and life might forever have remained in the sea.

In 1993, French scientists led by Jacques Laskar reported another lunar influence, one that stabilizes the tilt of the Earth's axis. This tilt is presently 23.4 degrees and accounts for the seasons: in July, for example, the Earth's north pole is tilted toward the Sun, warming the northern hemisphere and cooling the southern. However, the size of the tilt itself varies slightly—from 22.0 to 24.6 degrees and back again—every 41,000 years. Small though this oscillation is, it greatly affects the Earth's climate and is thought to be one cause of the ice ages. When the tilt is small, less sunlight reaches far northern latitudes in summer, and when northern summers are cool, ice accumulates and glaciers stampede south.

Without the Moon, though, things would be far worse, for the French found that the Earth's axial tilt would have varied wildly, from nearly 0 to 85 degrees. "Thus, had the planet not acquired the Moon, large variations in obliquity [axial tilt] resulting from its chaotic behaviour might have driven dramatic changes in climate," wrote the French. "In this sense one might consider the Moon to act as a potential climate regulator for the Earth." Whenever the Earth's axial tilt exceeded 54 degrees, the north and south poles would have gotten warmer than the equator, since they would have received more sunlight. On a world with such an erratic climate, life might not have evolved into intelligence, or even originated at all.

Disturbingly, other Earthlike worlds may lack the lunar benefits that Earth's inhabitants take for granted, because no other inner planet in this solar system has a satellite like the Moon. Mercury and Venus are both moonless, and the two moons of Mars are minuscule worlds just a few miles across. Current theory holds that the Moon formed when a large object hit the young Earth and splashed some of its mantle into orbit, where the material conglomerated into a large satellite. A large object also hit Mercury and created a basin bigger than Texas, so in their early lives, the terrestrial planets must have been wracked by enormous collisions.

Wetherill's simulations confirm this. "The way that I calculate the formation of terrestrial planets," said Wetherill, "impacts of the size that is necessary to make the Moon are fairly common; but whether or not they always make a Moon is another matter, because just having an impact doesn't necessarily make a Moon. The theory of making the Moon from an impact is not in very good shape, and the principal reason that most people believe the Moon was formed that way is because of a shortage of alternatives. The other approaches that have been taken have very serious problems." As a result, no one knows whether most Earthlike planets have such large satellites, but the Moon, like Jupiter and Saturn, may have been a key player in the development of intelligent life on Earth. In short, our existence may be profoundly dependent on our planet's having a large moon and on our solar system's having gas giants like Jupiter and Saturn.

The Anthropic Principle—Not

Wetherill's idea that intelligent life might not exist without Jupiter and Saturn is reminiscent of the so-called anthropic principle: that the existence of life demands that the universe have certain properties—those compatible with life. For example, gravity must be strong enough to form stars, galaxies, and planets, but not so strong that stars, which hold themselves up against gravity by burning fuel, would use up all their fuel before life evolves.

In one famous case, the anthropic principle even let a scientist predict an unknown property of nature. It actually occurred before the anthropic principle was christened. In the early 1950s, British astronomer Fred Hoyle was investigating the synthesis of elements inside stars. Other scientists had just figured out how stars fuse helium into carbon, but Hoyle saw that this carbon quickly reacted with the helium to form oxygen. That would leave the cosmos with little carbon, thereby preventing the development of carbon-based life, such as human beings. Hoyle therefore predicted that the carbon nucleus had a certain property, called a resonance, in order to facilitate its own creation. Although nuclear physicists were dubious, they agreed to examine the carbon nucleus; to their surprise, they found exactly the resonance Hoyle had foreseen.

Wetherill, though, said his idea is *not* the anthropic principle. "People always ask that question," he said. "It's entirely different; it's a much more straightforward thing. If what I proposed is actually true, there's nothing really very profound about it. It just means that we've selected a highly biased sample, if we choose *our* solar system as a model for the way that most solar systems should be. And there's a very straightforward cure for this: search for and study a large number of solar systems, and then you get out of this problem.

"Whereas the anthropic principle asserts that we're living in a highly biased universe. How you would go about making a search of universes so as to get an unbiased sample is hard to understand. So the one is sort of nuts-and-bolts-type stuff, and the other gets into rather mystical things." While our solar system may be just one of trillions in the cosmos, and we naturally inhabit one friendly to life, our universe may be the only one that exists. If so, its peculiar, life-giving properties suggest that it may have been consciously created for the purpose of giving rise to life—an idea that transcends science and enters the realm of philosophy and religion.

Where Are the Jupiters?

As Wetherill was advancing his idea that Jupiterlike planets might be rare, astronomers continued to search for such planets—and came up dry. One planet hunter was former planet debunker George Gatewood, the astrometrist at Allegheny Observatory who in the 1970s had challenged the alleged planets around Barnard's Star and other nearby stars. In 1986, he began observing about twenty nearby stars, including the solar-type stars Epsilon Eridani and Tau Ceti.

"When I started my project," said Gatewood, "I thought I was going to find a Jupiter around all twenty stars. I really did: 'This will be easy; this will be fun,' I thought. But it's not that easy; they're not that common. There are going to be stars that do not have Jupiters. I'm sure of that." Gatewood had invented a sophisticated device, the multichannel astrometric photometer (MAP), which uses electronic detectors to measure the positions of stars much more precisely than conventional photographic plates do. He hoped to find the stars wobbling a little from side to side in response to the pull of orbiting planets.

More bad news came from Canada. A team led by Bruce Campbell of the Dominion Astrophysical Observatory in Victoria and Gordon Walker of the University of British Columbia in Vancouver looked for stellar wobbles of a different sort. Whereas astrometrists like Gatewood sought wobbles *across* our line of sight to a star, the Canadian astronomers looked for wobbles *along* our line of sight. They did this by measuring the stars' Doppler shifts, the tiny blueshifts and redshifts that should result as a planet makes a star move toward and away from Earth. These movements along our line of sight are called radial velocities.

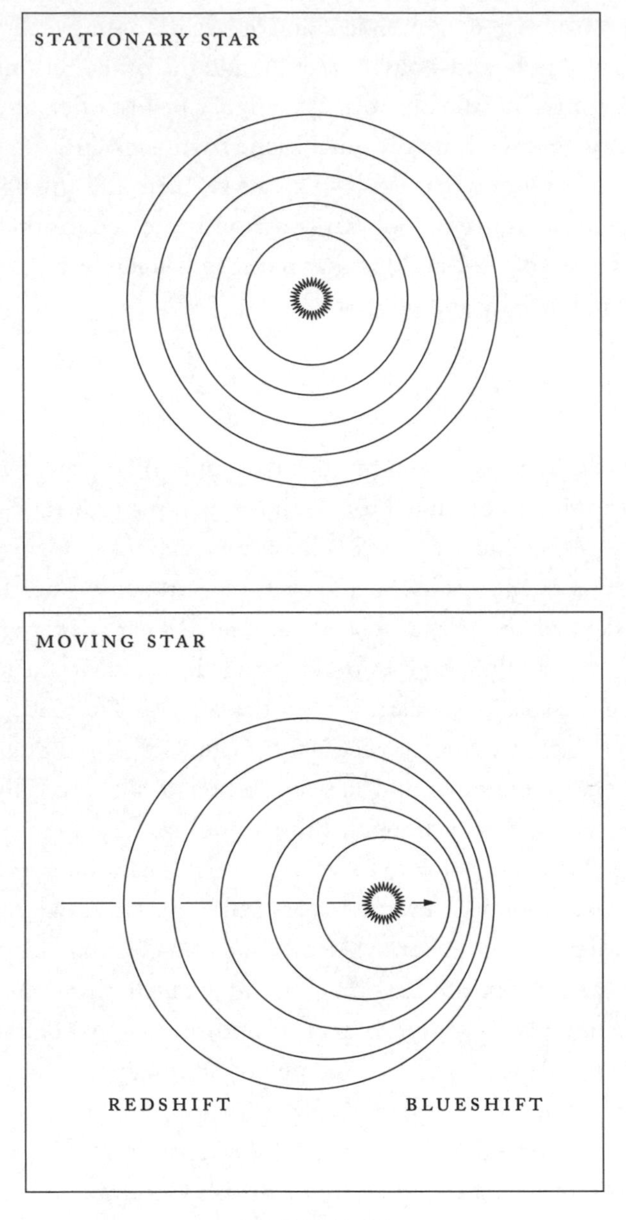

FIGURE 33

The Doppler shift. A star (top) emits regular light waves that get stretched out or scrunched up (bottom) if it moves. The stretching of light waves that occurs when the star moves away from an observer is called a redshift, because red light waves are long; the compression of light waves that occurs when the star moves toward an observer is called a blueshift, because blue light waves are short.

The Doppler technique nicely complements the astrometric approach. For obvious reasons, both techniques work best on big planets around small stars: the more massive the planet and the less massive the star, the more the planet perturbs the star. The astrometric technique best detects planets *far* from their star, such as Jupiter and Saturn, because a distant planet makes the star revolve around a center of mass that lies farther from the star, thereby forcing the star to execute a larger and easier-to-see wobble. In contrast, the Doppler technique works best for planets *close* to their star, because such planets revolve fast and force the star to do the same, and the Doppler technique relies solely on the star's radial velocity.

One key advantage of the Doppler technique is that a star need not be near the Sun in order to show the effect: a solar system just like the Sun's will exhibit exactly the same Doppler shift no matter how far away it is from Earth. In contrast, the astrometric technique works best on the closest stars. But one big *dis*advantage of the Doppler technique is its inability to extract the exact mass of the orbiting body. Instead, all the Doppler technique can establish is the object's *minimum* mass. This obtains only if astronomers view the system precisely edge-on, in which case the star swings toward and away from us to the maximum possible extent. If astronomers instead view the system exactly pole-on, the star will not move toward or away from us at all. For an intermediate case, the true mass is typically 1.27 times greater than the minimum mass; but for any particular system, it could be much greater—potentially dozens of times greater—which gives rise to a serious ambiguity that does not plague astrometric work.

In order to discover planets, the Canadians had to sharpen their Doppler precision enormously. Astronomers normally establish Doppler shifts by measuring the shift of a stellar spectrum relative to the spectrum of a stationary lamp. Switching from star to lamp, however, causes errors, and the resulting stellar velocities are accurate to about 1 kilometer per second, or 1000 meters per second. This precision won't do for planet hunters, since even mighty Jupiter induces a velocity in the Sun of just 12.5 meters per second. To obtain better precision, the Campbell-Walker team observed starlight through a container of hydrogen fluoride, a gas so lethal that other astronomers often refused to be anywhere around. Dangerous though it was, this technique super-

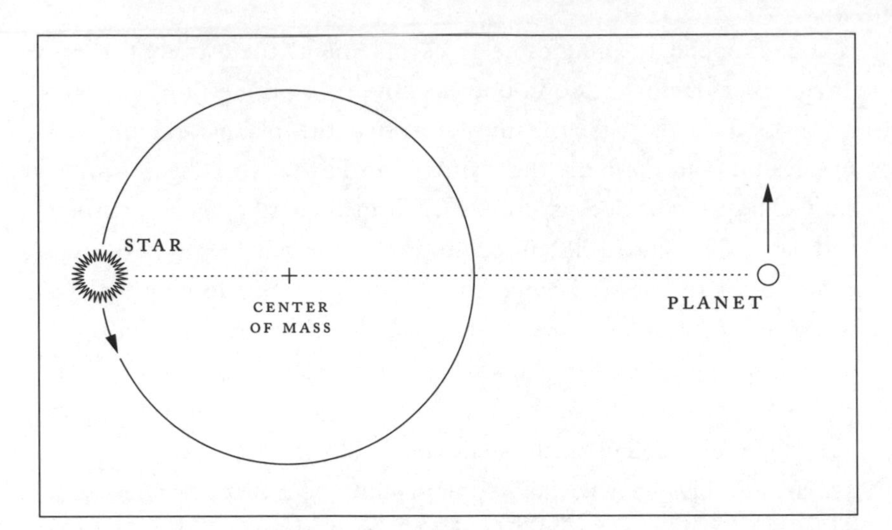

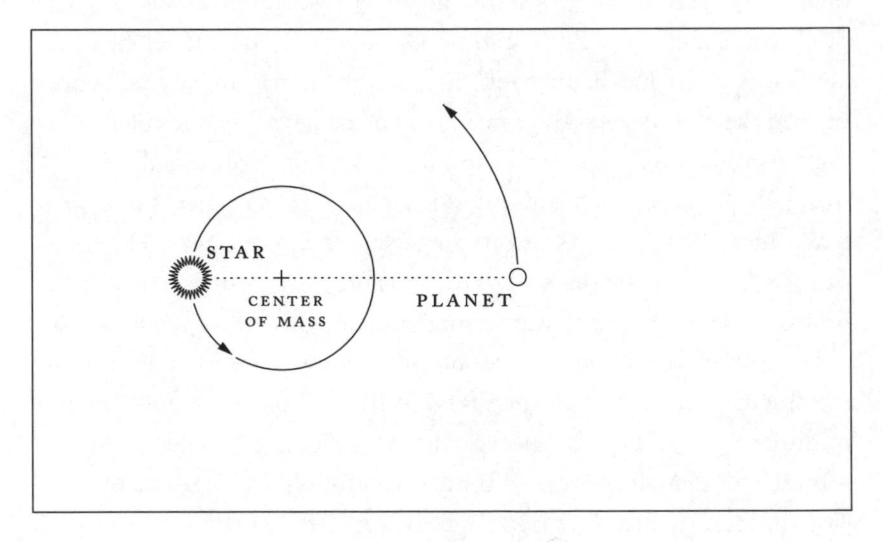

FIGURE 34

If a planet lies far from its star (top), the star will execute a large wobble around the center of mass, a wobble that astrometrists have a chance to detect; but the star will move only slowly, so Doppler observers, who measure stellar speeds, will fail. On the other hand, a star with a close-in planet (bottom) orbits around the center of mass fast, because the planet's period is short. This gives Doppler observers the chance to detect the wobble. But because the wobble is small, astrometrists, who measure the wobble's size, are out of luck.

imposed the spectrum of hydrogen fluoride directly onto the spectra of stars, thereby reducing errors and yielding the unprecedented precision of 15 meters per second—almost enough to detect a planet like Jupiter.

TABLE 9–1
Astrometry Versus Doppler

	Astrometry	Doppler
Easiest planetary mass to detect?	Large	Large
Easiest planetary distance from star?	Large	Small
Easiest orbital period of planet?	Long	Short
Best distance of star from Earth?	Nearby	Irrelevant
Can mass of planet be measured?	Yes	No; only a minimum

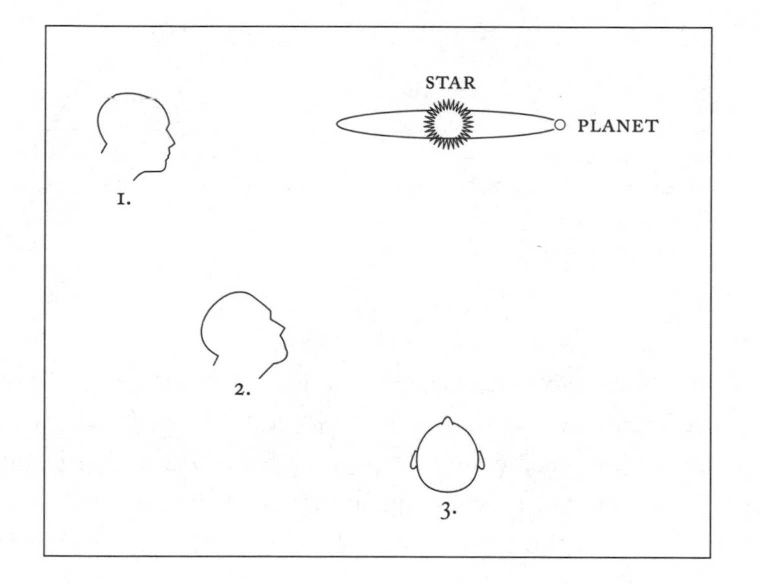

FIGURE 35

The Doppler shift that an observer detects in a star-planet system depends on the orientation of the system. If an observer (1) views the system edge-on, the star will move toward and away from the observer with full force; if an observer views the star-planet system from another perspective (2), the observed Doppler shift will be smaller; and an observer (3) who views the star exactly pole-on will see nothing, since the star does not move toward or away from the observer as it wobbles. For this reason, Doppler observers cannot normally determine a planet's mass: observer 1 will see a larger Doppler shift than will observers 2 or 3, even though all three are viewing exactly the same system.

TABLE 9–2

Astrometric Wobbles Versus Doppler Wobbles

Planet	Astrometric Wobble[a] (milliarcsecond)	Doppler Wobble[b] (meters per second)
Mercury	0.00002	0.008
Venus	0.00058	0.086
Earth	0.00098	0.089
Mars	0.00016	0.008
Jupiter	1.6	12.5
Saturn	0.89	2.8
Uranus	0.27	0.3
Neptune	0.51	0.3
Pluto	0.00008	0.00003

[a]For a Sunlike star viewed from a distance of ten light-years. A farther star would exhibit smaller astrometric wobbles. One milliarcsecond is 1/3,600,000 of a degree.

[b]For a Sunlike star whose planetary system is viewed exactly edge-on. A star with any other orientation would exhibit smaller Doppler wobbles.

Beginning in 1980, the Campbell-Walker team observed twenty-one stars. Some false alarms occurred later—in 1987, "probable planets" were reported around two of the stars, Epsilon Eridani and Gamma Cephei—but the final results, published in 1995, were depressing. "When we began this program over 14 years ago, we fully expected that, with sufficient precision, we would find several candidate giant planets," they wrote. "Not only is the Solar System dominated dynamically by Jupiter, but simulations . . . suggested that a Solar System-like distribution of planetary masses and . . . orbits would arise naturally around single stars. . . . The absence of giant planet detections . . . would seem to challenge these conclusions . . . and suggests that we shall now have to improve the precision of Doppler searches and look for lower-mass companions similar to Uranus." By 1995, other astronomers—in California, Texas, Arizona, and Switzerland—were also using precise

Doppler techniques to take aim at the stars, but these searches turned up nothing, either. "Our" Jupiter seemed to be more and more special, a planet that might indeed have fostered intelligent life here on Earth. And the paucity of alien Jupiters discouraged those who hoped for life elsewhere in the cosmos.

Then, in October 1995, the Swiss team announced a spectacular success: a genuine planet was circling a yellow star in the constellation Pegasus. It was the first planet ever detected around a Sunlike star outside the solar system.

10

Harvest of Planets

IT WAS a treasure that philosophers had foreseen millennia ago and that modern astronomers had awaited all their lives: in 1995, after five decades of false claims by others, two Swiss astronomers made one of the most spectacular discoveries of the twentieth century: the first genuine planet orbiting another normal star like the Sun. The star itself, obscure and barely visible to the naked eye, was number 51 in the large autumn constellation Pegasus, the Winged Horse; it lay just west of the constellation's trademark great square. Located fifty light-years from Earth, 51 Pegasi looked much as the Sun would if viewed from a similar distance. That is because 51 Pegasi, like Alpha Centauri A, is a virtual solar twin, a bright, yellow, main-sequence star of spectral type G.

Although 51 Pegasi resembles the star the Earth circles, its planet is a world completely alien to our own solar system: a huge gas giant lying so close to its sun that its temperature rockets well above those of Mercury and Venus.

The Swiss Discovery

Michel Mayor, of Geneva Observatory in Switzerland and one of the planet's discoverers, became enchanted with the night sky as a child. "I was a scout," said Mayor, "and when you are camping outside, in the mountains, the sky is clear, and you have a lot of stars around you. The scouts had different specialties, like cooking and topography, and I had the specialty of 'astronomer.' I had only to recognize a few stars and constellations and things like that." In graduate school, during the 1960s, Mayor investigated why some galaxies, like our own, exhibit beautiful spiral patterns that make them look like celestial pinwheels. Afterward, he developed instruments to measure Doppler shifts of stars. These instruments were not good enough to detect planets around the stars, but they did allow Mayor and his colleagues to discover many unseen stellar companions. In this way, Mayor helped confirm the brown dwarf candidate around the star HD 114762, a discovery that American astronomer David Latham had made in 1988.

Mayor had still greater goals in sight. With the help of graduate student Didier Queloz, he implemented a Doppler technique so precise it could reveal unseen planets. Mayor and Queloz used Haute-Provence Observatory in the south of France and attained a velocity precision of about 13 meters per second, similar to the velocity that Jupiter induces in the Sun and to the precision that Canadian astronomers Bruce Campbell and Gordon Walker had already achieved. Unlike the Canadians, however, Mayor and Queloz never messed with deadly hydrogen fluoride gas. Instead, they achieved their high precision by simultaneously observing a star and a lamp and comparing the two spectra.

The planet search started only in April 1994, and Mayor knew he was entering the competition late, long after several rivals. "All these teams had worked for at least five years without reporting anything," he said. "So our feeling was that if we wanted to discover something, we needed to have a large sample of stars." Mayor and Queloz selected 142 stars to observe, more than any other team. They had another crucial advantage over their greatest challengers, a team in California. The California astronomers had been observing since 1987, but they had been unable to reduce most of their data because of heavy computer requirements. In contrast, Queloz had designed a system that yielded Doppler shifts just minutes after the astronomers observed a star, literally while they were still in the observatory's dome. The Doppler shift appeared as a number on a computer screen.

All the stars that Mayor and Queloz chose resembled the Sun, being spectral type G (yellow) or K (orange), and the two astronomers first took aim at the fateful star 51 Pegasi in September 1994. "By around December," said Queloz, "we had a few points that were very strange. We decided to reobserve the star, and it was still unusual; and then we observed it all week long." The velocities of 51 Pegasi disagreed with one another by up to 60 meters per second, over four times worse than the precision of the spectrograph.

"My first reaction," said Mayor, "was that there was something wrong with the spectrograph. When you have initiated a program only a few months before with a new kind of instrument, the thing you are always afraid of is some problem with the instrument." Caution certainly made sense. After all, how could Mayor and Queloz succeed in mere months when other astronomers had searched for over a decade and failed to find a thing?

In January 1995, the astronomers first saw a pattern to 51 Pegasi's madness. The star's velocity was varying, all right, but it varied periodically, once every four days, just as it should if the star had a planet. The size of the variation implied that the planet had about half the mass of Jupiter. That was nothing strange, but the short period meant the planet circled the star in just four days. Since even Mercury takes eighty-eight days to revolve, the orbiting object seemed to be what could not be: a gas giant planet that lay closer to its star than Mercury does to the Sun.

Such a weird planet greatly worried the astronomers. "In March, we knew this could be very interesting, but we were not confident," said Queloz. "We just tried to keep cool and said, 'All right; this could be a very big discovery, and we have to be very careful not to make a mistake.' In the past, a lot of people had made mistakes, and we did not want to repeat that." To report a false planet would have been bad enough, but to report one so unusual that its own bizarre nature called into question its very existence would have been an even greater embarrassment.

Mayor and Queloz were forced to stop observing the star just before March, when it began to slip behind the Sun. Four long months followed, as the astronomers waited for the star to reemerge. During that time, they used their data to compute exactly what velocity the star should have upon its return to the night sky. In the first week of July, Mayor and Queloz took this list of predicted velocities, along with their families, into the dome of the observatory to await the

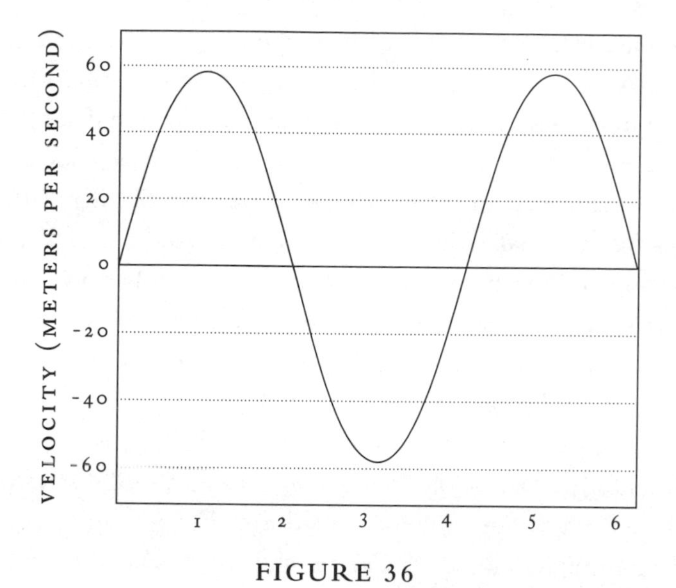

FIGURE 36

As 51 Pegasi wobbles in response to its planet, the star moves toward and away from us every 4.2 days.

star's verdict. Shortly before midnight, the head of the Winged Horse rose above the eastern horizon, and the astronomers pointed the telescope at 51 Pegasi. In front of them was the star's predicted velocity; above them was the star itself, sharing the light that would soon reveal the star's actual Doppler shift. A few minutes after the observation, the astronomers had the answer: the star's velocity matched the prediction that night, and every subsequent night as well.

"We both bought a bottle of champagne and a big cake," said Mayor. "We did a small celebration with the children. I have some children old enough to understand this kind of matter. Didier has a child who at the time was two years old. He probably enjoyed the cake, but not the discovery!"

Despite their celebration, the astronomers had yet to establish the planet's existence; all they had done was verify that the star's velocity was still oscillating periodically. "It was a crazy time," said Queloz. "We were excited, but we were also under very strong pressure. It was like a psychological step that you have to go through: this was the first planet around a normal star. And the planet was completely unusual—much closer to its sun than everybody thought it should be—so it was very, very difficult to be really convinced that this was a planet and not pulsation or rotation or something like that."

A planet was only one explanation for 51 Pegasi's velocity variation, and the astronomers spent July and August dealing with other possibilities. For example, the star might be pulsating, alternately contracting and expanding like a human heart. Pulsation also causes Doppler shifts, because when the star expands, its surface moves toward Earth and shows a slight blueshift; conversely, when the star contracts, its surface slips away from Earth and shows a slight redshift. As such a star expands and contracts, it also brightens and fades, so Mayor and Queloz had other astronomers measure the star's brightness. It did not fluctuate, which ruled out the pulsation hypothesis. But the star might have spots, like sunspots, that rotated in and out of view as the star did, giving rise to a periodic Doppler shift. If so, the star spun fast—every four days—and fast-spinning Sunlike stars generate strong magnetic fields that produce activity which 51 Pegasi lacked. The astronomers also eliminated this possibility.

A more troubling possibility still remained, though, one that bedevils all Doppler searches, including the one that had turned up the brown dwarf candidate around HD 114762 in 1988. Perhaps an object did indeed orbit 51 Pegasi, but it had much more than half a Jupiter mass—enough mass, in fact, to be a brown dwarf star rather than a planet. This would still cause a small Doppler shift if the astronomers viewed the system nearly pole-on, because the star would then move toward and away from Earth only slightly. Mayor and Queloz worked out the odds of this. The chance that the companion was more massive than 4 Jupiters was less than 1 percent, and the chance that it had enough mass to be a red dwarf star—80 Jupiter masses or more—was only 0.0025 percent. Nevertheless, because they had been observing 142 stars, the chance of catching one pole-on was considerably greater. Fortunately, the astronomers could use the star's own rotation to rule this out. A rotating star usually smears out its spectral lines, because as the star spins, one edge moves toward Earth and shows a small blueshift, while the other moves away and shows a small redshift. If a star is viewed pole-on, however, neither edge moves toward or away from Earth. In the case of 51 Pegasi, the spectral lines were somewhat smeared out, which meant that

Mayor and Queloz were *not* viewing the star pole-on; therefore, the companion had the mass of a genuine planet.

In early October, Mayor and Queloz traveled to Florence, Italy, to attend a conference about stars. They had earlier submitted a paper on the planet's discovery to the journal *Nature*, which sent it to three referees, scientists who would judge whether the paper should be accepted or rejected. During the conference, while in his hotel, Mayor received by fax the three referee reports—two said accept, one said reject—but none had raised what Mayor considered serious objections.

Buoyed by this news, Mayor announced the next day, October 6, the planet's discovery, to a room packed with nearly three hundred astronomers as well as a number of journalists. The new planet, said Mayor, revolved on a circular orbit every 4.2 days. It lay a mere 0.05 astronomical unit from its star—only one-eighth the distance between the Sun and Mercury—so 51 Pegasi must roast the planet to a temperature of some 1300 degrees Kelvin, or 1900 degrees Fahrenheit, far hotter than any planet in our solar system. Yet the new planet had a mass roughly half that of Jupiter and was presumably a gas giant, a planet that should have been far away from its star.

"It was very strange to consider the attitude of people facing something completely in disagreement with theory," said Mayor. "You had

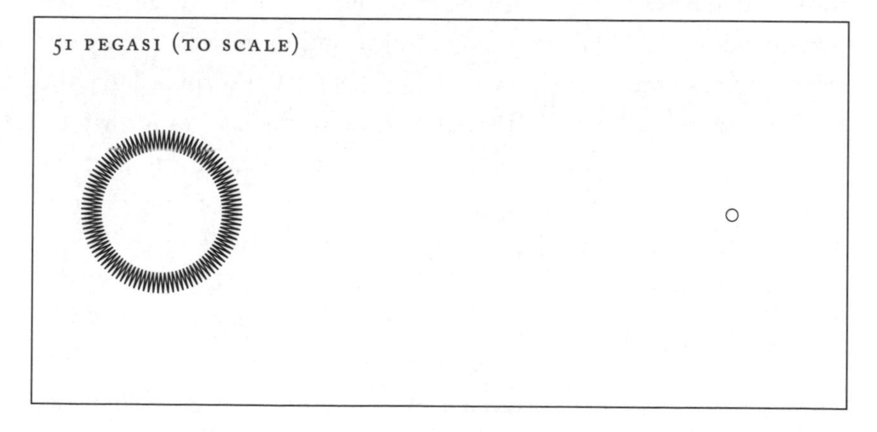

FIGURE 37

The planet around 51 Pegasi is only about five stellar diameters from the star. In contrast, Mercury is over forty solar diameters from the Sun.

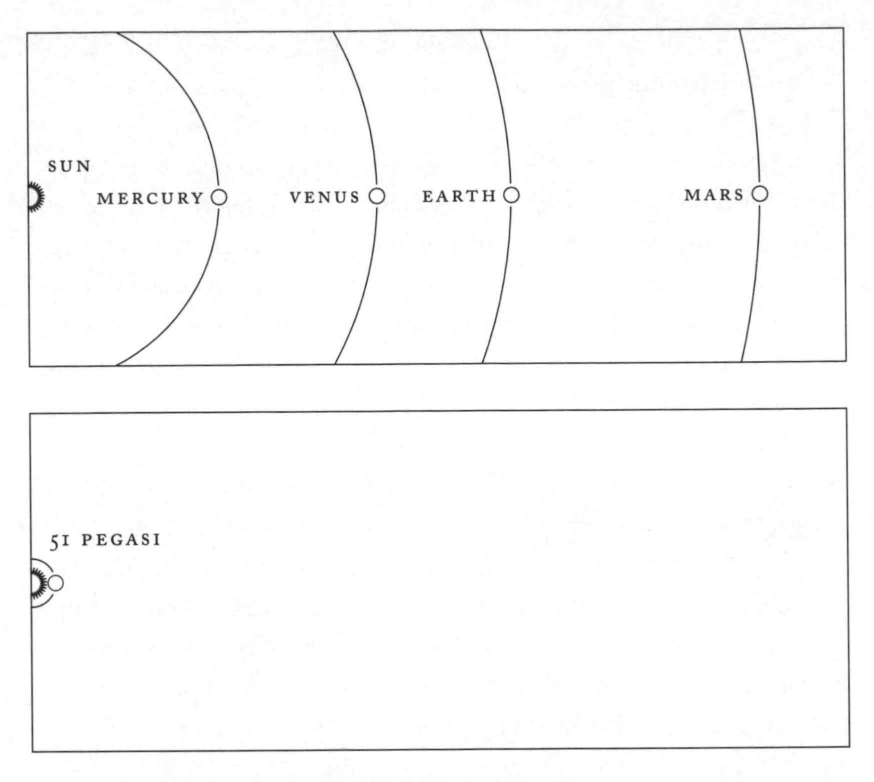

FIGURE 38

The 51 Pegasi planet is much closer to its star than Mercury is to the Sun.

some first-rank astrophysicists who said that the theory of the formation of planets was difficult, and we have only one example in our solar system, so they were absolutely not skeptical or surprised that the first planet around a normal star was quite different from what we were expecting. But you also had the more conservative astronomers, maybe not of the first rank of quality, saying, 'Oh, this is not a planet, because you cannot form Jupiterlike planets close to their stars.'"

The American Confirmation

Word of the new planet traveled fast. It soon reached California, home of the Swiss's greatest competitor, Geoffrey Marcy. A professor at San Francisco State University and a visiting scholar at the University of California at Berkeley, Marcy had been searching for extrasolar planets for eight years but had disproved so many false claims that he bore the nickname "Dr. Death."

"I, like almost everybody else, was greatly skeptical," said Marcy. "One of my colleagues was in Florence and said there were really cynical questions being asked: 'Come on, a four-day period is absurd for a planet. How can this be?' And I had to feel more or less the same way, because as Dr. Death, I certainly had more than my share of experience with fake detections. I remember saying, 'Give me a break. Are we going to have to waste telescope time debunking yet another false detection?'"

Marcy himself had become interested in astronomy when he was fourteen, after his parents gave him a telescope. "Saturn was my favorite," he said. "I loved the fact that you could look at Titan in a small telescope and come back the next night, and Titan had moved. I just thought that was the coolest thing."

In graduate school, Marcy observed not planets but stars, using their spectra to discern the strengths of their magnetic fields. The Sun's magnetic field causes sunspots and flares, and magnetic fields produce similar activity on other stars, sometimes with much more dramatic effects: spots and flares so large that they alter a star's entire brightness. But magnetic fields themselves affect stellar spectra so subtly that Marcy found the work difficult.

"I worried that I wasn't smart enough," he said. "I labored under this enormous weight, as do quite a few graduate students, that somehow I was not made of the right stuff. I really suffered. I wasn't suicidal, but I asked myself a few times, '*Am* I suicidal?' Sometimes the depression would get so great, and I thought, 'I'm such a failure; isn't it obvious to other people that I'm not smart enough?'" After earning his doctorate in 1982, Marcy found his work under attack. "In the dog-eat-dog world of science," he said, "it takes a special constitution to put up with critiques of your papers. It took me maybe ten years to learn that I can't be my own worst critic or else I'll just pummel myself to death."

Marcy knew that unlike most other postdocs, he could not easily continue the work he had done in graduate school. "I was feeling fairly depressed that my field of research had run up against this brick wall of difficulty," he said. "I remember one day being in the shower, feeling a little down about going into work and trudging through more magnetic fields. I was so depressed about it that I thought to myself: I've got to find a research topic that really grabs me—that's exciting to me on a personal level as well as a scientific level."

Marcy abandoned magnetic fields of stars and began thinking about their planets. Using the Doppler technique, he started to measure the velocities of sixty-five red dwarf stars. His precision was not yet good enough to detect genuine planets, but it would have revealed orbiting brown dwarfs, if any had existed. None did. During this time, however, the fiasco erupted over the nonexistent brown dwarf VB 8 B—the infamous object that astrometrists first "detected" from the wobble it produced and that infrared astronomers later "saw" directly, while earning front-page headlines by calling the brown dwarf a planet.

Marcy attended the 1985 conference on brown dwarfs that this "discovery" spawned. "I stood up and asked the question: 'How can the wobble that is being claimed by this one group be consistent with the direct infrared detection by this other group?' The wobble yielded a mass that was inconsistent with the infrared brightness. And I got some cockamamy answer that was entirely concocted. This VB 8 B showed up on the *cover* of the conference book, and it's just one of these real embarrassments where the whole astronomical community was living in complete, utter fantasy."

In 1987, Marcy started to search for planets. His approach resembled that of the Canadian team, which was then achieving a velocity precision of 15 meters per second by superimposing spectral lines of hydrogen fluoride onto the spectra of stars. But Marcy did not want anything to do with hydrogen fluoride. Aside from its lethal nature, hydrogen fluoride has spectral lines at infrared wavelengths, where stars have relatively few lines. This limited the Canadians' precision.

Marcy therefore asked R. Paul Butler, a graduate student with a degree in chemistry and an interest in astronomy, to find a better molecule. "About the time I was twelve or thirteen," said Butler, "I had read Bertrand Russell's books, and I stumbled upon the story of Tycho Brahe and Kepler and in particular Bruno. That certainly made astronomy a very exciting topic. It was the sort of stuff that only a few hundred years ago was so exciting that it would raise such emotions that they would go out and kill people over it."

Butler found a new molecule for Marcy: iodine. Unlike hydrogen fluoride, it's safe, and it has a huge number of spectral lines just where many stars do. By observing stars through a container of iodine gas, Marcy and Butler could superimpose iodine's spectral lines onto a star's spectrum and obtain a precise Doppler shift. Unfortunately, iodine almost had too many lines for its own good. To extract the stars' Doppler shifts required so much computer power that most of Marcy and Butler's data sat unreduced for years.

Worse, the few data that Marcy and Butler did reduce showed trouble. "There were a couple of big roller coaster rides in there that were quite frightening," said Marcy. "In the first four or five years of using the iodine cell, we were unable to achieve a Doppler precision better than about 30 or 40 meters per second. This was quite an embarrassment. Here we were using up telescope time at Lick Observatory that other people would have loved to have, and our precision was inferior by a factor of three compared with what the Canadians were doing. It was more than frightening. I lay in bed awake at midnight, wondering, 'What are we doing wrong?' I felt that maybe my career as an astronomer was being called into question." To make matters still worse, a competing group had arisen at the University of Texas at Austin. Led by William Cochran, these astronomers achieved a velocity precision similar to that of the Canadians, around 15 meters per second. Marcy and Butler were now in third place, behind both the Canadians and the Texans.

To achieve better precision, Marcy and Butler calculated how the spectrograph itself altered spectral lines, but this added yet more computer requirements to their data reduction routine. In 1994, the spectrograph's optics were refurbished, giving the astronomers the incredible precision of just 3 meters per second—better than anyone else in the world, more than good enough to detect a Jupiter, and nearly good enough to detect a Saturn. Because of the heavy computer demands, however, Marcy and Butler had reduced the data for only 25 of the 120 stars they were observing. None showed planets. Meanwhile, the Swiss astronomers Michel Mayor and Didier Queloz were unencumbered by heavy CPU needs, and they were racing ahead. It was they who had just announced the peculiar planet around 51 Pegasi.

As luck would have it, just five days after Mayor's announcement, Marcy and Butler had four nights scheduled at Lick Observatory, so they could observe 51 Pegasi and confirm or refute its velocity variation. By the third night of their observing run, it was obvious that the star's velocity was indeed varying, and Marcy and Butler soon verified that the period and amplitude matched the ones Mayor and Queloz had claimed. The weird planet, with a size rivaling Jupiter's and an orbit smaller than Mercury's, was more real than ever.

"This whole episode was a wonderful drama depicting how science should be done," said Marcy. "One team makes a claim that is spectacular and clearly of historic value; another team, us, being skeptical and even cynical, comes in, and what happens? We simply confirm what they found. It was science at its finest. If only the rest of society, from politics to sociology to religion, could operate in the same way."

On Tuesday, October 17, two days after returning from the mountain, Marcy and Butler put out a low-key press release announcing their result. It began, "Astronomers at the University of California at Berkeley and San Francisco State University have confirmed a recent report of a planet orbiting a nearby star—the first time a reported discovery of a planet around a normal star has withstood scrutiny."

"The next day," said Butler, "things went nuts. Geoff wasn't here for most of the day, and I was deluged with phone calls from *The New York Times*, *The Washington Post*—every decent paper—and by the following day, we were on the front pages of newspapers all over the world."

Ironically, in choosing stars to observe for their own program, Marcy and Butler had deliberately excluded 51 Pegasi, because they thought it was not a stable main-sequence star like the Sun but instead a subgiant, evolving from the main sequence to the giant stage. Subgiants can pulsate and create Doppler shifts that mimic orbiting planets. The incorrect classification of 51 Pegasi had appeared in the widely used Yale *Bright Star Catalogue*. Had the star been classified correctly, Marcy and Butler would have observed it; if they had reduced their data in time, they would have been the first to find a planet around another normal star.

"Sure, we were disappointed that we weren't first," said Butler. "On the other hand, we were extremely thrilled that there was a *signal*, because we'd spent all these years and seen all these stars doing nothing—flat lines. If they were hospital patients, they'd be dead. I actually couldn't sleep for about a week and a half or two weeks after this, because this has been my whole life."

The media remained keenly interested in the great planet story. On Thursday morning, October 19, Marcy taught an astronomy class and returned to his office. The time was eleven o'clock. The phone was ringing. It was ABC's *Nightline*.

Planet *Nightline*

"They said Ted Koppel is very interested in having a special program—they're going to cancel their planned show—but only if I can do it," said Marcy. "And I thought, 'My God, *Nightline*.' I remember walking into the physics department office: 'Well, you won't believe it, but *Nightline* has called,' and my department chair said, 'Oh, yeah, right,' and he just walked away. He thought I was kidding!"

Butler was then across the bay in Berkeley. "They don't teach you how to deal with the media in grad school," he said. "It was a mad dash: I had to get to San Francisco and not look like the complete bum that I normally look like. I usually wear shorts and a T-shirt, because the stars don't care." In the afternoon, a television crew filmed some footage with Marcy and Butler, and that evening Marcy would appear on the program live, with other American astronomers.

But neither Marcy nor Butler knew what was about to hit them. One of the other astronomers scheduled on the program was David Latham of the Harvard-Smithsonian Center for Astrophysics (CfA), who in 1988 had found the possible brown dwarf around HD 114762. Minutes before airtime, Latham told Marcy that he had discovered a *second* planet around 51 Pegasi—even though Latham's velocity precision was 500 meters per second, over a hundred times worse than Marcy's.

"I said, 'Wow, that's really great, Dave'—in that tone of voice," said Marcy. "It was like somebody was claiming that they had looked at a quasar with the naked eye and learned something that somebody else using the Palomar 200-inch telescope didn't learn. It was really quite a shock to hear him say that, and I knew Dave well enough to know that he sometimes—how shall I say it diplomatically?—exudes a certain enthusiasm about his results that maybe gets in front of him a little bit, if you can read between my lines. And immediately I thought to myself: there's something not right here. A precision of 500 meters per second is simply not good enough to find anything that you would call a planet. Well, I didn't say this to Dave, because the program was about to start, and poor old Ted Koppel gets on, and he quickly focused on the second planet."

On air, Latham and one of his colleagues described the new "planet," saying that it was much farther from the star than the one the Swiss astronomers had found. Whereas the inner planet was too hot for life, the outer one was too cold; but the "discovery" of the second planet meant that 51 Pegasi had a full-fledged solar system that might contain a planet in between, at just the right temperature for life.

Butler watched the entire spectacle from the studio. "I was aghast," said Butler. "I was embarrassed for Harvard; I was embarrassed for the CfA; I was embarrassed for American astronomy. What Dave did was extraordinarily bogus. There's already this really bad shimmer of snake oil associated with extrasolar planets, and Dave goes on in the most blatant and obvious way and—I don't know quite how to say what he did."

Not wanting to start a dispute on national television, Marcy said only that his own observations would be able to confirm or refute the planet. "Slowly but surely," said Marcy, "the clever Ted Koppel realized that something was wrong. Ted asked Dave Latham, 'When did you learn about this second planet?' [Koppel's exact words: 'How did you reach that conclusion? And what a coincidence that you reached it today.'] And Dave Latham said, 'Well, we really didn't look at our data too carefully until after *Nightline* called.' And Ted Koppel asked, 'How confident are you that there's really a second planet?' and Dave said, 'Well, we're not really 100 percent sure; we don't really know anything about it; we can't even tell what its orbital period or mass are; it might be there; it's probably there.' He was backpedaling faster than you can imagine, and Koppel was on to it. The brilliant nonscientist Koppel figured it out; but then the program ended, basically leaving the audience with this notion that there were two planets orbiting 51 Peg. It was a little unfortunate."

Naturally, Latham saw the incident differently. "I think my main motivation was to try to expose a national television audience to some of the excitement of an unfolding scientific story," he said. That afternoon, Latham had seen signs of the second planet in his data, signs that were even stronger than those which first alerted him to HD 114762 B. He then called Michel Mayor, who told him that he too saw a possible second planet in his data, but Mayor could reveal no details, because his paper was embargoed by the journal he had submitted it to. Having only this information, Latham then agonized over whether to mention the second planet on *Nightline*. "In the end," he said, "I decided to go ahead with the news, probably against my better judgment." During the broadcast, he noted, he said that the second planet needed to be confirmed and told Koppel, "I'm not going to bet my life on it, but I feel pretty good about it."

Nevertheless, that planet does not exist. Aside from the nonexistent planet, there were other problems with the broadcast. The star 51 Pegasi was repeatedly called "51 Pegasus," which is just as wrong as calling Alpha Centauri "Alpha Centaurus." It was an error that appeared in numerous publications, too, and arose from Marcy and Butler's press release. Far worse, though, was that no one mentioned the *first* planets discovered outside the solar system, those around the pulsar PSR B1257+12, which Alex Wolszczan and Dale Frail had found in 1991. Instead, viewers were told that 51 Pegasi's planet was the first extrasolar planet ever discovered.

Crimes Against *Nature*

Meanwhile, tempers were flaring in Switzerland. "We were angry," said Didier Queloz. "Some articles said that two Americans report the discovery of a planet, and just at the end, the articles said this has also been observed by a Swiss team."

Because Marcy and Butler had put out a press release, the media naturally contacted them; furthermore, the true discoverers, Mayor and Queloz, were not talking to the press—not because they didn't want to, but because they were *ordered* not to. Unlike most other scientific journals, the one to which Mayor and Queloz had submitted their paper, *Nature*, threatens to reject otherwise acceptable papers from scientists who dare talk to the media prior to a paper's publication. That way, the discovery is reported only when the paper appears in the journal and gives the journal prominent mention in the press. It does not take a great cynic to see that one of the chief beneficiaries of this policy is the journal itself.

"I don't blame Michel and Didier for being angry," said Marcy.

"They really should have gotten the bulk of the attention. I don't think they were angry at us, because clearly there was nothing we could do. What I think they were angry about was being hand tied by *Nature*. This discovery was a treasure given to humanity—the first planet around a solarlike star—and for that to be taken away by a publisher's edict was an abomination." In a free country, said Marcy, scientists should be free to report their results whenever they want.

"It was completely the fault of *Nature*," said Queloz. "We were in a very difficult situation, because we wanted to talk, we wanted to say what we did, but we could not do so, because of *Nature*'s embargo. We got a lot of phone calls from journalists, but we only said, 'Sorry, we cannot answer. Maybe ask somebody else.'"

"We are sure as hell not going to ever publish anything in *Nature*," said Butler. Marcy said, "With our planets, we've been counseled, 'Oh, you should publish in *Nature*; that's the place to publish really exciting results,' and my answer to that is: 'No.' As Scotty said on *Star Trek*, 'Fool me once, shame on you. Fool me twice, shame on me.' And I'm not going to be fooled twice by *Nature*. If they play games like that, forget it."

Michel Mayor, however, saw some justification for the general policy. "On the one hand," he said, "I don't know if making announcements in the press is the correct way to do science," because that can lead to errors and exaggerations. "But in the case of 51 Peg," he said, "everybody was already informed of most of the facts, and the embargo was a little bit stupid." Nevertheless, if he had it to do over again, Mayor said he does not know if he would choose to publish the discovery in *Nature*.

Also struggling with the embargo was Brian Marsden, head of the International Astronomical Union's Central Bureau for Astronomical Telegrams, which often reports fast-breaking astronomical news. Marsden was receiving requests from additional astronomers who had confirmed 51 Pegasi's velocity variation and naturally wanted to communicate that confirmation to others via an IAU circular.

"I said that we couldn't publish those observations," said Marsden, "because we knew that this object was discovered by these people in Switzerland. We have got to put that information on the IAU circular as well. The embargo on the *Nature* paper was the problem. Therefore, I wrote to Michel Mayor, saying, 'Look here; we've got to put the barest minimum of data on, to support your discovery. You've got to give some numbers.'"

Meanwhile, Mayor and Queloz were fighting with *Nature* for the same right that most people in the United States, and Switzerland, take for granted: the right to speak freely. That would not come until over a month later, when the paper was published, but Mayor was allowed to send a modicum of information to Marsden, and on October 25 Marsden issued an IAU circular on the discovery. Entitled simply "51 Pegasi," the first sentence read: "M. Mayor and D. Queloz, Geneva Observatory, have reported the discovery of a Jupiter-mass object in orbit around the solar-type star 51 Peg." A few numbers followed, along with reports of the discovery's confirmation.

As it turned out, the Swiss had their paper accepted the next week, exactly on Halloween—appropriate, since the peculiar nature of a giant planet in an orbit much smaller than Mercury's was already haunting the many theorists who had said such worlds could not exist. Just as appropriate, the paper appeared in the issue dated Thursday, November 23, which in America was Thanksgiving; and indeed, after decades of frustration, searchers of planets around normal stars finally had something to be thankful for: the first planet ever detected around another Sunlike star.

The Uninvited Planet

But not everyone was so thankful for the strange planet around 51 Pegasi. "It was a spectacular disproof of my prediction," said Alan Boss, a theorist at the Carnegie Institution. "I was surprised and shocked. It caused me a few sleepless nights, you might say." Just months before, Boss had said that extrasolar Jupiterlike planets must form far from their stars, as Jupiter and Saturn had. Furthermore, he had said, planet seekers who used the Doppler technique—such as Mayor and Queloz, and Marcy and Butler—were unlikely to succeed; astrometry, which is more sensitive to distant planets, was the way to go.

Boss himself had earned his degrees not in astronomy but in physics. "I was a science nerd in high school," said Boss. "I wanted to do physics, because it was the way to understand what the universe was about and because I viewed it as the toughest science to learn and to master. Physics was the thing that people really wanted to avoid in school, and I've always been a bit of a maverick. If I see the herd running in one direction, I tend to want to go in the opposite direction."

Boss actually learned of the new planet's discovery in early September, a month before almost everyone else, because he had been one of the referees of the Swiss paper. According to them, the planet had at least 0.47 Jupiter mass. This minimum mass held if the system was exactly edge-on. Statistically, the most probable mass was 1.27 times greater, or 0.60 Jupiter mass, which placed the planet's mass midway between those of the Sun's two gas giants, Jupiter and Saturn, the latter having 0.30 Jupiter mass. Although no one knew the planet's composition, the similarity in mass suggested the planet was also a gas giant: a big ball of hydrogen and helium surrounding a much smaller core of rock and ice.

Boss was hardly the only theorist who thought such planets must form far from their stars. In order for a gas giant planet to arise, a large core must form, whose gravity then attracts even larger amounts of the hydrogen and helium gas that dominate the disk orbiting a young sun. To construct that large core requires a lot of material—not just rock and iron, which exist throughout the disk, but also ice, which is several times more abundant but exists only far from the star, where the disk is cool. In addition, only the outer disk has a large enough volume to harbor large amounts of material. Thus, Boss had predicted that extrasolar Jupiters would exist at distances of around 5 astronomical units from their stars—that is, 5 times the distance of the Earth from the Sun—just as Jupiter does.

But 51 Pegasi's planet sits nearly on top of its star. The planet is only a twentieth as far as the Earth is from the Sun. It is much closer in than even Mercury. Whereas Mercury lies 0.39 astronomical unit from the Sun, 51 Pegasi's planet is a mere 0.05 astronomical unit from its star. Viewed from the planet, the star would loom some twenty times larger than the Sun looks in Earth's sky, and the star's gravity would raise tides thousands of times stronger than the Moon does on the Earth. These tides would force the same side of the planet to face the star forever, just as the same side of the Moon always faces the Earth.

"The missing element," said Boss, "was that I and others were thinking that planets wouldn't migrate much: wherever you make them, that's where they're going to stay. Perhaps the most important thing we've learned from 51 Pegasi is that this isn't necessarily so. Any responsible astrophysicist will argue that there's no way you can make the planet right there at 0.05 astronomical unit." If the planet had formed right where it now is, said Boss, the disk of orbiting material must have been outrageously massive to create a rock-iron core great enough to grab large amounts of hydrogen and helium and become a gas giant. This is because the inner disk had only a small volume and presumably was too hot to have ice. Another possibility, said Boss, is even more unlikely: that the world is an overgrown terrestrial planet, a giant hunk of pure rock and iron, because then the disk must have been even more massive in order to form a rock-iron object that big.

Instead, theorists believe that 51 Pegasi's planet formed farther out, at a distance around 5 astronomical units, and then spiraled closer in to the star. Even before the discovery of the odd planet, Douglas Lin at the University of California at Santa Cruz had proposed just this possibility. As a young planet circles its star in a disk of gas and dust, that material can drag the planet toward the star. Why the planet around 51 Pegasi did not go all the way and plunge into the star is not known. When it was young, however, 51 Pegasi may have spun fast, as many young stars do, and transferred some of its spin energy to the orbital energy of the planet, keeping the planet safely circling the star until the disk of gas and dust was gone. The planet's migration from 5 to 0.05 astronomical unit must have disrupted the inner part of the solar system, because the planet's gravity would have kicked Earthlike planets out of the system altogether. These may now be roaming interstellar space, cold, dark, and lifeless.

"51 Peg teaches us that you can do some real damage to a planetary system, if the gas is allowed to have free rein," said Boss. In contrast, and fortunately for Earth, the same fate did not befall Jupiter and Saturn; instead, the Sun's disk must have been thin enough or short-lived enough that these planets remained far from the Sun and never disturbed the Earth or the other inner planets. Thus, a planet like Jupiter may be a two-edged sword, on the one hand protecting small planets from the impacts of comets but on the other hand holding the potential to eject those same planets out into the void.

Some observers, however, are skeptical that 51 Pegasi's planet formed farther out. "If you want my personal opinion," said Geoffrey Marcy, "I think what's happening here is that we scientists are doing what we always do: hanging on to the paradigm until the last possible moment, the paradigm being that Jupiterlike planets form at 5 astronomical units. If I had to put my money on the craps table, I would bet that we're going to learn later that giant planets can form closer in and that we've been deluded by our own solar system. But the theorists really hate that idea." The theorists, though, do not have a great track record at predicting extrasolar planets. They did not foresee the planets circling the pulsar or the odd planet around 51 Pegasi.

A Genuine Brown Dwarf

Of course, theorists did foresee brown dwarfs, the stars with too little mass to ignite their hydrogen-1 fuel, but observers never found any definite examples. However, the same day of the same meeting in Florence where Michel Mayor announced 51 Pegasi's planet witnessed a second dramatic report: the direct detection of a star so cool and faint that it had to be a brown dwarf. In fact, the star's temperature was so low that its spectrum resembled Jupiter's.

Exciting though it was, this discovery garnered little publicity, in part because it was eclipsed by 51 Pegasi and in part because, to their credit, the discoverers never called the brown dwarf a planet. "That would have been ridiculous, to try to get some sort of hype out of that business," said Shrinivas Kulkarni of the California Institute of Technology. "I'm happy that our group has found the first object of a completely different sort, but I don't have strong opinions as to whether this should be publicized more. I don't want to sound as if I'm an old hand at this, but as a grad student, I had the greatest fortune of finding the very first millisecond pulsar, so I've been through enough of these things." Far more important than publicity, said Kulkarni, is that his grant get renewed.

Kulkarni's search for cool stars actually began with a search for a cool summer in his native India. "I had no interest in astronomy, apart from reading Isaac Asimov's books here and there," said Kulkarni, who was an applied physics major in college. "In my last year, I had one of these scholarships that I could take anywhere I wanted for the summer. I noticed there was a summer school in a very nice, very cool place, which in India in the summer does matter a lot!" His decision meant that he would study radio astronomy that summer, a subject he found a lot more fun than applied physics.

The brown dwarf hunt began years later, in 1992, after Kulkarni's group, including Tadashi Nakajima and Benjamin Oppenheimer of Caltech, obtained a coronagraph from two astronomers at Johns Hopkins University, Sam Durrance and David Golimowski. A coronagraph blocks a star's glare and reveals much fainter ones around it. Such a device had uncovered the disk of dust around Beta Pictoris back in 1984. The Caltech team also used a technique called adaptive optics, which attempts to undo the distortion that starlight suffers as it passes through the Earth's atmosphere. Armed with this equipment, the astronomers took aim at hundreds of stars near the Sun. Because young brown dwarfs shine brightest, Kulkarni wanted to examine stars that were roughly a billion years old, only a fifth the age of the Sun. Such stars stand out from others in two ways. First, they have circular orbits around the Galaxy, because stars are usually born on such orbits; as they age, however, stars encounter clouds of interstellar gas and dust whose gravity perturbs the stars and makes their orbits around the Galaxy more elliptical. Second, young stars often spin fast, which usually generates strong magnetic fields and activity that astronomers can see.

One star that fulfilled both of Kulkarni's criteria for youthfulness was Gliese 229, an apparently single red dwarf: a main-sequence star of spectral type M that burns hydrogen into helium at its center. It was the 229th star in a catalogue of nearby stars that German astronomer Wilhelm Gliese had compiled decades earlier. Gliese 229 resides just nineteen light-years from Earth. In October 1994, the astronomers discovered a faint and extremely red object beside the star. In September 1995, they acquired infrared data indicating that the star put out nearly all of its luminosity at infrared wavelengths rather than visible ones, which meant the star is cool. Right afterward, the astronomers obtained the object's infrared spectrum.

"The spectrum was something I'd never seen before," said Kulkarni, "and someone said, 'Gee, it must be a quasar.'" A distant quasar might happen to lie in nearly the same direction as Gliese 229 and have such a high redshift that unfamiliar spectral lines, normally at ultraviolet wavelengths, were being shifted into the infrared part of the spectrum. "So we are talking about it being some quasar, redshift seven or something," said Kulkarni, "and I was telling this guy, '*God*, all these years, and I don't *want* it to be a quasar; I want it to be a brown dwarf!' But fortunately Keith Matthews, who's our instrumentation person and is also the genuine planetary astronomer on our team, recognized those features as being methane."

Finding methane in a star is like seeing snow in a desert. No methane had ever been seen in a star before, because even red dwarfs are so hot that they would tear any methane (CH_4) to shreds. Instead, red stars often have their carbon bound up with oxygen in carbon monoxide (CO). At lower temperatures, however, chemical reactions convert carbon monoxide into methane. In fact, the solar system's giant planets all contain methane, and methane colors Uranus green and Neptune blue. After recognizing that Gliese 229 B had methane, Kulkarni's team pointed the telescope at Jupiter and saw the spectral similarity.

Despite the methane, Kulkarni hesitated to call the star a brown dwarf, because of all the false brown dwarfs reported before. "That's why this whole bloody field got a bad name," he said. "The number of retractions is equal to the number of discoveries. And I can tell you, we had a lot of trying moments in September, discussing whether we should publish this right away or not. I didn't want yet another brown dwarf going down the tubes.

"So everyone was trying to tell me that the methane does it. But to me, that's an indirect proof; it relies on chemistry." Instead, Kulkarni wanted an astronomical proof. That came three weeks later, in early October, when the astronomers observed Gliese 229 B at optical wavelengths and saw that it was in the same position, relative to Gliese 229 A, as a year earlier, which meant the two were moving through space together. Thus, the peculiar object was a true companion to Gliese 229 A and shared its distance of nineteen light-years. From this distance, and the star's apparent faintness, the astronomers calculated that Gliese 229 B was less than a tenth as luminous as a main-sequence star could possibly be. That convinced Kulkarni, and the team announced the detection of a genuine brown dwarf.

Other scientists hailed the discovery. "It's fantastic," said Alan Boss. "It's clear that that thing has got to be a brown dwarf star. The methane spectrum really nails it. And I think that discovery is equally as exciting as the 51 Pegasi discovery, because brown dwarfs are another key to our understanding of what's going on here. The fact that it was announced simultaneously with 51 Pegasi's planet is an incredible coincidence."

"Gliese 229 B is far and away the best brown dwarf candidate," said David Stevenson of the California Institute of Technology. "It is the only one I know of that I really feel quite comfortable about, because the temperature is clearly low enough that the star could not possibly be deriving its energy from nuclear reactions."

Especially pleased was Shiv Kumar, the astronomer who feels he has received inadequate recognition for his early work on brown dwarfs. "They apparently know the history," said Kumar, "because they gave me credit. So when I said people are not giving me credit, it's only some; others are. In the Gliese 229 B papers, they say that I predicted the existence of these objects in 1963." Both the paper on the brown dwarf's discovery and the one on its methane spectrum cited Kumar's 1963 paper before all others.

Gliese 229 B has between 40 and 55 times the mass of Jupiter, well below the main-sequence cutoff, first computed by Kumar, of around 80 Jupiter masses. The apparent distance between the brown dwarf and the red dwarf it orbits is 45 astronomical units, similar to the mean distance between the Sun and Pluto. However, the true separation between the two stars is greater, if they do not lie at precisely the same distance from Earth. During 1995, astronomers reported brown dwarf candidates in the Pleiades, a young star cluster. Because of their youth, these brown dwarf candidates have faded so little that they shine as brightly as red dwarfs. But they contain large amounts of lithium, which red dwarfs and even massive brown dwarfs—those having over 60 Jupiter masses—quickly consume. Thus, the high lithium levels suggest that these stars are brown dwarfs. Unfortunately, the exact rate at which a star destroys lithium is unknown, so these objects are not yet as definite. In contrast, the methane in Gliese 229 B demonstrates a temperature far too low for a main-sequence star. The coolest main-sequence stars are roughly 2000 degrees Kelvin, whereas Gliese 229 B's temperature is a mere 1000 degrees Kelvin.

"The theorists, who have been modeling these things, have been incredibly excited," said Kulkarni. "The only people who have been disappointed are people doing cosmology, when I tell them that we struggled for three years and looked through a few hundred stars, and we found only one brown dwarf. That doesn't solve the dark matter problem." If brown dwarfs contribute substantial quantities of dark matter to the Galaxy or universe, they must do so as lone stars. The work of Kulkarni's group, along with that of others, suggests that brown dwarfs are rare in orbit around other stars.

"To me," said Kulkarni, "the next goal is to find an object by reflected light. We've found a self-luminous object—Gliese 229 B—whereas Jupiter is primarily a reflector of sunlight. So I hope we can participate in a discovery showing direct detection of a gaseous object—I don't care whether it's a brown dwarf or a planet or whatever—but by *reflected* light, because that's a major challenge in our search for extrasolar planets."

If these objects can be seen directly, as Gliese 229 B was, astronomers can obtain their spectra. "We know what main-sequence stars look like," said Kulkarni, "and we know what cold planets—Jupiter, Saturn, Uranus, Neptune—look like. But we don't know what objects in between look like. Gliese 229 B tells you one very important thing: we now know what gaseous objects—I'm not using the word *planet*, you notice—look like at 1000 Kelvin." Gliese 229 B has much more mass than a planet and generates its own light, by converting gravitational energy into heat. But its temperature is similar to that of 51 Pegasi's planet, 1300 Kelvin, which owes its heat to the star it circles. Regardless of the heat source, it is primarily temperature, not mass, that determines the object's spectrum. Predicting a planet's spectrum is crucial, said Kulkarni, because then astronomers will know at which wavelengths the planet will be brightest and easiest to see.

New Worlds

Even after the spectacular discoveries of a planet and a brown dwarf around 51 Pegasi and Gliese 229, respectively, 1995 was about to confer two more prizes: another extrasolar planet, as well as what many astronomers regard as another brown dwarf. Both objects were found by Geoffrey Marcy and Paul Butler, the California astronomers who in October had confirmed the Swiss discovery of 51 Pegasi's planet.

"Here's what happened," said Marcy, "and I'm a little embarrassed about it, but the truth is the truth. Paul and I felt enormously excited that a planet had been discovered, but at the same time, put simply, we had been scooped." Marcy and Butler had spent over eight years refining their procedure, which gave them top-notch precision, but because their heavy computer demands prevented them from reducing most of their data, they had lost out to an upstart team operating at lower precision. Said Butler, "Mayor and Queloz don't have any huge CPU needs, so we imagined them just observing stars, getting velocities five minutes later, and finding all these planets, while we sat with eight years of data with higher precision. We were desperately frightened that they were going to scoop us object after object."

In early November, therefore, Marcy and Butler marshaled computers and worked overtime to march through all their data. The first discovery came a month later, when another obscure solar-type star—this one in the constellation Ursa Major, the Great Bear—showed a wobble. The star, named 47 Ursae Majoris, lies forty-six light-years away and is dimly visible to the naked eye, due south of the two pointer stars in the bowl of the Big Dipper.

Butler thought the wobble was real, but Marcy was skeptical. "I harbor a sincere trepidation about making a false claim," said Marcy, "because once your credibility is dashed, it's virtually impossible to get it back. I still have Barnard's Star and VB 8 B in the back of my mind." But subsequent observations soon verified that 47 Ursae Majoris was wobbling.

The shape of the planet's orbit suggests that it is indeed a planet and not a brown dwarf. "The companion of 47 Ursae Majoris is likely to be a gas giant planet, by virtue of its circular orbit," said Boss. "Planets form inside a disk, which has fairly small eccentricity. You should form it on a circular orbit, and it should stay circular as it interacts with the disk. In contrast, a brown dwarf star in orbit around another star forms when an interstellar cloud comes screaming down from interstellar densities to collapse and brutally fragment into two clumps. The very nature of that process means that when the clumps are formed, they're going to be moving on highly elliptical orbits."

Of all the discoveries of 1995, this was perhaps the most interesting, because it was so normal. Like the three planets around PSR B1257+12, which uncannily mirror Mercury, Venus, and Earth, the planet around 47 Ursae Majoris felt like a world that those bound to the Sun would find familiar, for it resembles an overgrown Jupiter. The planet has a minimum mass of 2.4 Jupiters and a most probable mass of 3.0 Jupiters. It lies closer to its star than Jupiter does to the Sun, having a distance of 2.1 astronomical units and an orbital period of three years, so if the planet circled the Sun it would lie in the asteroid belt between the orbits of Mars (1.5 astronomical units) and Jupiter (5.2 astronomical units). Because of this, the planet's gravity probably prevented the formation of a planet at the right distance from its star to support life, just as Jupiter prevented a planet from forming in the asteroid belt. But the discovery suggested that additional giant planets, like Saturn, might lie beyond it.

More controversial was an object that Marcy and Butler found a few weeks later, on December 30. "It was a Saturday morning," said Butler. "I came into the office at 8 A.M.—I'm a loser; I come to the office all the time—and the computer had finished with 70 Virginis, so I ran the code and put all the velocities together, and the thing looked bizarre." The star's velocity varied by several hundred meters a second, and Butler soon determined that the period of this variation was 117 days. "I was knocked out," he said. "I was sitting in my chair, speechless. I just stared at my computer for about an hour. Then I called Geoff. He was worried that I had called him at home, and he was also quite worried because I was still speechless. All I could do was mumble, 'Get here! Come to the office now!' He thought that the building was on fire, and we were losing all of our data."

The companion of 70 Virginis, another solar-type star, has a most likely mass of 8.4 Jupiters and a high orbital eccentricity, 0.40, so it resembles the object around HD 114762, the brown dwarf candidate that David Latham had found back in 1988. This object has a most probable mass of 12 Jupiters and an orbital eccentricity of 0.38. At the time, though, many astronomers rejected that discovery because of fears that the star was being viewed pole-on, in which case the companion was so massive that it was a *red* dwarf, not a brown dwarf. But the 70 Virginis discovery changed the thinking, since it is unlikely that astronomers would view two such systems exactly pole-on. HD 114762's velocity variation had been confirmed by Michel Mayor himself.

"At that point," said Butler, "we were very worried that Mayor could have 70 Virginis any day. We were going to the AAS meeting in San Antonio about two weeks later, and it was the obvious place to make this announcement. So we basically sat on it for about two weeks, just frightened to death the whole time that Mayor was going to make an announcement."

He did not, though, and in January 1996 the companions to 47 Ursae Majoris and 70 Virginis earned front-page headlines. Marcy and Butler called both objects planets. No one disputed this for the companion of 47 Ursae Majoris, but the one around 70 Virginis had such a high mass and elliptical orbit it seemed to be a brown dwarf. In 1967, Shiv Kumar had even proposed that orbital eccentricity could distinguish between brown dwarfs and planets in just this way. If one instead considers 70 Virginis's companion a planet, Marcy and Butler argue that one should also call the companion to HD 114762 a planet too, since the two objects seem similar. Ironically, that would mean the first extrasolar planet was discovered way back in 1988 by Latham, the same astronomer whom Marcy and Butler blasted for announcing 51 Pegasi's infamous second planet on *Nightline*.

Calling the objects around HD 114762 and 70 Virginis planets is contentious, however. "That you found something with 10 times the mass of Jupiter does not entitle you to say it's a planet," said David Stevenson. "But it's inevitable that observers will do this. It makes for a nice headline." In contrast, as Gliese 229 B demonstrated, no one gets prominent headlines for finding a mere brown dwarf.

Alan Boss said that the companions of HD 114762 and 70 Virginis are brown dwarfs, not planets. "This is much more than semantics," said Boss. "If you think that the object is actually a giant planet, then by implication you expect when it was formed that there were also some terrestrial planets. But if 70 Virginis's companion formed as a brown dwarf, before planets had a chance to form, you have these two fairly massive bodies orbiting around each other. We have to admit that we don't understand very well how to form planets in binary star systems, but the fear is that it isn't going to help. If a brown dwarf star was there from the very beginning, then it's going to be pretty tough to grow planets." Thus, in Boss's view, the discovery of the brown dwarf around 70 Virginis means that this star system probably has no planets at all.

Marcy admitted, when pressed, that he was not entirely comfortable labeling the companions to HD 114762 and 70 Virginis planets, but he was adamant that they were *not* brown dwarfs. "There's not even a prayer that they're brown dwarfs," said Marcy. "It's a very ironclad argument. There is a giant gulf between 10 and 40 Jupiter masses, where we don't have any companions, and bear in mind that any companions to our stars that had 10 to 40 Jupiter masses would have stood out like a sore thumb—they produce whopping velocity variations. Brown dwarfs were conjured up thirty years ago by Kumar to be an extension of the bottom of the main sequence, at 80 Jupiter masses. HD 114762 B and 70 Virginis B are no extension of any main sequence." That is because a gap—Marcy calls it the "brown dwarf desert"—exists between them and higher-mass brown dwarfs, like Gliese 229 B.

Marcy's argument did not sway another planet discoverer. "The companion of 70 Virginis is a brown dwarf," said Michel Mayor, noting the object's high orbital eccentricity. "When you have a very small number of objects, you can always see gaps. So in my mind, there's no reason to believe that we have a gap." Brown dwarfs of *any* mass are rare, at least in orbit around other stars. Since only three firm or reasonably firm brown dwarfs had been reported by early 1996—Gliese 229 B (40 to 55 Jupiters), HD 114762 B (12 Jupiters), and 70 Virginis B (8 Jupiters)—gaps in the mass distribution will inevitably appear, said Mayor, but these will vanish as more brown dwarfs are discovered. On the other hand, if Marcy is right and a mass gap does exist, then the gap will remain empty, even as more objects are found.

The astronomer who stands the most to gain by classifying HD 114762 B and 70 Virginis B as planets is surely David Latham, since he discovered the former long before the crop of recent discoveries; but he considers the objects brown dwarfs. "It is kind of interesting to see how Marcy and Butler ignored the companion of HD 114762 until they discovered the companion of 70 Virginis," said Latham. "Then, because these two systems are so similar, I guess they could no longer ignore HD 114762 B, and they started mentioning it, but kind of in passing." Nor is Latham pleased with Marcy and Butler's comments concerning his appearance on *Nightline*.

Adam Burrows, of the University of Arizona, took yet another view of the brown-dwarf-versus-planet question: "I think it's premature to type things so rigidly in the beginning of this rather new field. We don't know what to call these things. They could be planets; they could be brown dwarfs. We shouldn't be fixated too much on the words. For years, people have been telling us exactly *what* planets are, *where* they're formed, *what* their properties are—and they've defined a parameter space that has completely exploded with these observations. The observations should tell us what the characteristics of these objects are, and we don't want to give them names or classify them now, because if we do, we're going to have to change all that. There's a whole tradition in astronomy for typing things incorrectly, only to have to go back and change the typing—to the confusion of future generations." One example is the once alphabetical, and now unalphabetical, line-up of spectral types: O, B, A, F, G, K, and M.

More than just the classification of 70 Virginis's companion provoked controversy. When Marcy and Butler put out a press release on the discoveries, the very first sentence mentioned the possibility that the planet around 47 Ursae Majoris and the object around 70 Virginis had liquid water. This emphasized that these companions, unlike 51 Pegasi's, lay farther from their stars and so were cooler. But it also fueled speculation that the objects harbored life, causing other astronomers to groan.

Michel Mayor cited this as the danger of announcing discoveries in the press rather than through a refereed scientific journal. "I'm sure that no referee would accept this kind of thing," said Mayor, "because it's absolutely clear that no water has been discovered." Said Allegheny Observatory's long-time planet debunker George Gatewood, "Come on, Geoff; you're probably one of the most conservative guys I know. What did they do to make you say that? These are Jupiter-sized bodies, and the fact that they might have water—which hasn't even been shown yet—probably has nothing to do with life. Where's the water going to be: down in the metallic hydrogen? And then somebody said, 'Oh yeah, they might have moons,' and this got published, too. We don't even know there's a planet there for sure, and now we've got moons and life on them."

Later in 1996, Marcy and Butler reported planets around the solar-type stars Rho^1 Cancri A, Tau Boötis A, and Upsilon Andromedae. The first two stars have red dwarf companions a large distance away, showing that at least wide double star systems can have planets. The discoveries received relatively little attention, however, because Marcy and Butler were burned out on the media and never actually issued press releases. Nevertheless, the planets are just as real as the others. All three resemble the strange world around 51 Pegasi. The discoveries proved that 51 Pegasi was not unique but that a whole flock of "Pegasian" planets exist—giant planets located right next to their stars. Of course, these are exactly the planets that the Doppler technique can best pick up, which is why so many of these presumably peculiar planets are now known.

Also in 1996, Gatewood did a first: he himself announced a new planet, around Lalande 21185, the nearby red dwarf around which he had earlier disproved alleged planets. "This is very scary," said Gatewood. "I spent about three months trying to make it go away. You just know you're going to get shot down." Because Gatewood uses astrometry, the apparent planet he found lies much farther from its star than do those detected by the Doppler technique. The planet's mass is roughly Jupiter's, and its orbital period is about thirty-five years, which means the planet is almost as far from Lalande 21185 as Saturn is from the Sun. Lalande 21185 itself is the fourth nearest star system to the Sun, lying just 8.3 light-years away, which is much closer than 51 Pegasi and the other solar-type stars having planets, and much, much closer than PSR B1257+12. Thus, this would be the nearest known extrasolar planet yet found—if it is real. Said Gatewood, "One of the cosmologists once said, 'If I'm wrong, you won't know it till I'm dead and gone.' Well, this isn't quite that bad, but it is insurance on a long-term position. You're not going to confirm a thirty-five-year period in a short period of time. I think we're going to be looking at this for quite a while."

The latter half of 1996 saw discoveries that bore on the controversial classification of the eccentric, high-mass objects around HD 114762 and 70 Virginis, the objects that some astronomers claim to be brown dwarfs and other astronomers consider planets. In July, the Swiss team produced evidence favoring the brown dwarf interpretation, reporting the discovery of several brown dwarfs with masses between 10 and 40 times that of Jupiter. These brown dwarfs fill the alleged desert that Marcy had invoked to argue for the planetary nature of the companions to HD 114762 and 70 Virginis.

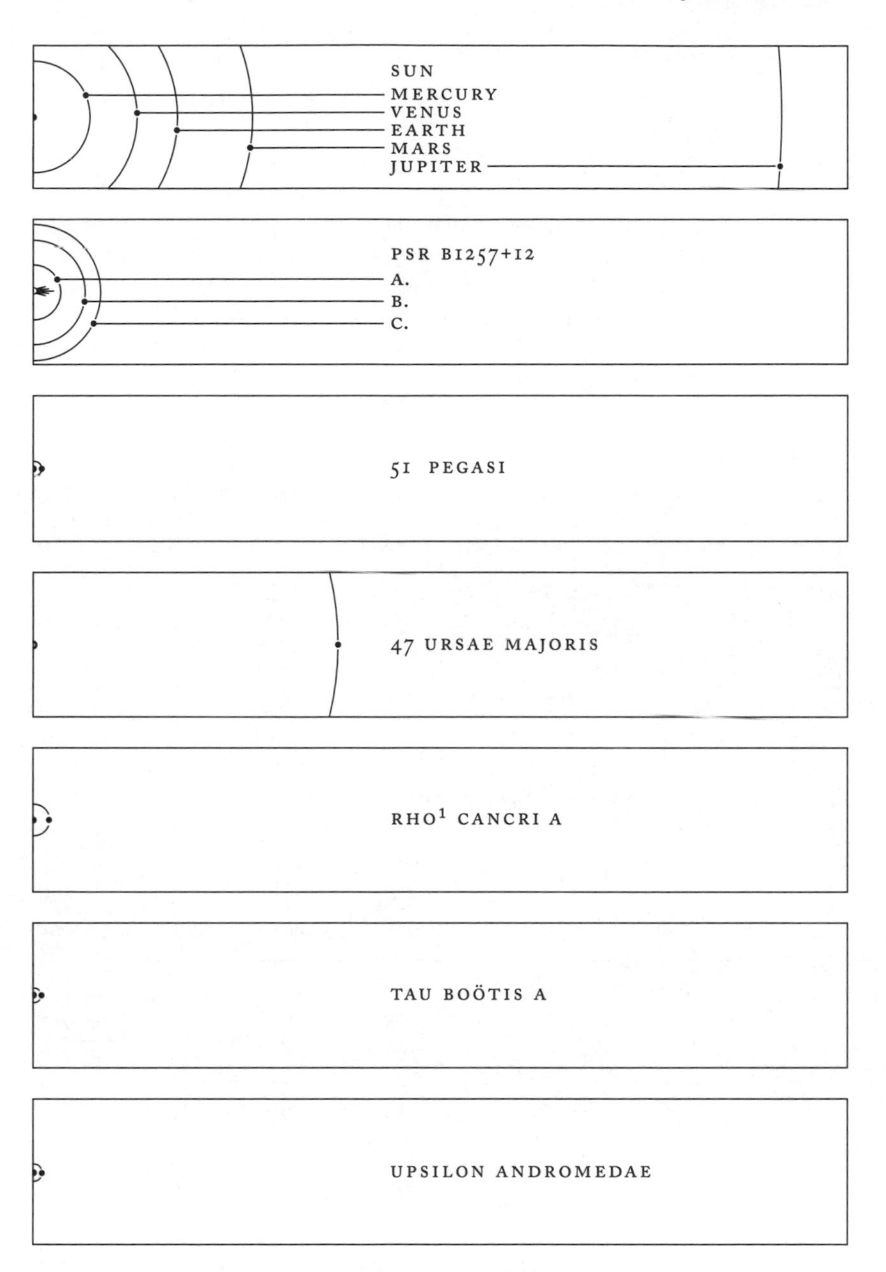

FIGURE 39

A line-up of the first extrasolar planets discovered shows some similarities but even greater differences with those of our solar system.

TABLE 10–1
The First Planets

TERRESTRIAL PLANETS

	Discovery		Distance			
	Year	Discoverers	from Star (AU)	Period (days)	Mass[a] (Earth = 1)	Orbital Eccentricity
Mercury	—	—	0.39	88	0.055	0.206
Venus	—	—	0.72	225	0.815	0.007
Earth	—	—	1.00	365	1.000	0.017
Mars	—	—	1.52	687	0.107	0.093
PSR B1257+12 a	1993	Wolszczan	0.19	25	0.019	0.0
PSR B1257+12 b	1991	Wolszczan and Frail	0.36	67	4.3	0.018
PSR B1257+12 c	1991	Wolszczan and Frail	0.47	98	3.6	0.026

PEGASIAN PLANETS

	Discovery		Distance			
	Year	Discoverers	from Star (AU)	Period (days)	Mass[a] (Jupiter = 1)	Orbital Eccentricity
51 Pegasi b	1995	Mayor and Queloz	0.05	4.2	0.6	0.0
Rho[1] Cancri A b	1996	Marcy and Butler	0.11	14.6	1.0	0.0
Tau Boötis A b	1996	Marcy and Butler	0.05	3.3	4.9	0.0
Upsilon Andromedae b	1996	Marcy and Butler	0.06	4.6	0.8	0.0
Rho Coronae Borealis b	1997	Noyes et al.	0.23	39.6	1.5	0.0

continued

TABLE 10–1

The First Planets (continued)

CLASSICAL GIANT PLANETS

	Discovery		Distance from Star	Period	Mass[a]	Orbital
	Year	Discoverers	(AU)	(years)	(Jupiter = 1)	Eccentricity
Jupiter	—	—	5.2	12	1.0	0.05
Saturn	—	—	9.5	29.5	0.3	0.05
Uranus	1781	William Herschel	19.2	84	0.05	0.05
Neptune	1846	Galle and d'Arrest	30.1	165	0.05	0.01
47 Ursae Majoris b	1995	Marcy and Butler	2.1	3.0	3.0	0.03

ICY PLANETS

	Discovery		Distance from Star	Period	Mass	Orbital
	Year	Discoverer	(AU)	(years)	(Earth = 1)	Eccentricity
Pluto	1930	Tombaugh	39.5	248	0.002	0.25

[a]For the pulsar planets and the planets detected with the Doppler technique—the Pegasian planets and the planet around 47 Ursae Majoris—these are the most likely masses. To obtain the minimum masses, divide by 1.27.

Three months later, however, another announcement bolstered those who consider the companions around HD 114762 and 70 Virginis to be planets: Marcy and Butler, and independently William Cochran and Artie Hatzes of the University of Texas at Austin, discovered an object that had a high orbital eccentricity—around 0.60—but a *low* mass, only about twice Jupiter's. Just as remarkable is the system in which this object lies, for 16 Cygni is a double star whose two components are both yellow and nearly identical to the Sun. The two stars lie several hundred astronomical units apart, allowing plenty of room for orbiting bodies. The new object was discovered around the fainter star, 16 Cygni B. If the object belonged to our solar system, it would swing from the orbit of Venus to beyond the orbit of Mars, and its gravity would mercilessly eject all four of the Sun's inner planets, including the Earth, out of the solar system and into the cold depths of interstellar space. Either way one looks at it, 16 Cygni B's companion muddies the waters: if it is a brown dwarf, because of its high orbital eccentricity, then some brown dwarfs are less massive than some planets; if it is a planet, then some planets have the highly eccentric orbits that are characteristic of brown dwarfs.

The next year, 1997, started badly, with a near-sacred extrasolar planet under fire. In a paper originally titled "The Planet Around the Star 51 Peg Does Not Exist," Canadian astronomer David Gray asserted that a pulsation, not a planet, causes the star's Doppler shifts. Although the editor toned down the title, Gray said he had detected periodic variations in the spectral lines' *shapes*, which a planet should not alter but which pulsation does. As a pulsating star expands and shrinks, it normally brightens and dims; but 51 Pegasi is special, said Gray, because as one region of the star rises, another falls, and the brightness barely changes. Gray's pulsation hypothesis could explain periods a few days long, jeopardizing the other Pegasian planets, too.

Even before Gray's paper appeared, Mayor, Queloz, Marcy, and Butler posted a stinging rebuttal on the World Wide Web. They questioned his data, noted the star's steadfast brightness—constant to an impressive 0.02 percent—and said the period is perfectly stable, unlike the wavering periods of many pulsating stars. Furthermore, although the Sun vibrates, it does so every 5 minutes, not every 4 days. Similar stars should vibrate at similar frequencies, just as a trumpet should sound like a trumpet and not a tuba. Moreover, this pulsation should produce additional frequencies, akin to musical overtones, yet the star has none. Unfortunately, just as battle erupted, 51 Pegasi slipped behind the Sun, so astronomers could not quickly confirm or refute Gray's data.

A few months later, the author polled nine astronomers not affiliated with either Gray or any planet-hunting team—one astronomer for each of the Sun's planets. The result was decisive: planet, 8 votes; pulsation, 1 vote. Again and again, the astronomers cited the planet as the simpler and better hypothesis. But one astronomer who favored the planet nevertheless excoriated its promoters for publishing their attack on the Internet rather than in a refereed journal.

Spring saw an exciting discovery that indirectly bolstered the planet. Announced in April and published in July 1997, the new find featured Rho Coronae Borealis, a Sunlike star 57 light-years distant, around which Robert Noyes of the Harvard-Smithsonian Center for Astrophysics and his colleagues discovered a Jupiter-sized planet. The planet is Pegasian, but it is farther from its star than the previously reported Pegasian planets. Whereas those planets orbit in less than 15 days, Rho Coronae Borealis's planet revolves every 40 days. This still means its orbit is smaller than Mercury's, but it carries two important implications. First, and crucial to the 51 Pegasi controversy, Gray's pulsation hypothesis cannot account for such a long period, so the new discovery is almost certainly a true planet, proving that some stars do have giant planets close to themselves. Second, the object has a circular orbit, consistent with its being a planet rather than a brown dwarf. The other Pegasian planets do too, but at least three of them lie so near their stars that they could have been born on elliptical orbits that got transformed into circular ones by stellar tides. Rho Coronae Borealis's planet is too far from its star for that to have happened. Thus this planet, and presumably the other Pegasian planets, arose on the circular orbits characteristic of genuine planets.

Still more discoveries are in the offing. A large number of extrasolar planets will allow planet hunters to achieve some of their ultimate goals: classify other planets; see what patterns other solar systems exhibit; and determine just how typical, or atypical, our own solar system, which has spawned life, really is.

Extrasolar planets are already falling into several distinct categories. The first three extrasolar planets found, around PSR B1257+12, have masses and orbits similar to those of the inner planets of the Sun. They are probably terrestrial planets, albeit circling a most unusual star. Circling normal stars are the very abnormal Pegasian planets, which seem to exist around a few percent of stars, including 51 Pegasi, Rho^1 Cancri A, Tau Boötis A, and Upsilon Andromedae. Of course, astronomers never expected to find either pulsar planets or Pegasian planets; they had instead anticipated what might be called "classical" giant planets, so named because four of them belong to our solar system. These, astronomers now know, also exist outside the solar system, for 47 Ursae Majoris has one and Lalande 21185 might, too. In addition, Marcy and Butler have seen tentative signs of such a planet around Rho[1] Cancri A, much farther than that star's Pegasian planet.

Back in 1992, George Wetherill had speculated that Jupiter and Saturn may be unusually large planets, responsible for intelligent life on Earth. Said Alan Boss, "I think it's still true that Jupiter and Saturn are prerequisites for life—George's argument still holds: we'd be bombarded by comets incessantly if those planets didn't exist—but the fact that we are finding other Jupiters implies that there are going to be other Earth-like planets there, and since they have a Jupiter that can protect them from their comets, too, then I think that really increases the chances that there'll be life in these other solar systems." Neither Michel Mayor nor Geoffrey Marcy, however, could yet say just how common or rare "real" Jupiters are—that is, Jupiter-mass planets at 5 astronomical units—because such distant planets are difficult for the Doppler technique to detect. Over the coming years, both Mayor and Marcy look forward to addressing that crucial question. Only George Gatewood said firmly that some stars lack Jupiters, a result that poses major barriers to the development of intelligent life elsewhere in the Galaxy.

Bruno's Dream

The discovery of planets around Sunlike stars greatly excited the public. "For me, it's really amazing to consider that the reaction was so huge," said Michel Mayor. "This is very fascinating to see—that the word *planet* carries some kind of magic. But if we are to be honest, it's still not life. We have only discovered a few Jupiter-type planets, and these are probably big and full of gas. I don't like to push things beyond what we have discovered, and say that we are just on the border of discovering life. No! We have discovered good examples of Jupiterlike planets, and now we have to work harder and find terrestrial planets."

When Mayor announced the discovery of 51 Pegasi's planet in 1995, many people asked him to name it. He said no, that would have been premature, and cited the "planet" seen inside the orbit of Mercury during the nineteenth century. That planet, named Vulcan, did not exist, and Mayor did not want to confer a name on his planet until it was confirmed. Now that this has happened, Mayor would like to christen the world *Epicurus*, for the Greek philosopher had postulated the existence of extrasolar planets over two millennia earlier. Mayor preferred this name to mythological ones, such as Bellerophon, who rode the winged horse Pegasus.

Geoffrey Marcy faced a great challenge dealing with the unrelenting media interest. "Clearly, the public is enthralled," he said. "I get letters from children, I get letters from eighty-five-year-old people in the hospital, and I feel a certain obligation to communicate our results to the public—they're paying the bill. The public wants to know. You have found some things: so what? And the 'so what' is quite profound. Is our Earth unique? Are there other Earths out there? Did Jesus have to go to all of those planets? People saw something they rarely get to see: how science can enrich their lives and have meaning beyond just a bottom-line profit. That's a wonderful thing, and I'm happy to contribute.

"On the other hand," said Marcy, "we ain't getting no work done no more. And if it weren't for my graduate students, and Paul Butler, who works much harder than I do, we really would have gotten almost nothing done. So I'm torn. We all have challenges in our personal lives, our love lives, our lives with our friends, and this is a challenge that I'm happy to take on. But it's a challenge. Sometimes, to be totally honest with you, I have to take a deep breath when, like today, the BBC said they would be here for two hours and then they ended up staying for seven."

Successful though it ultimately was, the long search for the first planets around normal stars was an enormous struggle. "Much of that time," said Paul Butler, "I was greatly depressed. I was working extremely hard, I wasn't producing papers, and nobody had ever heard of me. I developed most of this computer code while I was a grad student at the University of Maryland, and about every three months, I would get hate email from the Maryland astronomy computer committee, because I was using so much CPU time. I felt like people had no clue about the magnitude of the thing I was attempting to do—not just that I was personally struggling with my own demons, because I didn't know if I was ever going to get an answer, but I was furthermore struggling against this bureaucracy, which didn't give a damn. I'm making no money—like $10,000 a year; complete grad student poverty—and I'm getting bitched at regularly by the department for doing my research. There were a lot of temptations to bail out." At one point, Butler was offered a lucrative job in the computer industry, which he declined.

The failure of other planet seekers hardly helped matters. "We knew Campbell and Walker's results," said Butler, "and at that point we were despairing that there even *were* planets around other solar-type stars. The thing that always kept me going, even in moments of deepest despair, was back when I was a kid, I'd read about Kepler and Bruno—the fact that Kepler lived in fear of his life because of what he did and that Bruno was burned at the stake. These guys had put in years and years of work to come up with their results, and here I was, four hundred years down the road, trying to pull off Bruno's dream. And no matter how depressed I got, that, as much as anything, kept me plugging away."

Said Marcy, "The most important thing to me is that the human species, with its glorious intellect and curiosity, survive. And I think we as scientists can help to enhance the chances, and one way we can help has to do with planets. If we can someday venture to another planet, that would provide the kind of geographical diversity for the human species that I think would promote our chances of survival. We might blow ourselves up here, or maybe the colony on Alpha Centauri might blow themselves up; but there would be other colonies still surviving, with the human species pushing forward, and we would learn from the failures of our brethren who are on the other ill-fated planet. If we can make it for another fifty or hundred years, I think these planets we are finding may serve as safe harbors, within which the seafaring ships of the human genome can rest and proliferate and protect themselves."

Meanwhile, the hunt for extrasolar planets has only just begun. "People say we're in the golden age," said Marcy, "but we're not. We're in the bronze age of planet searching. Our velocity precision, 3 meters per second, is better than ever, but it's not good enough. We really need to be at 1 meter per second, so that we can definitively find Jupiters at 5 AU and Saturns at 10 AU." Someday astronomers will eventually reach planets with even lower masses, like Uranus and Neptune, which may be more common than Jupiters and Saturns.

For many planet hunters, though, the ultimate goal is still greater—or actually, smaller—prey: terrestrial planets, like Earth, circling a star like the Sun. Astronomers already know that three such planets orbit at least one pulsar. But planet hunters will not rest until they are in sight of a small blue world, warm and wet, in whose azure skies and upon whose wind-whipped oceans shines a bright yellow star like our own.

11

Future Worlds

A NEW ERA in the exploration of the universe has commenced. Although astronomers have long known of other stars and galaxies, only in the 1990s did observers unveil a third major component of the cosmos: planets circling other stars. What was once a dubious field in search of its subject has become an actual discipline that has already begun to study and classify new worlds far stranger than anyone had imagined. PSR B1257+12 and 51 Pegasi will forever bask in fame, but these stars' planets were only the first. Within the next few years, astronomers will have dozens, then hundreds, of extrasolar planets, some even odder than the ones known today. Not only will additional Jupiter-sized planets emerge, but new techniques will soon uncover less massive planets, like Uranus and Neptune, and eventually planets as small as Earth. Furthermore, astronomers will actually see some of these worlds *directly*, by glimpsing their light rather than just the wobbles they induce in their stars. Then the spectra of these planets will reveal their atmospheric composition and even indicate whether they harbor life.

Traditionally, of course, planet searchers hunted for their prey astrometrically, through planet-induced wobbles in a star's motion *across* our line of sight; but for decades astrometry produced only false detections, the most famous being the "planets" of Barnard's Star. The first true extrasolar planets were revealed instead by the Doppler technique, which looks for wobbles in a star's motion *along* our line of sight. This is essentially how the first pulsar planets were discovered in 1991, because they tugged the pulsar toward and away from Earth and altered the intervals between the observed pulses. The 1995 discovery of the planet around the Sunlike star 51 Pegasi also resulted from the Doppler technique, as did most of the other planets reported in the mid-1990s.

The Doppler technique won the planet-searching race in part because it was more sensitive to short-period planets—which can be quickly confirmed—and in part because it had become more sensitive than astrometry. In July 1996, Geoffrey Marcy and Paul Butler took their technique to the Keck telescope in Hawaii, the largest optical telescope in the world, where they aim to attain a Doppler precision of just 2 meters per second, sufficient to reveal both a Jupiter and a Saturn around a Sunlike star. In contrast, the best astrometry, which George Gatewood was doing electronically using his multichannel astrometric photometer (MAP) at Allegheny Observatory, had a precision of 1 milliarcsecond—the apparent thickness of a human hair seen from two miles and sufficient to detect a Jupiter around a Sunlike star ten light-years from Earth, but not a Saturn, or even a Jupiter around much farther stars. Gatewood's precision, astrometrists thought, was the best they could do from the ground, because turbulence in the atmosphere prevented still sharper views. Only an expensive satellite could do better.

In the early 1990s, however, scientists at the Jet Propulsion Laboratory realized that without launching a satellite, they could achieve a precision far superior. "What they're doing at JPL is really spectacular stuff," said Geoffrey Marcy. "I can't tell you how good a scientist Mike Shao is—he's really sharp, hard-working, clever, and insightful. He and his people are going to blow us out of the water. I look forward to the day, and I know it's coming very soon, when the whole Doppler technique is rendered obsolete, and I'm going to be out on Telegraph Avenue with a tin cup." Telegraph is Berkeley's haven for bums. "George Gatewood," said Marcy, "may be one of the people next to me with the tin cup." That is because the new technique will provide an astrometric precision a hundred times better than Gatewood's MAP—so good that it should detect planets as small as Uranus and Neptune.

Astrometry's Revenge: Interferometry

Ironically, the weapon behind astrometry's reemergence as a planet-hunting tool will be a technique that optical astronomers have long scorned: interferometry. This exploits the wave nature of light and combines the light received by two or more separate telescopes to yield incredibly sharp views of the heavens. Each telescope receives the peaks and troughs in a star's light waves, and comparing those from one telescope with those from the other reveals the star's exact position on the sky. Radio astronomers have long employed interferometry to obtain crisp views of radio sources. For example, in 1991 Dale Frail used the Very Large Array in New Mexico to measure a precise position for the first pulsar with planets, PSR B1257+12. The VLA is an interferometer with twenty-seven separate radio antennas spread over some 15 miles. It attains the same resolution—that is, the same sharpness—as would a single radio telescope whose dish was 15 miles across.

But interferometry at optical wavelengths is much more difficult. "Optical interferometry had a bad name," said Michael Shao. "In the early days, we would write proposals, and the optical people would come back and say, 'This is stupid; it'll never work.'" As a graduate student at the Massachusetts Institute of Technology in the 1970s, Shao had found most of his support coming from radio astronomers, not optical astronomers. Optical interferometry is tougher than radio interferometry because light waves are much shorter than radio waves. The latter may be inches, feet, or even miles long, so a radio interferometer need not be much more exact than this. But because light waves are microscopic, the components of an optical interferometer must be nearly perfect, and many of Shao's predecessors had met with failure.

"George Gatewood was actually a pretty good guy," said Shao. "He was skeptical in the beginning, but he was properly skeptical. Any time people come up with a claim that they can do orders of magnitude better, someone should be skeptical. There were others, though, who had a gut emotional reaction, and George had a good story for that. He was one of the first to go from photographic measurements to photoelectric ones, and he met with the same sort of thing. What he said was that in astrometry, progress is made by the old generation's dying off."

During the 1980s, Shao's group successfully built an optical interferometer, but he still thought he could do no better than 1 milliarcsecond, the same precision Gatewood was already achieving with the much simpler and cheaper MAP. This limit was thought to exist because the Earth's atmosphere distorts light waves. To see how, imagine a vast ocean that sends a large wave of water toward shore. If the wave meets no obstacles, it will crash on the shore uniformly. This is analogous to the perfect way in which light waves travel through space. Now imagine that before it reaches shore, the water wave encounters several giant boulders that stick out of the ocean floor. They will mess up the wave so that different sections of the shore receive the wave at different times. The Earth's atmosphere does the same thing to a light wave, and astronomers suffer the consequences.

One consequence is that two telescopes in an optical interferometer have a hard time piecing together the original wavefront from a star, because the wave may strike one telescope at some unknown time before it strikes the other. Ultimately, it was thought, this would limit astrometrists to a precision of 1 milliarcsecond, no matter how good their instrument, and Shao's group found that this was indeed the case for many of their observations. But when he observed double stars—two stars very close together—the interferometer gave positions for the two stars relative to each other that were ten times better than this supposed limit. The astronomers then realized that the atmosphere was corrupting the light waves from both stars, but it was doing so equally for both, because their light waves took nearly the same path through the atmosphere. Thus, the interferometer could measure a star's position precisely by comparing it with a nearby reference star, which is all that one needs to discern tiny planet-induced wobbles. Shao also found that he could achieve the best results if he observed at infrared wavelengths, somewhat longer than the eye can see.

To take advantage of these new findings, Shao's group spent several years building a powerful infrared interferometer atop Palomar Mountain in California. The new interferometer began searching for planets in 1997. Its two telescopes are separated by a hundred meters. The interferometer's precision is so exquisite that Shao measures it not in milliarcseconds but in *micro*arcseconds, which are a thousand times smaller. One microarcsecond is 1/3,600,000,000 of a degree—

the apparent thickness of a human hair as viewed from two thousand miles. A Jupiterlike planet orbiting a Sunlike star ten light-years from Earth would induce an astrometric wobble of 1600 microarcseconds; Neptune, 510 microarcseconds; and Earth, 1 microarcsecond. The Palomar interferometer has an expected astrometric precision of 50 microarcseconds, so it can find a Jupiter, Saturn, Uranus, and Neptune around a Sunlike star ten light-years away.

One of the first extrasolar planets that Shao hopes to detect is the one that Geoffrey Marcy and Paul Butler discovered around 47 Ursae Majoris. This is the planet that has a few times more mass than Jupiter and revolves around the star every three years on a circular orbit. Because the planet lies fairly far from its star—2.1 astronomical units—its astrometric signature is fairly large, around 400 microarcseconds.

"Astrometry has no ambiguity about the planet's mass," said Shao, so the astrometric wobble will determine the planet's true mass—rather than just a lower limit, as Marcy and Butler did. "Another advantage of astrometry is that the accuracy in radial velocities will ultimately be limited by the star itself, because stellar activity can change the shape of the spectral lines." Spots on stars produce artificial Doppler shifts and will prevent the Doppler technique from measuring genuine wobbles as small as astrometry can.

Meanwhile, George Gatewood began to use the giant Keck telescope to combine astrometry and the Doppler technique, simultaneously measuring a star's wobble both across and along our line of sight. Gatewood started observations in 1997, with a new instrument called the MAPS—multichannel astrometric photometer with spectrograph. The spectrograph, which measures the Doppler shifts, was developed by Robert McMillan of the University of Arizona and aims for an accuracy of 2 or 3 meters per second. Although Gatewood's astrometric precision will not be as good as Shao's, Gatewood expects to do ten times better from Keck than from Allegheny Observatory.

Pure Doppler searches will continue to turn up new planets, too. In fact, the Pegasian planets will likely keep Marcy and other Doppler users off the streets, because their technique is most sensitive to planets close to stars, which Shao's astrometric approach cannot detect. For example, the planet around 51 Pegasi induces an astrometric wobble of only about 2 microarcseconds, and even smaller Pegasian planets, like Neptune, would be even further beyond Shao's reach. In contrast, Marcy and Butler's Doppler technique can establish whether Uranus- and Neptune-sized Pegasian planets are as common as Jupiter- and Saturn-sized ones.

Nevertheless, Shao even plans to detect the Pegasian planets, not through astrometry, but directly through the planets' own heat, by linking the two 10-meter Keck telescopes in Hawaii into an awesome optical-infrared interferometer. Said Shao, "Originally, back in 1991 when we proposed this technique, we said we would have enough sensitivity, just barely, to see a Jupiter-sized planet that's at a temperature of 300 Kelvin. And of course, what they said was, 'But the *real* Jupiter is at 125 Kelvin.'" That mattered a lot, because a planet's infrared output plummets with decreasing temperature. "So we couldn't see a real Jupiter," said Shao, "and the then-current theory said that Jupiters could form only beyond the ice condensation zone; therefore, it wouldn't make any sense to try to do direct detection from the Keck."

The Pegasian planets changed all that. These giant planets lie so close to their stars that their temperatures are around 1000 Kelvin, which causes the planets to glow so brightly in the infrared that the Keck interferometer will easily see them, even though they lie right next to their stars.

Flickerings of Planets

The Pegasian planets also boost the prospects of another method of planetary detection. Like astrometry and the Doppler technique, this method is indirect, because it does not show the planet's actual image. Instead, astronomers would discover the planet as it passed in front of its star and blocked part of its light. In order for the planet to do this, the system must be one of the relatively few that happens to be viewed edge-on from Earth, so astronomers would have to monitor hundreds or thousands of stars simply to detect a single planet. Once a dip in a star's light was discovered, astronomers could observe the star for a repeat performance and deduce the planet's period. Prior to the discovery of the Pegasian planets, however, this method seemed much more daunting. If a solar system like the Sun's were viewed edge-on, Jupiter would dim the Sun by only 1 percent, and that for only a day once every twelve years. Earth would cause a mere 0.01 percent dip. Disturbances on the star itself, such as spots, could easily produce changes larger than that.

Enter the Pegasian planets. Because the planet is so close to the star, the chance of such a transit's occurring is much better, just as the chance that two leaves on a tree happen to appear superimposed on each other is much better if they are close together. Also, because the planet revolves every few days, astronomers would not have to wait long to see it again and confirm it. The best stars to observe are red dwarfs, because they are small and a planet would block a large fraction of the star's light.

Also promising are double star systems in which the two stars eclipse each other. In these cases, astronomers know they view the systems edge-on; otherwise, the two stars would not eclipse. The most famous eclipsing binary is Algol, which is normally a second-magnitude star in the constellation Perseus but frequently fades when the fainter star in the system eclipses the brighter. Any planets presumably orbit the stars in nearly the same plane as the stars themselves do, so astronomers can examine these systems for transits of planets. The bad news, though, is that double stars may not easily form planets, if the gravity of one star prevented planets from developing around the other. Of course, astronomers also once thought that planets could not exist around pulsars, and astronomers could be just as wrong about planets around binaries. In any case, the best eclipsing binaries to observe are those which consist of two red dwarfs, because a passing planet would dim a greater fraction of the light of such small stars. An ideal system is CM Draconis, an eclipsing M-type binary forty-seven light-years away in the constellation Draco. A Neptune-sized planet around CM Draconis would cause a greater decrease in brightness than would a Jupiter-sized planet around a Sunlike star.

Another method for the indirect detection of planets exploits what astronomers call gravitational microlensing. When a star passes directly in front of another star, the gravity of the nearer star magnifies, or "microlenses," the farther's light. In a 1986 paper, Princeton astronomer Bohdan Paczyński proposed that microlensing could detect not planets but the Galaxy's dark matter, which is at least as elusive. If the dark matter surrounding the Milky Way's disk is composed of dim stars, said Paczyński, they will occasionally pass in front of bright stars in other galaxies and microlense those stars' light. A few years later, astronomers implemented Paczyński's plan and in 1993 reported the first successes. The exact number of these events, together with the masses of the objects responsible, will eventually reveal the composition of the Galaxy's dark matter.

In 1991, Paczyński and Shude Mao, then a graduate student, realized that if the star causing the microlensing event had a planet, the planet's gravity could also distort the more distant star's brightness for a few hours, giving astronomers a new way to detect planets. "This is doable," said Paczyński, "but the problem is you need hardware and software that is vastly more powerful than what we have right now. The reason it is so difficult is that planets are low-mass objects, so the microlensing events caused by them will be much shorter in time and much less probable. Even if there is a planet around the star, in only a very few cases will the geometry be just right to notice the planet in the light curve, and that blip in the light curve will last for just a few hours." To detect the dark matter events, astronomers are currently monitoring tens of millions of stars every night. To have a realistic hope of detecting a planet, said Paczyński, astronomers must observe hundreds of millions or perhaps billions of stars per night; therefore, the best place to look is toward the center of the Galaxy, in the constellation Sagittarius, where stars are most numerous. Even if they find a planet, astronomers may not be able to follow up the discovery, since the star-planet system that caused the microlensing event may be too faint to see.

The microlensing technique, however, is the only one that monitors enough stars to stand a chance of detecting interstellar planets—planets that have been cast off by their stars. Billions of such planets may exist in the Galaxy, because a supernova explosion should liberate the planets of a dying star, and vast numbers of supernovae have exploded during the Milky Way's life. In addition, as a Pegasian planet migrates from the outer part of its solar system to the inner, the planet's gravity kicks terrestrial planets into space. Most interstellar planets must be cold and dark, for no star keeps them warm. But an interstellar planet could stay warm if it has a large supply of radioactive material that gives off heat, or if it has a moon that raises intense tides which warm the planet. Whatever the planet's temperature, if such a world passed directly in front of a star, its gravity would microlense the star's light just as a star would, allowing astronomers to discover planets roaming between the stars.

Adaptive Optics

Meanwhile, astronomers are also devising new ways to detect planets *directly*. "In the direct methods," said J. Roger Angel of the University of Arizona, "you have the aesthetic or public appeal advantage of making an actual picture where you see a planet there. You also have the possibility of sorting out multiple planets, which I think is a real plus. If we get an adaptive optics system running, one picture exposed over a night should resolve all the giant planets in a solar system."

Adaptive optics is a technique that untwinkles the stars, showing images as sharp as those from a telescope in space. The idea of adaptive optics was first proposed in 1953 by an American astronomer, Horace Babcock, but it was then adopted by the military, which wanted to view objects like Soviet satellites through the Earth's murky atmosphere. For years this information was classified, but in 1991 the military released it to astronomers. Adaptive optics undoes the damage a light wave suffers as it navigates through the atmosphere. To know the exact damage done, the system observes a bright reference star. Then, to correct this damage, a "rubber" mirror bends and flexes in such a way that it continually undistorts the light and gives a sharp view of the bright object and any objects near it. This technique was partially responsible for the discovery of the brown dwarf orbiting Gliese 229, and Michael Shao also plans to use adaptive optics when he links up the two Keck telescopes into an infrared interferometer.

Much more powerful adaptive optics systems may someday reveal extrasolar planets at visible wavelengths. Extrasolar planets are made to order for adaptive optics, since the planets' own star provides the bright reference that the system needs. Because of this need, the best target stars are those bright enough for the naked eye to see. Nearby stars, such as Vega and Altair, are ideal; then the apparent separation between the star and its planets is greatest. For example, if our solar system were viewed from a distance of ten light-years, Jupiter would appear up to 1.7 arcseconds from the Sun, which is far enough for even a backyard telescope to separate. The problem, of course, is that the star is a billion times brighter, and its glare washes out the planet. By removing the blur caused by the Earth's atmosphere, however, adaptive optics transforms the fuzzy star into a pinpoint of light. Someday this technique may take actual pictures of extrasolar planets.

"The planets that are going to be discovered by astrometry will be the best targets," said Angel, "because astrometry finds the planets that have enough angular separation that we can hope to resolve them. In contrast, the Doppler technique favors those planets which are close in—the ones that are most difficult for adaptive optics to see. For instance, 51 Pegasi's planet is hopelessly close in. I take the detection of these planets as super-encouraging—there *are* Jupiterlike planets to look at—but I think we have to wait a little bit to find the ones that are going to be the really good candidates."

The direct images of these planets will complement those which Michael Shao plans to make of the Pegasian planets. Whereas the latter are hot and bright, emitting heat that Shao plans to detect, the giant planets that Angel hopes to see reside in the opposite corners of their solar systems: far from their stars, where the planets are cold and faint, and will be detected by reflected light rather than by their heat.

Angel said that adaptive optics will photograph only giant planets, not terrestrial ones. "If you regard the imaging of Earthlike planets as an important goal," said Angel, "then the kind of technologies and problems that you face in imaging Jupiterlike planets are really testing out the things you're going to have to do to look for Earthlike planets." Photographing extrasolar Earthlike planets, however, will also be possible—for those who venture into space.

Extrasolar Life

To reach still more planets, by both indirect and direct means, astronomers want to place interferometers above the atmosphere. In the first decade of the twenty-first century, NASA may launch an astrometric interferometer that will achieve an accuracy of about 1 microarcsecond, almost enough to sense the wobble induced by an Earthlike planet around a Sunlike star ten light-years away. "Whether NASA will have money to actually pull this off," said Shao, "God knows."

Actually, God does not decide such things; Congress does. Shao estimated that the first space-borne interferometer would cost half a billion dollars, considerably less than the Hubble Space Telescope cost to build (and fix). Planet detection would not be the mission's main goal, however. Instead, the exquisite astrometric accuracy of the space-borne interferometer would let astronomers measure parallaxes and thus distances of distant stars in our Galaxy and even those in other galaxies, too. These distances would help establish the distance scale and age of the universe.

NASA's recent history supplies plenty of ammunition for both supporters and detractors of this or any other NASA mission. On the one hand, this is the same government agency that in 1986 launched a space shuttle which exploded and killed seven people; in 1990 launched an expensive telescope with bungled optics and then put out numerous press releases exaggerating the telescope's accomplishments; in 1993 lost a spacecraft around Mars; and in 1995 watched a spacecraft reach Jupiter unable to transmit most of its data to Earth. On the other hand, NASA has also had recent successes: the 1986 and 1989 Voyager encounters with Uranus and Neptune, respectively; the 1990 Magellan mission to Venus; the 1991 launch of the Compton Gamma Ray Observatory; and the 1992 discovery, by the Cosmic Background Explorer, of ripples in the big bang's afterglow.

If the proposed astrometric interferometer is launched, later ones, sensitive to infrared wavelengths, could yield still greater planetary prizes: direct images of Earthlike planets around Sunlike stars, along with a determination of whether these planets have life. A space-borne infrared interferometer would see the planets both by attaining extremely sharp views and by canceling out the light waves of the central star. The interferometer can cancel a star's light by adding the crests of its light waves received by one telescope to the troughs received by the other telescope. Off-center objects, like a planet, will generally not be canceled. This idea for a so-called nulling interferometer had first been suggested by Stanford University's Ronald Bracewell in 1978. A somewhat similar trick can eliminate vocals from compact disks. Loosening the input cords that lead from the compact disk player to the stereo receiver causes any sounds that are the same on the left and right channels to cancel, while other sounds continue to appear. Since vocals are usually equal left and right, they get eliminated, whereas many accompanying musical instruments are biased to either the left or right and remain. The vocals are like stars, and the musical instruments like planets.

The best wavelengths at which to see planets directly are infrared. At visible wavelengths, Jupiter and Earth are, respectively, 1 and 10 billion times fainter than the Sun, because planets emit no light of their own but simply reflect some of the little light they intercept from their star. At infrared wavelengths, however, the story changes: planets emit their own radiation, because they are warm. In fact, at some infrared wavelengths, the small Earth is actually brighter than giant Jupiter, because Earth is much warmer. Nevertheless, even at these wavelengths, the Sun is some 10 million times brighter than the Earth. To achieve the best results, astronomers have proposed placing the interferometer far from the Sun, at the distance of Jupiter, where both interplanetary dust and the interferometer itself would be so cool that they would emit little infrared radiation to ruin the observations. In addition, astronomers need to know more about how infrared-emitting dust in other solar systems might obscure sought-after planets.

The infrared section of the spectrum offers the best chance to see not only Earthlike planets but also extraterrestrial life. "You're driven to make the direct detection of the planet in the infrared," said Roger Angel, "because that's where the contrast with the star is best, and then you notice that the infrared spectrum of the Earth has three enormous features in it, due to the greenhouse gases. The same gases that warm the Earth do it by blocking the escape of infrared radiation, and so you see these loud and clear because chunks of infrared emission are missing." All three gases carry important information. "This is what nature has provided for us," said Angel, "and what is so remarkable and encouraging is that the three molecules we can look for are so interesting: carbon dioxide is common in the atmospheres of Venus, Earth, and Mars, and it tells you that the planet does indeed have an atmosphere; water vapor, which is unique to the Earth, tells you that the planet has oceans; and ozone, in the case of the Earth, tells you the planet has life."

Ozone (O_3) is the most crucial, because on Earth, at least, oxygen signifies life. It did not exist in Earth's air before life arose, and if all terrestrial life suddenly died, the oxygen in the air would eventually

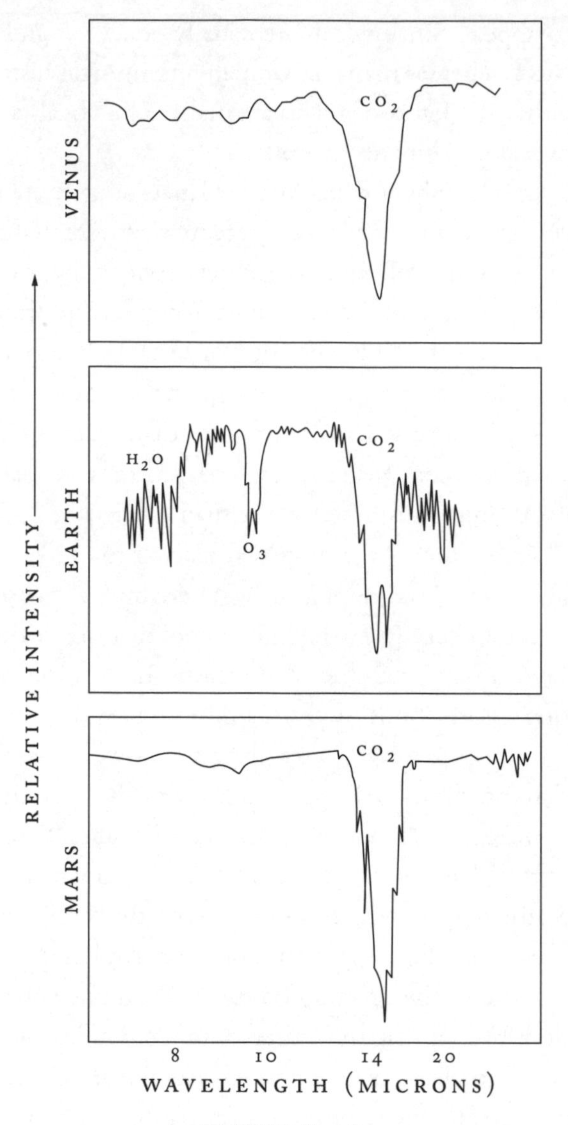

FIGURE 40

Three key molecules—carbon dioxide (CO_2), water (H_2O), and ozone (O_3)—may someday be detected in the spectra of extrasolar Earth-mass planets, signifying, respectively, atmospheres, oceans, and life.

vanish, because oxygen is a reactive element that readily combines with other elements. Most oxygen in Earth's atmosphere is diatomic (O_2), but that molecule is much harder to detect on another planet than ozone, and ozone is an excellent tracer of O_2.

TABLE 11–1
Atmospheres of Venus, Earth, and Mars
(abundance by volume in percent)

Gas	Venus	Earth	Mars
Carbon dioxide (CO_2)	96.5	0.035	95.32
Nitrogen (N_2)	3.5	78.08	2.7
Oxygen (O_2)	—	20.95	0.13
Water vapor (H_2O)	0.003	0 to 4	0.03
Argon (Ar)	0.007	0.93	1.6
Carbon monoxide (CO)	0.002	—	0.07
Neon (Ne)	0.0007	0.0018	0.00025
Helium (He)	0.001	0.0005	—
Sulfur dioxide (SO_2)	0.01	—	—
Methane (CH_4)	—	0.00017	—
Hydrogen (H_2)	—	0.00005	—
Nitrous oxide (N_2O)	—	0.00003	—
Xenon (Xe)	—	0.000009	0.000008
Ozone (O_3)	—	0.00004	0.00001

Oxygen itself is a good tracer of life, but it may not be perfect. Can a planet have oxygen without life? And can a planet have life without oxygen? In some cases, the unfortunate answer to the first question is yes. For example, if a planet with oceans gets too hot, those oceans will boil away, and the resulting water vapor (H_2O) will split into hydrogen and oxygen, giving the atmosphere a temporary whiff of oxygen, even if it lacks life. This may have happened to Venus, billions of years ago. Such planetary tragedies should be rare, at least at any one time, so astronomers are unlikely to run across them. In the 1990s, astronomers did find oxygen around Jupiter's presumably lifeless moons Europa and Ganymede, suggesting again that atmospheric oxygen can exist without life, but the oxygen around Europa and

Ganymede is so tenuous that it could never have been seen around an extrasolar planet. Thus, in most cases, an extrasolar planet with detectable oxygen should also be a planet with life.

But is a planet with life a planet with oxygen? Life as we know it produces oxygen, but what about life as we do not know it? "The argument that life is likely to produce oxygen sounds pretty good," said Angel. "But before we discovered extrasolar Jupiterlike planets, everybody thought these planets were going to be far away from their stars, and all these poor people spent a decade looking far from the stars, only to find out that the planets were really easy to detect once you look close in. So the message is that our imaginations may not be smart enough."

Despite the possible ambiguities, atmospheric oxygen holds great promise to distinguish between life-bearing worlds and their lifeless brethren. In fact, it might well be a much better test than the ongoing search for radio waves from extraterrestrial intelligence, since that search can detect only those life-bearing planets whose life is sufficiently intelligent to be able to broadcast radio waves into space and sufficiently stupid to do so deliberately, for such a broadcast would reveal their presence to a Galaxy that might teem with hostile races ready to gobble them up. Some claim this is a ridiculous concern—that any advanced civilization would necessarily be benevolent—but there are no data whatsoever to support this assertion, and Earth history suggests caution. In the 1930s, one of the most scientifically advanced civilizations on the Earth was the less-than-benevolent Nazi Germany. A further problem with the search for extraterrestrial intelligence is that intelligence may be extremely rare, even if life is common. After all, of the millions of species living on Earth, only one can send and receive radio signals.

In contrast, even primitive life produces oxygen, so the direct detection of Earthlike planets, along with their atmospheric supply of carbon dioxide, water vapor, and ozone, could answer one of the chief questions of astronomy: how common is life in the universe? This is actually more a question of biology than of astronomy, because astronomers know that most of the astronomical requirements have been fulfilled. Our Galaxy has billions of stars like the Sun, and astronomers now know that Jupiter-sized planets exist around at least some of them. The chief remaining astronomical unknown is how common Earthlike planets are around Sunlike stars.

But the biological unknowns are far greater. Suppose a warm, wet planet like Earth circles a good, bright star like the Sun. Will the planet develop life? Will that primitive life evolve into complex life? And will some of that complex life acquire intelligence? On Earth, the answers to these three questions were yes, yes, and yes; but the only *honest* answers to these questions in general are: no one knows, no one knows, and no one knows.

A search for Earthlike planets around other Sunlike stars would remove the remaining astronomical unknown in the question of extraterrestrial life, because such a search would determine whether terrestrial planets exist around stars like our own. Furthermore, spectra of any such planets that are discovered would address the biological questions. Some of the Earthlike planets might show nothing atmospheric, so they would be nearly airless worlds like Mercury, unable to support oceans or life. Other planets would probably show strong carbon dioxide lines, indicating the presence of an atmosphere. A few of these planets would also show water vapor, because they lie at the right distance from their stars to have mild temperatures, liquid water, and thus the possibility of life. And some of these planets might also show ozone, the spectroscopic hallmark of life.

It is with the ozone that the biology begins. Suppose astronomers discovered a hundred Earthlike planets bearing water, and every single one also had ozone. Then life must be a common phenomenon, one that arises whenever conditions are right: warm seas under a bright star. Astronomers could search these planets for radio signals of intelligent origin. On the other hand, every single one of those water-bearing worlds could just as easily lack ozone. Life, even primitive life, would then be seen to be incredibly rare, perhaps miraculous.

"One of the most significant results of the whole NASA enterprise was to find that life within our own solar system was probably restricted to the Earth," said Angel. "Ever since the invention of the telescope and Galileo and the discovery of the New World, there was almost an expectation of the plurality of worlds and life, and then in the twentieth century we've seen that expectation pretty much disappear, certainly in our solar system." Now, though, astronomers can search for life elsewhere. "By good fortune," he said, "because nature placed these water and ozone lines right in a place where they're most accessible to us, here's a way that we can look for life in an even more definitive way than by taking a picture. To me, that's been the incredible excitement of the last year: here's this visionary thing which is actually completely consistent with what's technically feasible."

The future, said Angel, will be tremendously exciting. "This is one of those moments in science when a whole new field is opening up," he said. "For technical reasons, we had been constrained to look at big bright things, like stars and galaxies. But now the technology is at a point where we're not forced to look only at big bright things; we can look at *planets*, which have so much human interest."

From their platform on Earth and armed with increasingly powerful instruments, astronomers will continue to scrutinize the planets they have discovered and look for still more. They do so across a gulf of trillions of miles. That huge gulf is the greatest reason for, and the greatest obstacle to, exploring these systems by sending spacecraft to photograph them, investigate possible life forms, and assess how intelligent that life might be. But vast though it is, that gulf is one which scientists and engineers of the twenty-first century may just be able to bridge.

12

Into the Cosmos

THE STARS—red, yellow, orange, blue, and white—shine from across a void of trillions of miles, far beyond the farthest planets of the solar system. At least some of these stars, astronomers now know, warm orbiting planets, and future years may see the discovery of planets that harbor water and even life. From afar, astronomers can try to study these planets and deduce basic properties, but the only way to scrutinize new worlds in detail is by sending spacecraft to them. Before spacecraft flew past the planets of the Sun, astronomers trapped on Earth knew little about them: some thought Venus had oceans, Mars canals, and no one knew much at all about distant worlds like Uranus and Neptune.

Targets for interstellar missions abound, because the space within just a dozen light-years of the Sun boasts a rich variety of stars and possible planets. The nearest target is the famous triple star Alpha Centauri, 4.3 light-years away, whose two brightest members resemble the Sun. Alpha Centauri A is a yellow G-type main-sequence star, and Alpha Centauri B is an orange K-type main-sequence star. On average, the two lie a bit farther apart than the Sun and Uranus, and they never get closer together than the Sun and Saturn. Each star could therefore have four planets spaced like Mercury, Venus, Earth, and Mars, because the orbits of such planets would be little disturbed by the other star's gravity. Billions of years ago, however, when Alpha Centauri was forming, one star's gravity might have perturbed the material attempting to form planets around the other so that planets never arose around either star. Far from these two suns is the dim red dwarf Proxima Centauri, which revolves around them roughly once every million years and may have its own retinue of planets.

At greater distances than Alpha Centauri is the red dwarf Barnard's Star, 6.0 light-years away; Wolf 359, another red dwarf, 7.8 light-years away; and Lalande 21185, which is 8.3 light-years away. A bit farther lies white-hot Sirius, the brightest nighttime star, and beyond Sirius the solar-type stars Epsilon Eridani and Tau Ceti, which astronomers in 1960 searched for radio waves of intelligent origin. Today astronomers know that Epsilon Eridani is only around a billion years old, probably too young to have intelligent life, and a journey to this star might reveal simple life just now originating on one of its young planets. Tau Ceti is older, but it has a smaller supply of planet-forming, and life-forming, elements like carbon and oxygen. Nevertheless, this star ranks high on any space traveler's list of interstellar destinations. Whether Tau Ceti has planets, no one knows. Astrometrists have found nothing, and Doppler searchers will probably fail too, because the star's pole points nearly at the Sun, so any planets would pull the star only slightly toward and away from Earth.

Interstellar Travel: Prospects and Problems

Despite the appeal of voyaging to the stars, some scientists believe such missions will never occur, because even the nearest stars are so distant. In 1960, Harvard astronomer Edward Purcell wrote, "All this stuff about traveling around the universe . . . belongs back where it came from, on the cereal box." The energy and expense of a journey to Alpha Centauri would be enormous, problems that even advocates of interstellar travel concede. In a 1996 paper, longtime starship proponent Robert Forward wrote, "interstellar flight is difficult and expensive, but not impossible." As Eugene Mallove and Gregory Matloff's generally optimistic *Starflight Handbook* admitted, "Starflight is not just very hard, it is very, very, very hard!" Proponents like to point to amusing statements denying the possibility of things everyone now takes for granted. For example, early in the twentieth century, some scientists claimed that airplanes would never fly across the Atlantic. Opponents of starflight can point with equal glee to past statements that have erred in the opposite extreme, such as predictions that people in the 1990s would commute to work in their own personal helicopters.

Today some astronomers actually have a vested interest in ridiculing interstellar travel: they believe, without any evidence, that intelligent life is common in the Galaxy. Yet none of this life has ever visited Earth, a fact that prompted physicist Enrico Fermi in 1950 to ask the famous question, "Where are they?" To explain this apparent contradiction, now called the Fermi paradox, astronomers who believe that extraterrestrial intelligence exists argue that interstellar travel is so difficult and expensive that no civilization would undertake it. Thus, we Earthlings will never succeed at it, either.

To date, terrestrial spacecraft have explored all the Sun's planets from Mercury to Neptune; only Pluto has been left out in the cold. The first interplanetary spacecraft focused on the two planets closest to Earth, Venus and Mars, but in 1972 the United States launched Pioneer 10, which bravely crossed the asteroid belt and encountered Jupiter. A second spacecraft, Pioneer 11, sped past Jupiter in 1974 and greeted Saturn in 1979. In 1977, the two Voyager spacecraft were launched. Voyager 1 passed Jupiter in 1979 and Saturn in 1980, and Voyager 2 met Jupiter in 1979, Saturn in 1981, Uranus in 1986, and Neptune in 1989.

In a sense, all four of these far-flying spacecraft—Pioneer 10, Pioneer 11, Voyager 1, and Voyager 2—are humankind's first interstellar craft, for all four are leaving the solar system at high speed, bound for the stars. The Pioneer spacecraft cover about 2.3 astronomical units every year, and the faster Voyagers about 3.4 astronomical units per year. But the stars are so distant that even the Voyager spacecraft will require 80,000 years just to reach the present distance of Alpha Centauri. With luck, though, they will never make it that far: spacefarers of future centuries will overtake and capture those spacecraft and preserve them in a space museum.

The greatest difficulty confronting voyagers to the stars stems not from Einstein's special theory of relativity, which decrees a maximum speed on interstellar voyages—the speed of light, or 671 million miles per hour—but from the enormous distances of the stars themselves. Indeed, compared with that basic barrier, the relativistic speed limit is a mere nuisance.

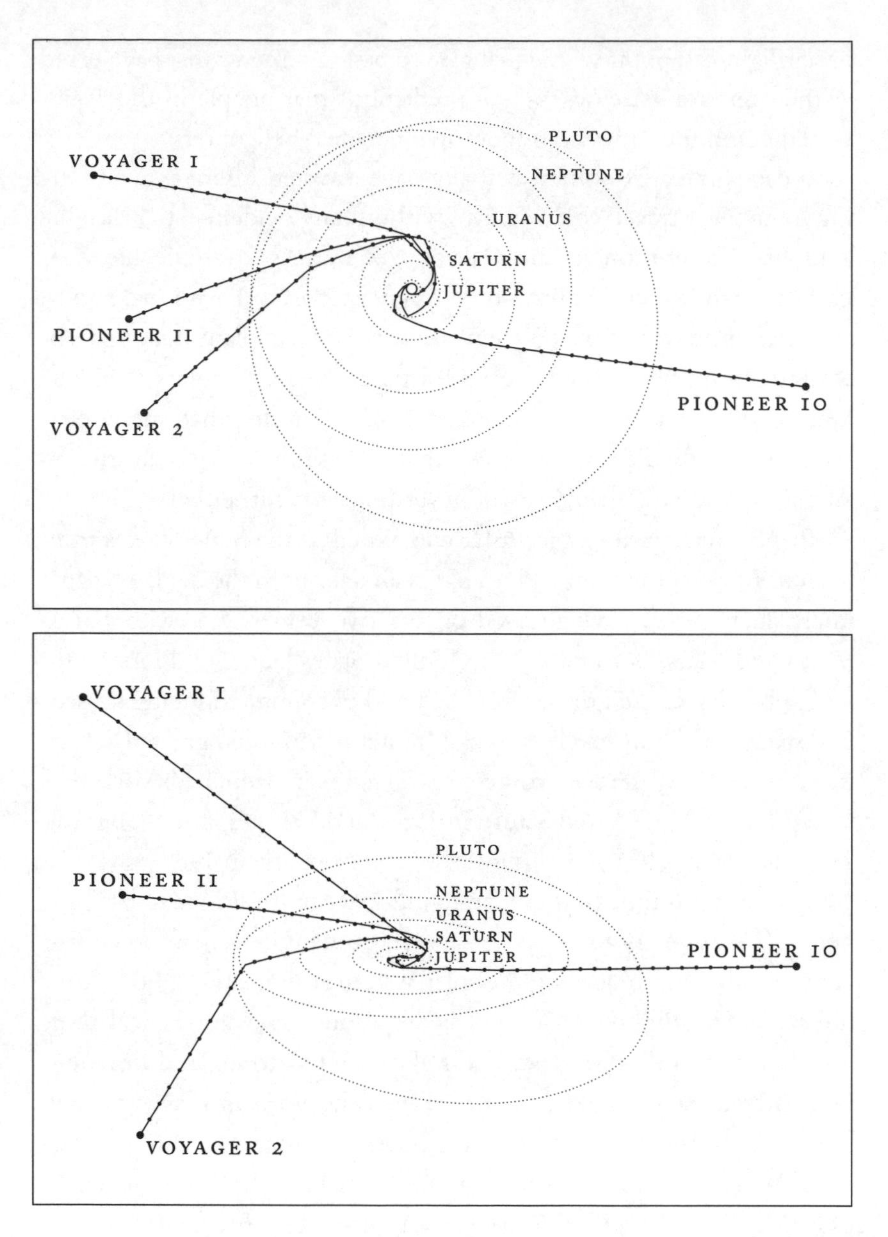

FIGURE 41

Four spacecraft are already heading to the stars: Pioneers 10 and 11, and Voyagers 1 and 2. None will get there any time soon.

Astronomy enthusiasts toss around distances in light-years so often that they sometimes forget just how incredibly large a light-year is. Light is swift: in a single second, a light beam could whip around the Earth seven and a half times, so the distance that light travels in an entire year is gargantuan. To visualize how far even the nearest stars are, imagine that the Galaxy shrank and the Earth and Sun were an inch apart. Then Jupiter would be just five inches from the Sun and distant Neptune only thirty inches. On this same scale, though, a light-year would equal a full *mile*, so Alpha Centauri would be 4.3 miles away. All other stars, such as 51 Pegasi and 47 Ursae Majoris, would be even farther and harder to reach. And the Milky Way Galaxy itself is so immense that if its disk shrank still further, to the size of a dime, the entire observable universe, from the Earth to the farthest known quasar, would be only two miles wide. We live in an enormous Galaxy whose stars are separated by enormous distances.

To cross those distances, starships must travel fast. The Voyager spacecraft have 0.005 percent of the speed of light, but true starships must attain at least 10 percent of the speed of light in order to reach Alpha Centauri in less than half a century. By contrast, a ship that left the solar system with "only" 1 percent of light speed would require 430 years to get to Alpha Centauri, and during that lengthy time technological improvements on Earth might have enabled the construction of faster starships. Imagine if a long-lived Christopher Columbus had taken five hundred years to sail the Atlantic. He would have watched faster ships pass him by, and then airplanes zipping between Europe and America, all before he ever reached America. When he finally got there, the New World would have looked pretty old.

Traveling fast, however, is difficult, because it costs great amounts of energy—and money—to attain high speeds. For example, a one-ton ship traveling at a third the speed of light would have as much energy as the United States consumes in three weeks. By solar standards, however, this is not much, since every second the Sun radiates into space millions of times more energy. So the energy is there, if human beings can figure out how to harness it. The cost of such a mission might be more than a trillion dollars, another impediment that critics cite. But a trillion dollars is loose change, compared with what the American gross domestic product will be in future centuries. If the

American economy grows at the modest rate of 2 percent per year after inflation, then in two centuries the GDP will be fifty-two times larger than it is now, or about 350 trillion dollars in current dollars. What is outrageously expensive today may be cheap centuries hence. After all, a moonflight would have been both technically impossible and ridiculously expensive to the American colonists, yet their descendants successfully put a man on the Moon. And if we can put a man on the Moon, why can't we put one around Alpha Centauri?

Undoubtedly, though, the first interstellar voyages will carry only machines, not people. Human beings have reached no farther than the Moon, whereas robotic spacecraft have already made it past Neptune. Human beings require air, water, and food; machines do not. Furthermore, in coming decades, computers and other instruments will become smaller, lighter, and more powerful, thereby reducing the weight that a spacecraft must carry, which in turn will reduce the energy and cost of a trip to a star.

In space, for better or worse, once a craft attains a particular speed, it generally retains it. Thus, a craft traveling at, say, half the speed of light will keep going at that speed without expending any additional energy. The hard part, of course, is achieving that high speed. By contrast, in ordinary life, one must consume fuel not only to reach a certain speed but also to maintain it. This interstellar blessing is mixed, however, because a spacecraft arriving at another star cannot easily stop and look around unless it carries a lot of fuel to decelerate itself, fuel that would add weight and cost to the mission. Even spacecraft in the solar system often fly past their targets rather than orbit them, an option that would cost too much energy and money.

An interstellar voyage must confront Einstein's special theory of relativity, which affects objects that travel at an appreciable fraction of the speed of light. The best-known relativistic barrier is the speed of light itself, so Earth's residents will always have to wait at least 4.3 years for a ship to get to Alpha Centauri and then another 4.3 years for the ship's data to reach the Earth.

Special relativity also affects mass and time. As a ship's speed increases, so does its mass, which is bad because it makes it harder to accelerate the ship to still greater speeds. But to any passengers aboard the ship, time proceeds more slowly, which is good because they can travel even farther. These two relativistic effects, involving mass and time, are small at low speeds and grow greatly as a ship nears the speed of light. At the speed of light, the mass becomes infinite, which is why no material body can actually travel that fast.

To gauge the severity of these relativistic effects, scientists employ the Lorentz factor, named for Dutch physicist Hendrik Lorentz. The Lorentz factor depends on speed: it is 1.00 at zero velocity and increases at higher ones, becoming infinite at the speed of light. At 20 percent of the speed of light, the Lorentz factor is only 1.02, which means that a starship this fast is 2 percent more massive than it was at rest and time slows so that the crew sees 1 hour elapse in the time that 1.02 hours pass on Earth. At 50 percent of the speed of light, the Lorentz factor is 1.15, which is still fairly low: the mass is now 15 percent greater than it was at rest, and 1 ship-hour equals 1.15 hours on Earth. Only at speeds over 80 percent of the speed of light does the Lorentz factor start to rise greatly. It reaches 2.00 at 87 percent of light speed, which means that mass doubles and time halves; it reaches 3.00 at 94 percent of the speed of light, 4.00 at 97 percent, 5.00 at 98 percent, 7.00 at 99 percent, and even higher values at still greater speeds.

The relativistic increase in mass leads to a popular but fallacious argument against interstellar travel. This argument proposes that a starship continually accelerate at the comfortable rate of the Earth's gravitational acceleration. The ship's speed then steadily increases, soon attaining 99 percent of the speed of light, and then increases further, growing closer and closer to the speed of light. Because the Lorentz factor is large, time slows so much for the passengers that during their lifetimes they can actually travel not just to the nearest stars but to the center of the Galaxy, 27,000 light-years from Earth, and even to other galaxies; but also because of the Lorentz factor, the ship's mass has increased so much that its kinetic energy exceeds all the energy the United States produces in a century. Thus, the argument goes, interstellar travel is impossible.

The mathematics is right, but the argument is wrong. To people on Earth, who would presumably pay for the voyage, there is little advantage to sending a ship at near light speed. A craft that zooms away at 99.999 percent of the speed of light will reach Alpha Centauri only a year before one traveling at 80 percent of that speed, but the Lorentz

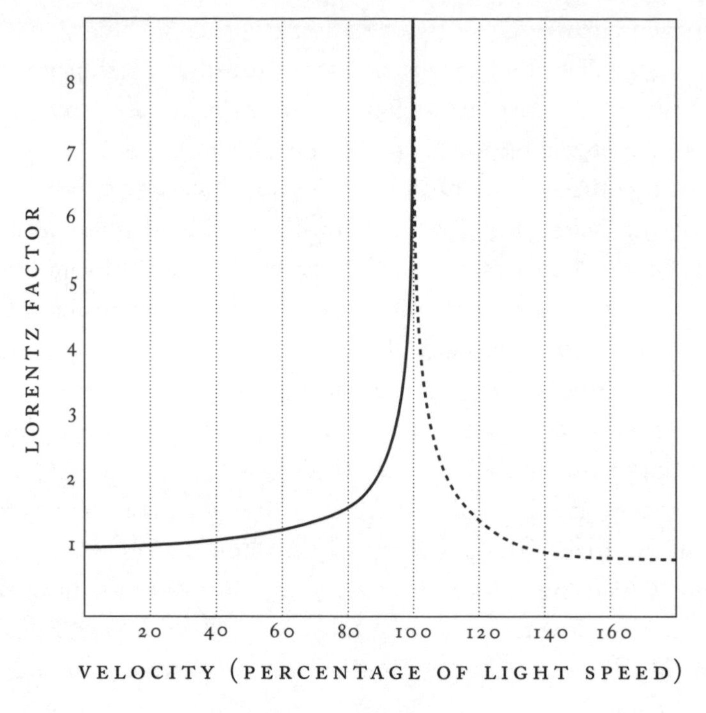

FIGURE 42

The Lorentz factor, which quantifies the effects of special relativity, is small at low speeds but shoots up at high ones, becoming infinite at the speed of light. At speeds greater than light, the Lorentz factor is imaginary and decreases with increasing speed.

factor for the faster ship is prohibitively large—224 versus 1.67. The moral for future starship captains: travel fast, all right, but not so fast that the Lorentz factor burdens you by raising your mass.

The real problem for proponents of interstellar travel is not special relativity but how to attain the high speeds at which it becomes a concern. Even 10 percent of the speed of light vastly exceeds the fastest spacecraft ever launched. Ordinary chemical rockets, which power present spacecraft, will not do, because they release little energy. Somewhat better is nuclear fission, in which heavy atomic nuclei split into lighter ones, but though this powers nuclear reactors, it is also too wimpy for interstellar spacecraft. Nuclear fusion, which occurs when light nuclei join to create heavier ones, is more powerful and heats the Sun and most other stars. However, physicists have yet to sustain nuclear fusion on Earth, let alone in space, and even fusion releases only a tiny fraction of the energy locked in matter. For example, when the Sun converts hydrogen into helium, only 0.7 percent of the mass becomes energy.

In principle, the best rocket fuel is the one Captain Kirk uses: antimatter, which mirrors ordinary matter. In ordinary matter, the atomic nucleus is positively charged and the orbiting electrons are negatively charged, but in antimatter, the atomic nucleus is negatively charged and the orbiting "electrons," called positrons, are positively charged. When matter meets antimatter, they completely annihilate each other, converting all mass into energy. Matter and antimatter are therefore potent fuels, because even a small amount of mass m contains an energy E equal to mc^2. Since c, the speed of light, is so large, the amount of energy in even a small piece of matter, or antimatter, is huge. If you shook hands with your antimatter twin, the resulting energy could power the entire United States for the next two months—or send a small ship to Alpha Centauri.

Unfortunately, antimatter does not exist naturally on Earth; if it did, it would have exploded when it touched ordinary matter. And astronomers know of no sources of antimatter in the solar system, such as antimatter asteroids. Antimatter does arise in nuclear reactions, but only in minuscule quantities, and to produce even the relatively small amount of antimatter needed to power a starship would be extremely expensive. Furthermore, antimatter is dangerous to handle, because it must not touch any ordinary matter. The good thing about antimatter, though, is that it packs a lot of punch per pound.

Any rocket, even one powered by a matter-antimatter drive, suffers a basic curse: it must accelerate not only the spacecraft but also the fuel itself. Thus, the rocket must carry more fuel just to accelerate that fuel, but the more fuel the rocket carries, the more additional fuel it must carry to accelerate that extra fuel. Scientists have therefore devised schemes to accelerate starships without rockets. It sounds impossible, but ideas do exist. In 1960, Robert Bussard suggested using the fuel in space itself. Seemingly empty space contains a trace of hydrogen atoms. If a starship could scoop these up and fuse them together in a nuclear reactor, the resulting energy could power the ship. Unfortunately, interstellar space contains only about one atom per cubic centimeter, so the ship would have to scoop up atoms over a radius of hundreds or thousands of miles. Furthermore, the most promising reactions require the heavy hydrogen isotope deuterium, or hydrogen-2, which is much rarer than ordinary hydrogen, hydrogen-1. In addition, nuclear fusion demands high temperatures, which is why stars can do it much more easily than physicists can.

Another scheme that gets rid of the rocket has been advocated by Robert Forward, who proposes that lasers push ships through space. This works because light itself exerts pressure, albeit one so weak that no one has ever been knocked over by a sunbeam. Because photon pressure is weak, the lasers must be large and powerful, covering hundreds of miles, and their beam must remain collimated over trillions of miles of space. In addition, if the spacecraft carries people, they would have little control over their voyage. Instead, they would be at the mercy of the laser station, light-years behind them.

Such ideas may sound farfetched and suggest that starflight is indeed as impractical as its critics allege, but at least these proposals employ known physics. Even more speculative schemes invoke new and unproved physics. For example, shortcuts called wormholes may exist in space that would let ships travel from the Sun to Alpha Centauri without passing through the entire 4.3 light-years of space between the two. It would be like tunneling through the Earth from the United States to China instead of treading the longer route over the Earth's surface. Also in the domain of speculation is faster-than-light travel. Technically, Einstein's special theory of relativity does not rule this out. At the speed of light, the Lorentz factor is infinite, but above this speed, it becomes what mathematicians call imaginary and actually decreases as the ship's speed increases. How one could cross the light-speed barrier, at which the Lorentz factor becomes infinite, is unclear, though, and jumping back from faster-than-light speed to slower-than-light speed may be equally difficult. Particles that exceed the speed of light are called tachyons, but no one has ever seen any, which suggests that even nature itself does not know how to do it. Perhaps, though, a parallel universe exists in which everything is forced to travel faster than light, and people there long to lead a slower life. Maybe we can make a deal with them.

Until they hear from such a universe, scientists must struggle with the one they know. As a first step toward actual interstellar travel, sci-

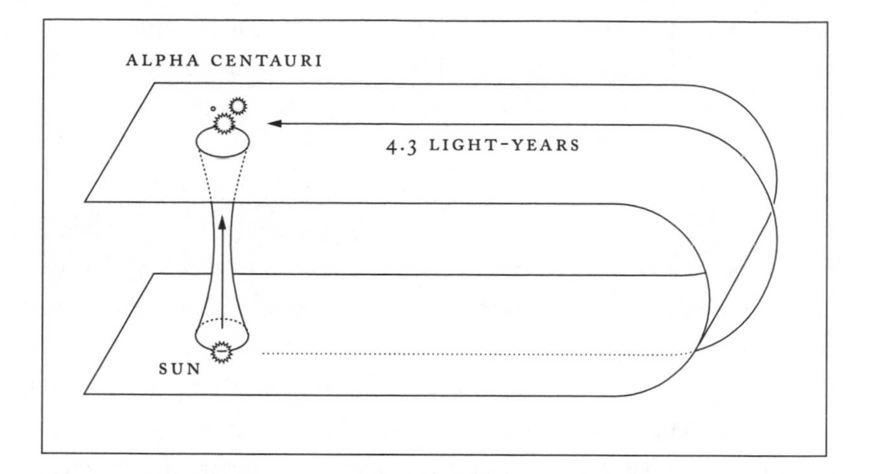

FIGURE 43

A friendly wormhole through space will get you to Alpha Centauri sooner.

entists have proposed launching a robotic mission that would travel fast and far enough to test some of the concepts of interstellar flight without actually reaching the nearest stars. The proposed craft is named TAU, because it would travel a thousand astronomical units from the Sun, or twenty-five times the mean distance of Pluto. The ship would take about a century to achieve that distance, which is only a fraction of 1 percent of the distance to Alpha Centauri. Still, the TAU spacecraft could probe the solar system far beyond Neptune and Pluto, use parallax to measure the distances of distant stars in the Galaxy, and serve as a trailblazer for faster craft.

The challenges of interstellar travel are enormous—perhaps so enormous that its critics are right, and no civilization will ever be able to achieve it, thereby explaining why our Galaxy could abound with intelligent species we have never met. Yet it would be folly to dismiss the capability of terrestrial civilization a few centuries from now, let alone that of other civilizations which may be millions or billions of years more advanced. Furthermore, if a genuine twin of Earth were discovered around a nearby star, such as Alpha Centauri, Lalande 21185, or Tau Ceti, the urge to explore that world directly would be irresistible. Conceivably, such a mission could be launched during the twenty-first or twenty-second century. If so, those who believe that intelligent life is common in the cosmos would be forced to explain why none of that life has done the same and launched a mission toward a most promising star system: the Sun.

Cosmic Symphony

During the 1990s, astronomers for the first time established that the universe at large does indeed possess the four basic astronomical ingredients for life. Prior to that decade, astronomers already knew that stars forge life-giving elements like carbon and oxygen; that giant galaxies such as the Milky Way sculpt these life-giving elements and recycle them into new star systems; and that many stars generate the abundant light and warmth which life requires. Now astronomers know that the fourth and final ingredient also exists: planets around other stars, planets that can provide stable platforms on which life might arise, develop, and thrive. At least three planets comparable in mass to Earth exist around a pulsar in Virgo, and planets like Jupiter circle Sunlike stars in Pegasus, Boötes, Ursa Major, and elsewhere. Someday astronomers will also be able to discover extrasolar Earth-mass planets around Sunlike stars and determine which have water and life.

Even before these exciting discoveries occur, astronomers know that the universe is an amazing place, one that has the right physical properties to allow for the formation of the material objects—stars, galaxies, and planets—that life, which many view as a spiritual entity, requires. Although the universe is studded with spiral and elliptical galaxies, stars burning red and yellow and blue, and planets large and small, the most incredible feature of the cosmos, life, is so subtle that astronomers have yet to ascertain its existence anywhere else.

Life could not exist if the universe did not have the specific properties that it does. If the gravitational force were weaker, matter would never have conglomerated into galaxies, stars, and planets; if the primordial universe had possessed equal quantities of matter and antimatter, they would have completely annihilated each other; if the carbon atom did not have the specific resonance that Fred Hoyle had predicted, carbon would be too rare for terrestrial life to exist; and so on: change the charge of the electron, the mass of the proton, or the speed of light, and the life-supporting universe collapses.

Is it mere coincidence that the universe happens to possess just those properties which allow part of it to be alive? Some people say yes; it was simply good luck that the universe was born with the particular characteristics that it has. Others say no; our universe is only one of many universes, each with different properties, and we naturally inhabit one of the few that can have life, just as we inhabit one of the few planets that has a mild climate. Still others, of a more spiritual persuasion, see the universe's remarkable offspring as a sign that an intelligent creator wrote a tremendous symphony whose melodies the stars, galaxies, and planets now play with beauty and precision, and we living beings are one of the fortunate resulting chords, perhaps the climatic chord in that symphony's greatest movement. Whatever the case, and vast and complex though the universe is, its most astonishing features are two of the simplest: it exists, and so do we.

Catalogue of Planets

Stellar Data

Star	Constellation	Distance (light-years)	Spectral Type	Color	Magnitudes	
					Apparent Visual	Absolute Visual
Sun	—	0	G2	Yellow	−26.74	4.83
PSR B1257+12	Virgo	1300	Pulsar	Black	—	—
51 Pegasi	Pegasus	50	G2.5	Yellow	5.49	4.56
47 Ursae Majoris	Ursa Major	46	G0	Yellow	5.05	4.31
Rho[1] Cancri A	Cancer	42	G8	Yellow	5.95	5.38
Rho[1] Cancri B		42	M3.5	Red	13.14	12.57
Tau Boötis A	Boötes	60	F7	Yellow-white	4.50	3.18
Tau Boötis B		60	M2	Red	11.0	9.7
Upsilon Andromedae	Andromeda	57	F8	Yellow-white	4.09	2.86
Rho Coronae Borealis	Corona Borealis	57	G0	Yellow	5.41	4.19

Planetary Data

Planet	Discovery		Distance from Star (AU)	Period	Mass[a] (Jupiter = 1)	Orbital Eccentricity	Temperature		
	Year	Discoverers					Fahrenheit	Celsius	Kelvin
Mercury	—	—	0.39	88 d	0.00017	0.206	+250	+120	390
Venus	—	—	0.72	225 d	0.0026	0.007	+860	+460	730
Earth	—	—	1.00	365 d	0.0031	0.017	+60	+16	289
Mars	—	—	1.52	687 d	0.00034	0.093	–67	–55	218
Jupiter	—	—	5.2	12 y	1.00	0.048	–236	–149	124
Saturn	—	—	9.5	29.5 y	0.30	0.054	–289	–178	95
Uranus	1781	William Herschel	19.2	84 y	0.046	0.047	–353	–214	59
Neptune	1846	Galle and d'Arrest	30.1	165 y	0.054	0.009	–353	–214	59
Pluto	1930	Tombaugh	39.5	248 y	0.000006	0.25	–400?	–240?	30?
PSR B1257+12 a	1993	Wolszczan	0.19	25 d	0.000060	0.0	?	?	?
PSR B1257+12 b	1991	Wolszczan and Frail	0.36	67 d	0.014	0.018	?	?	?
PSR B1257+12 c	1991	Wolszczan and Frail	0.47	98 d	0.011	0.026	?	?	?
51 Pegasi b	1995	Mayor and Queloz	0.05	4.2 d	0.6	0.0	1900	1000	1300
47 Ursae Majoris b	1995	Marcy and Butler	2.1	3.0 y	3.0	0.03	–100	–70	200
Rho[1] Cancri A b	1996	Marcy and Butler	0.11	14.6 d	1.0	0.0	1300	700	1000
Tau Boötis A b	1996	Marcy and Butler	0.05	3.3 d	4.9	0.0	2200	1200	1500
Upsilon Andromedae b	1996	Marcy and Butler	0.06	4.6 d	0.8	0.0	2000	1100	1400
Rho Coronae Borealis b	1997	Noyes et al.	0.23	39.6 d	1.5	0.0	600	300	600

[a]For the pulsar planets and the planets detected with the Doppler technique—the Pegasian planets and the planet around 47 Ursae Majoris—these are the most likely masses. To obtain the minimum masses, divide by 1.27.

Glossary

A Spectral type for stars that are hot and white, such as Sirius, Altair, Fomalhaut, Vega, and Deneb. The most luminous stars within thirty light-years of the Sun are spectral type A. A-type stars make a great contribution to the sky, and five of the twenty brightest stars in the night sky are spectral type A. It was around A-type stars that in 1983 the Infrared Astronomical Satellite detected the first rings of dust orbiting other main-sequence stars.

Absolute Magnitude A measure of a star's true, or intrinsic, brightness—how much light the star actually casts into space. In contrast, a star's *apparent* magnitude measures how bright the star looks and depends on both how much light the star emits and how far the star is from Earth. *Absolute* magnitude removes the effect of distance, because it is the apparent magnitude a star would have if viewed from a distance of 32.6 light-years (10 parsecs). As with all other magnitudes, the lower the number, the brighter the star. For example, the Sun's absolute magnitude is +4.83, whereas Sirius, which is more luminous, has an absolute magnitude of +1.45. See also **Apparent Magnitude**.

Adaptive Optics A technique that yields sharp views of objects by employing a "rubber mirror" to correct the distortion which light suffers as it travels through the Earth's atmosphere.

Aldebaran An orange K-type giant that is the brightest star in the constellation Taurus the Bull. Aldebaran lies 67 light-years from the Earth and is accompanied by a red dwarf star.

Algol A famous eclipsing binary in the constellation Perseus. Algol consists of a hot, bright star and a cool, fainter one that revolve every 2.87 days and periodically pass in front of and behind each other. Normally, Algol is apparent magnitude 2.1, but when the fainter star eclipses the brighter, Algol fades to magnitude 3.4.

Alpha Centauri The nearest star system to the Sun, Alpha Centauri actually consists of three separate stars—two bright and one faint. The two bright members, Alpha Centauri A and B, resemble the Sun, being main-sequence stars of spectral types G2 and K1, respectively. They are 4.35 light-years from Earth. Their distance from each other varies from 11 to 35 astronomical units, and they revolve around each other every eighty years. The third member of the system, Alpha Centauri C or Proxima Centauri, is a red dwarf that lies some 13,000 astronomical units from the other two and 4.24 light-years from the Earth. Alpha Centauri is the third brightest star in the night sky, but lies so far south that it is not visible except below latitudes of 25 degrees north.

Altair The brightest star in the constellation Aquila the Eagle, Altair is a white A-type star 16.6 light-years away, or about twice the distance of Sirius, the nearest A-type star to Earth.

Ammonia One of the three main "ices" that existed in the outer solar system during its formation. Along with water and methane ice, ammonia ice was responsible for giving birth to the large cores that formed the giant planets Jupiter, Saturn, Uranus, and Neptune. Ammonia (NH_3) consists of nitrogen and hydrogen, two of the most common elements in the cosmos.

Angular Momentum A measure of how much spin momentum an object has. A spinning star or planet has angular momentum, as does a star or planet that orbits a star. The amount of angular momentum depends on the mass of the object (the more mass, the more angular momentum) and on how fast the object spins or revolves (the faster, the more angular momentum); it also depends on how extended a spinning object is (the more extended, the more angular momentum) and on how far an object lies from the star around which it revolves (the farther, the more angular momentum). Angular momentum is conserved. For example, if a spinning object contracts, it must spin faster in order to retain the same amount of angular momentum; conversely, if a spinning object expands, it must spin more slowly.

Antares A red supergiant star in the constellation Scorpius the Scorpion, Antares is a prominent sight in the southern sky during summers in the northern hemisphere. The star lies about 500 light-years from the Earth.

Antimatter The mirror of normal matter: in antimatter, "protons" are negatively charged and "electrons" are positively charged. When matter and antimatter meet, they annihilate each other in a burst of energy.

Apparent Magnitude A measure of how bright a star looks; the brighter the star, the lower the apparent magnitude. Two consecutive apparent magnitudes differ by a factor of 2.5, so a magnitude +1 star looks 2.5 times brighter than a magnitude +2 star, which in turn looks 2.5 times brighter than a magnitude +3 star, and so on. Most of the brightest stars have apparent magnitudes around 0 and +1. To convert a star's apparent magnitude into *absolute* magnitude, which measures the star's intrinsic brightness, one must know the star's distance. See also **Absolute Magnitude**.

Arcsecond A unit of angular measure equal to 1/3600 of a degree, or the apparent thickness of a human hair viewed from ten feet. Planets induce astrometric wobbles in their stars that are much smaller than this, so astronomers also speak of milliarcseconds (1/1000 of an arcsecond) and microarcseconds (1/1,000,000 of an arcsecond).

Arcturus The fourth brightest star in the night sky, Arcturus is a beautiful orange K-type giant in the constellation Boötes that lies 37 light-years from Earth.

Asteroid A small rocky body. In our solar system, most asteroids revolve around the Sun between the orbits of Mars and Jupiter. The first four asteroids discovered were Ceres, Pallas, Juno, and Vesta. Ceres is by far the largest asteroid, having a diameter of about 900 kilometers.

Astrometry The branch of astronomy that deals with measuring the positions of celestial objects. In astrometry's domain are *parallax*, the tiny shift in a star's position induced by the orbital motion of the Earth, which reveals the star's distance; *proper motion*, the change in a star's position year after year as the star travels through space; and *extrasolar plan-*

ets, whose gravity perturbs a star's position. Astrometry is most sensitive to massive planets that lie *far* from their stars. In contrast, the Doppler technique is most sensitive to massive planets that lie *near* their stars.

Astronomical Unit (AU) The mean distance between the Sun and Earth, or about 93 million miles. Jupiter's mean distance from the Sun is 5.2 AU, and Pluto's is 39.5 AU. A light-year is a much larger unit of distance, equal to 63,240 AU.

Axial Tilt The angle that a planet's rotation axis makes with the line perpendicular to the plane of its orbit. The Earth's axial tilt is 23.4 degrees and causes the seasons. When the Earth's northern hemisphere leans toward the Sun, spring and summer occur in the northern hemisphere and fall and winter in the southern. Axial tilt is sometimes called obliquity.

B Spectral type for stars that are hot and blue, such as Regulus, a B-type main-sequence star, and Rigel, a B-type supergiant star. Although B-type main-sequence, giant, and supergiant stars are rare, they make a great contribution to the night sky, since they are so luminous that they can be seen across great distances. Five of the twenty brightest stars in the night sky are spectral type B. The nearest B-type main-sequence star to the Earth is Regulus, 74 light-years away.

Barnard's Star The second nearest star system to the Sun, after the triple star Alpha Centauri. Barnard's Star is a red dwarf that lies 6.0 light-years from the Sun in the constellation Ophiuchus. It was discovered in 1916 by Edward Emerson Barnard, who found that the star had the greatest proper motion ever seen: 10.3 arcseconds per year. Today, Barnard's Star is most famous, or infamous, for the planets that Peter van de Kamp claimed around it, but modern observations indicate that van de Kamp's planets do not exist.

Beta Pictoris An A-type star about fifty light-years from the Earth, Beta Pictoris became famous in 1984 when Bradford Smith and Richard Terrile photographed a disk of dust around the star. Some believe that Beta Pictoris is a main-sequence star, like the Sun, while others think it is so young that it has not yet reached the main sequence.

Betelgeuse The bright red star in the constellation Orion, Betelgeuse is an M-type supergiant and the brightest red supergiant in Earth's sky. Someday Betelgeuse will explode in a supernova. All the bright blue stars that now shine in Orion will become red supergiants, too, and then explode.

Binary Star A system of two stars that orbit each other. Binary stars are common. Of the nineteen star systems within a dozen light-years of the Sun, six are binary (Sirius, Luyten 726-8, 61 Cygni, Procyon, Struve 2398, and Groombridge 34), and two are triple (Alpha Centauri and Luyten 789-6). If the two stars in a binary are close together, then planets may orbit both; if the two stars are far apart, then each star may have its own planetary system; but if the two stars fall between these two extremes, planets may not be able to form at all.

Black Hole An object with such intense gravity that not even light, the fastest thing in the universe, can escape. Black holes form when the most massive stars—those born with more than about forty times the mass of the Sun—run out of fuel and collapse. The most famous black hole candidate, Cygnus X-1, was discovered in 1971.

Black Widow Pulsar A once-dead pulsar that has been revived by a companion star and is now tearing that companion apart. The companion dumped material onto the pulsar and made it spin fast; the pulsar's beam then began to cut through the companion star and destroy it. The first black widow pulsar to be found was PSR B1957+20, whose strange nature was discovered in 1988.

Blueshift The shift of an object's spectrum caused when the object moves toward us. This movement scrunches up the light waves, reducing their wavelength. Blue light has a short wavelength, which is why this shift is called a blueshift. The greater the blueshift is, the faster the object is moving toward us.

Bode's Law A basic rule that most planetary distances in the solar system obey. See **Titius-Bode Law**.

Brown Dwarf A star with too little mass to ignite its hydrogen-1 fuel and achieve a long life on the main sequence. Despite their name, brown dwarfs are not brown. When young, brown dwarfs glow red as they convert gravitational energy into heat; as they age, they fade, cool, and turn black. Brown dwarfs have less than 8 percent of the mass of the Sun. The first definitive brown dwarf was reported in 1995, orbiting the nearby red dwarf Gliese 229.

Callisto The outermost of the four large moons of Jupiter, Callisto has a dark, cratered surface and a diameter of 4806 kilometers, making it the third largest moon in the solar system, after Ganymede and Titan. Callisto revolves around Jupiter every 16.689 days.

Canopus The second brightest star in the night sky, Canopus lies in the southern constellation Carina. It is a yellow-white F-type star roughly 200 light-years from Earth.

Capella The sixth brightest star in the night sky. Capella lies 41 light-years from Earth and consists of two bright yellow G-type giants, which are the nearest such stars to Earth and revolve around each other every 104 days. Two faint red M-type dwarfs lie quite far from the main stars.

Carbon Element with atomic number six, having six protons in its nucleus, and the element on which terrestrial life is based. Most carbon originates in stars that do not explode. After these stars become red giants, they convert helium into carbon, which can seep into the star's atmosphere and get ejected when the star casts off its atmosphere to form a planetary nebula.

Carbon Dioxide The gas (CO_2) that makes up most of the atmospheres of Venus and Mars but only a fraction of Earth's. Carbon dioxide is a greenhouse gas that traps solar energy and warms the Earth.

Castor The second brightest star, after Pollux, in the constellation Gemini the Twins. Castor is actually a sextuple star, containing four white A-type stars and two red M-type dwarfs. The system lies 44 light-years from the Earth.

Center of Mass The point around which two stars, or a star and a planet, revolve. The center of mass lies on a line joining the two and is closer to the more massive object. For example, the Sun is a thousand times more massive than Jupiter, so the center of mass of the

Sun-Jupiter system lies a thousand times closer to the Sun. Jupiter revolves around the center of mass every twelve years, but so does the Sun. The Sun's wobble gives extraterrestrial astronomers the chance to discover Jupiter's presence.

Ceres The largest asteroid, Ceres has a diameter of about 900 kilometers and lies between the orbits of Mars and Jupiter, as most other asteroids do. Ceres is 2.77 astronomical units from the Sun and revolves every 4.6 years. Its orbital eccentricity is 0.097, and its orbital inclination is 10.6 degrees.

Charon Pluto's moon, which James Christy discovered in 1978. Because Charon revolves around Pluto, its motion allowed astronomers to determine Pluto's mass. This turned out to be much too small to perturb Uranus or Neptune.

Chiron An object that American astronomer Charles Kowal discovered between the orbits of Saturn and Uranus in 1977. Although Chiron was first classified as an asteroid, astronomers now regard it as a large comet, a likely interloper from the Kuiper belt.

CM Draconis An eclipsing binary in the constellation Draco that consists of two red dwarfs. It is an ideal system in which to hunt for a planet passing in front of the stars, because they are small and a planet would therefore block a large fraction of their light.

CO_2 See **Carbon Dioxide**.

Comet A small, icy body. In our solar system, most comets go around the Sun on highly elliptical orbits. The most famous comet is Halley's, which revolves every 76 years and will next return to Earth's vicinity in 2061. Two bright comets recently passed Earth: Comet Hyakutake, in 1996, and Comet Hale-Bopp, in 1997. Astronomers divide comets into those of the short-period variety, which revolve in less than two hundred years, and those of the long-period variety, which take over two hundred years to do so. The former originate in the Kuiper belt and the latter in the Oort cloud. Halley is a short-period comet, whereas Hyakutake and Hale-Bopp are long-period comets.

Constellation A region of the sky as viewed from Earth. There are eighty-eight constellations. The five largest are Hydra, Virgo, Ursa

Major, Cetus, and Hercules; the smallest is Crux, the Southern Cross. Two objects in the same constellation do not necessarily have anything to do with each other, since they may lie at greatly different distances from the Earth, and thus from each other.

Coronagraph A device that blocks the light of a star or planet, allowing astronomers to see faint objects nearby. Coronagraphs helped astronomers spot the dust disk around Beta Pictoris and the brown dwarf around Gliese 229.

Crab Nebula The remains of a massive star that exploded in the constellation Taurus. The explosion was witnessed in 1054 A.D.

Dark Halo A huge cloak of material that surrounds the Milky Way's luminous disk, emits little if any light, but contains most of the Galaxy's mass.

Dark Matter Material that emits little if any light. Most of the Galaxy, and the universe, consists of dark matter. Candidates for dark matter include planets, brown dwarfs, white dwarfs that have cooled and faded, neutron stars, and black holes, as well as exotic subatomic particles.

Day The length of time that a planet takes to spin once. In our solar system, most planets spin fast and days are short.

Degenerate The state of matter in which pressure among subatomic particles, such as electrons or neutrons, supports a star against the inward pull of its gravity. White dwarfs and faded brown dwarfs are degenerate, because the inward force of their gravity is balanced by the outward pressure of electrons. Neutron stars are also degenerate, supported by the pressure of neutrons.

Deimos The smaller of the two moons of Mars. Its mean diameter is only 12 kilometers.

Deneb A white A-type supergiant in the constellation Cygnus and the most distant first-magnitude star visible from Earth. Deneb lies roughly 1500 light-years from the Sun. It will evolve into a red supergiant and then explode in a supernova.

Density Mass divided by volume. In our solar system, giant planets have low densities and terrestrial planets high densities.

Deuterium A hydrogen isotope, hydrogen-2, in which the nucleus contains one proton and one neutron; in contrast, ordinary hydrogen, or hydrogen-1, contains one proton and no neutrons. Deuterium readily burns, even in most brown dwarfs.

Diameter A measure of an object's physical size, the diameter is the length of an imaginary straight line passing through the object's center. In our solar system, the planet with the largest diameter is Jupiter.

Doppler Shift The shift of an object's spectrum because it moves toward or away from Earth. This phenomenon was first described by Austrian physicist Christian Doppler in 1842 but not detected for light waves until 1888. When an object moves toward us, its light waves get scrunched up and shifted to shorter, or bluer, wavelengths, producing a blueshift; when the object moves away from us, the light waves get stretched out and shifted to longer, or redder, wavelengths, producing a redshift. When a planet orbits a star, the star wobbles around the center of mass between itself and its planet; this wobble causes the star to move toward and away from Earth with the same period as the planet, producing small Doppler shifts in the star's spectrum that can reveal the planet's existence. The Doppler technique is most sensitive to massive planets near their stars, such as the Pegasian planets. In contrast, the astrometric technique is most sensitive to massive planets far from their stars, such as Jupiter and Saturn.

Double Star A star system containing two stars. Many stars in the Galaxy are double.

Dwarf A star on the main sequence. By itself, the term does not refer to white dwarfs or brown dwarfs.

Earth The third planet from the Sun and an oasis of life in the cosmos.

Eccentricity Describes how circular or elongated an orbit is. An orbit that is perfectly circular has an orbital eccentricity of 0.00; the most elliptical orbit possible has an eccentricity of just under 1.00. An orbit with an eccentricity of 1.00 is parabolic, and one with an eccentricity of over 1.00 is hyperbolic; neither represents the paths of bound objects. In our solar system, planets, and especially giant planets, have orbits with low eccentricities. In contrast, binary stars usually have orbits with high eccentricities.

Eclipsing Binary A double star in which at least one star blocks the light of the other as they revolve. Because any planets would likely lie in the same plane as the two stars, these planets might also block the stars' light, giving astronomers the chance to discover the planets. The most famous eclipsing binary is Algol; the most promising for planetary detection is CM Draconis, because the two stars are both small red dwarfs, and a passing planet would therefore block a noticeable fraction of their light.

Electron The negatively charged subatomic particle that lies outside the positively charged nucleus of an atom. A neutral atom has as many electrons as protons. For example, neutral hydrogen has one electron, neutral helium two electrons, and neutral oxygen eight electrons.

Ephemeris A list of predicted positions of a planet or other celestial object. Deviations between the ephemeris and the actual observed positions of a planet might indicate that another planet's gravity is tugging on the observed planet and pulling it off course.

Epsilon Eridani A nearby orange K-type dwarf that lies 10.7 light-years from Earth. In 1960, astronomers searched Epsilon Eridani for signs of intelligent life and found none. Today astronomers know Epsilon Eridani is probably too young to support such life, having an age of only around 1 billion years.

Epsilon Indi A nearby orange K-type dwarf 11.3 light-years from Earth. It is an old star, probably older than the Sun.

Europa The second of the four large moons of Jupiter. Europa is 3130 kilometers in diameter and revolves around Jupiter every 3.55 days. It bears an icy white surface.

Extrasolar Planet A planet that lies beyond the Sun's solar system. Extrasolar planets may orbit other stars or drift through the Galaxy alone, unbound to any star.

F Spectral type for yellow-white stars, which are somewhat warmer than the Sun. Some F-type stars could support life on orbiting planets. The two brightest F-type stars are Procyon, which lies just 11.4 light-years away, and Canopus, which is much farther. Polaris, the North Star, is also spectral type F.

51 Pegasi A yellow G-type main-sequence star 50 light-years from Earth in the constellation Pegasus and the site of the first planet discovered around a Sunlike star. The discovery was made by Swiss astronomers Michel Mayor and Didier Queloz in 1995.

Fission The splitting apart of a heavy atomic nucleus, a process that releases energy. Fission occurs in nuclear reactors and is less efficient than fusion, which powers the Sun and most other stars.

Fomalhaut A bright white A-type main-sequence star in the constellation Piscis Austrinus. Fomalhaut lies 22 light-years away and is the second most luminous star within that distance of the Sun, after Sirius. It was one of the first stars around which the Infrared Astronomical Satellite detected a ring of dust.

47 Ursae Majoris A yellow G-type main-sequence star, 46 light-years away, around which a planet was discovered in 1995 by American astronomers Geoffrey Marcy and Paul Butler. The planet has a circular orbit, revolves every three years, and resembles a somewhat overgrown Jupiter.

Fusion The joining together of light atomic nuclei to create energy. Fusion powers most stars, including the Sun, but physicists have yet to sustain fusion on Earth.

G Spectral type for yellow stars, such as the Sun, Alpha Centauri A, Tau Ceti, 51 Pegasi, and Capella. Because they resemble the Sun, G-type main-sequence stars are excellent candidates for harboring life on orbiting planets.

Galaxy A conglomeration of millions, billions, or trillions of stars that are held to one another by the force of gravity. Our Galaxy is the

Milky Way, a giant that has at least ten lesser galaxies revolving around it. The nearest giant galaxy to our Galaxy is the Andromeda Galaxy.

Galilean Satellite One of the four large moons of Jupiter: Io, Europa, Ganymede, and Callisto. All were discovered by Galileo in 1610.

Ganymede The third of the four big satellites of Jupiter, Ganymede is the solar system's largest moon, with a diameter of 5268 kilometers—larger than Mercury and Pluto. Ganymede has a gray, cratered surface and revolves around Jupiter every 7.15 days.

Gas Giant A planet consisting mostly of hydrogen and helium. Our solar system has two gas giants: Jupiter and Saturn. Uranus and Neptune are sometimes erroneously called gas giants, too, but they are not, since the hydrogen and helium constitute only a fraction of the planets' masses.

GD 165 A white dwarf in the constellation Boötes around which a possible brown dwarf was found in 1988. The status of this brown dwarf candidate is unclear.

Giant Planet A large planet. Our solar system has four giant planets: Jupiter, Saturn, Uranus, and Neptune. The first two are gas giants, and the other two are sometimes called ice giants.

Giant Star A large star, not on the main sequence, that shines roughly a hundred times more brightly than the Sun. Giant stars no longer fuse hydrogen into helium at their cores. Instead, they may fuse hydrogen into helium outside their cores, or engage in other nuclear reactions, or do both. Giants evolve from stars that are born with less than eight times the mass of the Sun, so someday the Sun itself will become a giant star. Most giants are yellow (spectral type G), orange (K), or red (M).

Giclas 29-38 A white dwarf in the constellation Pisces around which a brown dwarf was reported in 1987. Later work, however, suggested that the infrared excess which indicated a brown dwarf actually arose from a cloud of orbiting dust.

Gliese 229 A red dwarf 19 light-years from Earth around which the first definite brown dwarf was reported, in 1995. Gliese 229 B is so cool that methane appears in its atmosphere.

Greenhouse Effect The warming of a planet by certain gases in the atmosphere—such as carbon dioxide, water vapor, methane, and ozone—that allow visible light from a star to reach and warm the planet's surface but prevent the surface from radiating some of its heat back into space. Without the greenhouse effect, Earth would be an ice-covered world hostile to life.

Halley's Comet This most famous of comets revolves around the Sun on a highly elliptical orbit (eccentricity 0.967) that takes the comet from just inside Venus's orbit to beyond Neptune's. Halley's Comet revolves once every 76 years and will next appear in Earth's sky in 2061.

HD *Henry Draper Catalogue*, a star catalogue named for an amateur astronomer whose widow donated money to Harvard University. The catalogue lists over 200,000 stars and was published between 1918 and 1924.

HD 114762 An F-type star in the constellation Coma Berenices around which a possible brown dwarf was discovered in 1988.

Helium The second lightest and second most common element in the universe. Gas giant planets contain large supplies of this element. Stars produce helium by fusing hydrogen; helium itself fuses into carbon and oxygen.

Hydrogen The lightest and most common element in the universe. Gas giant planets, such as Jupiter and Saturn, consist mostly of hydrogen. Main-sequence stars, such as the Sun, generate energy by fusing hydrogen into helium at their cores. The atomic nucleus of hydrogen contains one proton. Hydrogen has three chief isotopes: hydrogen-1, the most common isotope, contains one proton and no neutrons; hydrogen-2, or deuterium, contains one proton and one neutron, and it readily burns, even in most brown dwarfs; and hydrogen-3, or tritium, contains one proton and two neutrons and is radioactive.

Ice Giant The term occasionally used for Uranus and Neptune, giant planets that are not gas giants. These planets consist mostly of "ices"—water, methane, and ammonia—surrounded by thick atmospheres of hydrogen and helium.

Inclination The angle between the plane of an object's orbit around the Sun and the plane of the Earth's orbit. Most planets have orbital inclinations of only a few degrees; the largest is Pluto's, at 17.1 degrees.

Infrared The form of radiation whose wavelength is somewhat longer than that of visible light. Stars and planets, whether hot, warm, or cool, emit infrared radiation, but the cooler the object, the greater the percentage of energy that emerges in the infrared. In particular, brown dwarfs glow brightly in the infrared, as do planets and dust surrounding stars.

Infrared Excess If a star emits more infrared radiation than it should, based on its temperature, it is said to exhibit an infrared excess. Such an excess signals a cool companion or orbiting dust.

Interferometry Combining two or more telescopes so that light waves from one interfere with those from the other. Interferometry yields much sharper views of the heavens than do single telescopes.

Interstellar Cloud A mass of gas and dust in the space between the stars. Some interstellar clouds give birth to new stars and planets.

Interstellar Comet A comet that roams the Galaxy at large, unbound to any star. When the Sun was young, Jupiter and Saturn ejected large numbers of comets into interstellar space, and if other stars have gas giant planets, they should have done the same. An interstellar comet that entered the solar system would have a hyperbolic orbit, distinguishing it from the elliptical orbits of the Sun's own comets. Astronomers have never seen a single interstellar comet.

Interstellar Planet A planet that drifts through the Galaxy on its own, free of any star. Billions of interstellar planets probably exist in the Milky Way. When a star explodes in a supernova, it should cast its planets into space. In addition, the Pegasian planets are thought to have originated farther from their stars and migrated inward. As they did so, their gravity should have ejected terrestrial planets into the void.

Io The innermost of the four large Jovian satellites, Io sports active volcanoes and a brilliantly colored surface. With a diameter of 3643 kilometers, Io is similar in size to the Moon. Io circles Jupiter once every 1.769 days.

IRAS The Infrared Astronomical Satellite, launched in 1983. It discovered rings of dust around Vega, Fomalhaut, Beta Pictoris, and Epsilon Eridani.

Iron Element with atomic number 26, which makes up much of the Earth's core and a vital part of human blood.

Juno The third asteroid to be discovered. Its diameter is around 240 kilometers, its mean distance from the Sun is 2.67 astronomical units, and its orbital period is 4.4 years.

Jupiter The largest planet in the solar system, Jupiter has 318 Earth masses, over twice as much as all the other planets of the solar system combined. Jupiter is a gas giant, made mostly of hydrogen and helium. When the solar system was young, Jupiter and its partner Saturn helped pave the way for intelligent life on Earth by casting large numbers of deadly comets out of the solar system, so that few remain to hit the Earth today.

K Spectral type for orange stars, which are somewhat cooler than the Sun. K-type stars split into two main groups: modest stars on the main sequence, such as the nearby stars Epsilon Eridani and Epsilon Indi, and bright giant stars, such as Arcturus and Aldebaran. Although the former are more numerous, the latter are more prominent because of their much greater luminosity.

Kelvin A temperature scale on which the coldest possible temperature is 0 degrees, which corresponds to -460° Fahrenheit. Room temperature is about 295 Kelvin, and the Sun's surface is 5800 Kelvin, or 10,000° Fahrenheit.

Kilometer Per Second A unit of speed that astronomers often employ. One kilometer per second equals 2237 miles per hour. In recent years, as astronomers have developed more sensitive methods for measuring stellar velocities, they have begun to express velocities in meters per second instead.

Kuiper Belt The ring of icy bodies that lie just beyond Neptune's orbit. The Kuiper belt supplies short-period comets to the solar system.

Lalande 21185 The fourth nearest star system to the Sun, after the triple star Alpha Centauri, Barnard's Star, and Wolf 359. Lalande 21185 is a red dwarf 8.3 light-years from the Sun in the constellation Ursa Major.

Light-Year The distance that light travels in one year, or 5.88 trillion miles. Alpha Centauri, the nearest star system to the Sun, is 4.3 light-years from Earth.

Lorentz Factor Quantifies the severity of the effects of special relativity. Mathematically, the Lorentz factor is

$$\frac{1}{\sqrt{1-\frac{v^2}{c^2}}}$$

where v is the object's velocity and c is the speed of light. The Lorentz factor is 1.00 at zero speed, where the effects of special relativity do not exist, and grows greatly as the velocity approaches the speed of light and relativistic effects become important. These include the relativistic increase of mass and dilation of time.

Luminosity The total amount of energy that a star emits into space. This is different from how bright or faint a star *looks*, since a luminous star will look faint if it is far away. See also **Absolute Magnitude**.

M Spectral type for cool, red stars. M-type stars come in two main varieties: dim main-sequence stars, called red dwarfs, which emit far less light than the Sun, and giants and supergiants, which are hundreds or thousands of times brighter than the Sun. Red dwarfs are the most common type of star in the Galaxy, accounting for about 80 percent of all stars, including Proxima Centauri, Barnard's Star, Wolf 359, and Lalande 21185, but red dwarfs are so faint that not a single one is visible to the naked eye. Much less common, but far more prominent, are red giants, like Mira, and red supergiants, like Antares and Betelgeuse.

Magnitude A measure of a star's brightness, either intrinsic or apparent. See **Absolute Magnitude** for the former and **Apparent Magnitude** for the latter.

Main-Sequence Star A star that fuses hydrogen into helium at its core. The Sun and most other stars in the Galaxy are main-sequence stars.

Mars The fourth planet in the solar system, Mars is a small world half the size of Earth. Because of its greater distance from the Sun, Mars is cold. Its atmosphere consists primarily of carbon dioxide but is so thin that it raises the temperature by only 10 degrees Fahrenheit.

Mass The amount of "stuff" in an object. The more mass an object has, the stronger is its gravitational pull on other objects.

Mercury The innermost of the Sun's planets, Mercury is a hostile world with an extreme climate and a cratered landscape.

Meter Per Second The unit of speed that planet searchers employing the Doppler technique use. One meter per second is 2.2 miles per hour. Jupiter and Saturn induce a velocity in the Sun of 12.5 and 2.8 meters per second, respectively.

Methane One of the three "ices," together with water and ammonia, that prevailed in the cool outer solar system and helped form the cores of the giant planets. Methane (CH_4) consists of two of the most common elements in the cosmos, carbon and hydrogen. Until 1995, methane had never been seen in a star, but astronomers that year detected it in the brown dwarf Gliese 229 B.

Microarcsecond An extremely fine unit of angular measure, equal to 1/3,600,000,000 of a degree, or one millionth of an arcsecond—the apparent thickness of a human hair as viewed from a distance of two thousand miles. This is the size of the wobble that an Earthlike planet would induce in a Sunlike star that lies ten light-years away.

Microlensing The gravitational distortion of light caused by an object, such as a planet or star, passing in front of a star. Astronomers may someday detect planets this way.

Micron The unit used by infrared astronomers to measure wavelengths, a micron equals a millionth of a meter, or 10,000 angstroms. Visible light ranges in wavelength from 0.4 to 0.7 microns.

Milky Way Our Galaxy. The Milky Way is a giant among galaxies, having hundreds of billions of stars and roughly a trillion times more mass than the Sun. It is home to the only intelligent civilization known to exist in the cosmos. The Milky Way's brightest component is a luminous disk 130,000 light-years across that contains the Sun. A dark halo surrounds the disk, emits little if any light, but outweighs the rest of the Galaxy. Orbiting the Milky Way are at least ten satellite galaxies.

Milliarcsecond A unit of angular measure, equal to 1/3,600,000 of a degree, or one thousandth of an arcsecond—the apparent thickness of a human hair as viewed from two miles. A Jupiterlike planet in orbit around a Sunlike star that lies ten light-years away would induce an astrometric wobble of 1.6 milliarcseconds.

Millisecond Pulsar A pulsar that spins so fast its period is measured in mere milliseconds. The first millisecond pulsar was discovered in 1982. The first extrasolar planets were detected around the millisecond pulsar PSR B1257+12 in 1991.

Moon A natural object that revolves around a planet. The planets of the solar system have over five dozen moons. Saturn has the most, with eighteen, followed by Jupiter, with sixteen, and Uranus, with fifteen. Of all the moons in the solar system, however, only seven are large. One of these seven large moons is the Earth's own, which is simply called "the Moon" (capitalized). The Moon is 3475 kilometers across, and it revolves around the Earth once every 27.32 days.

Nebula A cloud of gas and dust in space. Some nebulae, such as the Orion Nebula, give birth to new stars and planets; other nebulae, such as the Crab Nebula, represent the material cast into space when a star dies.

Neptune The eighth planet from the Sun, Neptune is a giant blue planet with seventeen times the mass of the Earth. It has eight moons. Neptune resembles Uranus, having a similar color, mass, size, and composition.

Nereid A small moon of Neptune discovered in 1949. It has the most elliptical orbit of any moon in the solar system, with an eccentricity of 0.75. Nereid revolves around Neptune every 360 days.

Neutron An uncharged subatomic particle that appears in every atomic nucleus except hydrogen-1. Light atomic nuclei usually have about as many neutrons as protons.

Neutron Star A collapsed star that forms when a massive star explodes and dies. A typical neutron star has 1.4 times the mass of the Sun but measures only about ten miles across. Its gravity is supported by the pressure of its neutrons. After it has just been born, a neutron star can radiate as a pulsar. See **Pulsar**.

Nitrogen The element with atomic number seven, the chief gas in Earth's atmosphere, and a vital component of terrestrial life. Nearly all nitrogen on Earth was produced in stars that did not explode; when these stars grew old, they expanded into red giants and cast their atmospheres into space, forming planetary nebulae and enriching the Galaxy with nitrogen.

North Star The star toward which the Earth's northern axis points. At the present time, this star is Polaris.

Nova A double star system in which one star transfers mass to the other, usually a white dwarf, and causes an explosion. Unlike a supernova, a nova does not destroy either star.

Nuclear Fission The breaking apart of heavy atomic nuclei into lighter ones, to release energy. This process powers nuclear reactors.

Nuclear Fusion The joining together of light atomic nuclei into heavier ones, to release energy. This process powers most stars, including the Sun, but physicists have yet to succeed in sustaining fusion on Earth.

O Spectral type for stars that are hot and blue—hotter even than B-type stars.

Obliquity Axial tilt. See **Axial Tilt**.

Oort Cloud The vast spherical swarm of icy bodies that surround the solar system at large distances from the Sun, far beyond the Kuiper belt. Whereas the Kuiper belt supplies the solar system's short-period comets, which revolve around the Sun in fewer than two hundred years, the Oort cloud contributes the long-period comets.

Orange Dwarf A main-sequence star that is orange and spectral type K. Orange dwarfs are common, and five lie within a dozen light-years of the Sun: Alpha Centauri B, 4.35 light-years away; Epsilon Eridani, 10.7 light-years away; Epsilon Indi, 11.3 light-years away; and 61 Cygni A and B, 11.4 light-years away.

Orbit The path that one body follows around another.

Orbital Eccentricity A measure of an orbit's shape. See **Eccentricity**.

Orbital Inclination A measure of how tilted an orbit is. See **Inclination**.

Orbital Period The time a body takes to revolve around another. In general, the farther apart the two bodies are, the longer the orbital period: Pluto lies a hundred times farther from the Sun than Mercury, and its orbital period around the Sun is a thousand times longer.

Orion Nebula A huge cloud of interstellar gas and dust, visible to the naked eye, that is giving birth to new stars in the sword of the constellation Orion.

Oxygen Element with atomic number eight, an important part of water and of Earth's atmosphere, and the element that fuels terrestrial animal life. It is produced primarily by high-mass stars when they fuse helium into oxygen (and carbon); this oxygen gets ejected into the Galaxy when the stars explode as supernovae.

Ozone A molecule containing three oxygen atoms (O_3) that shields Earth's surface from deadly ultraviolet radiation. It is also a greenhouse gas that absorbs some of the Earth's warmth and bounces it back to the surface. Detectable ozone on an extrasolar planet would suggest the presence of life, since Earth's atmospheric oxygen is produced by life and would disappear if life perished.

Pallas The second asteroid to be discovered, Pallas has a diameter of about 500 kilometers, a mean distance from the Sun of 2.77 astronomical units, and an orbital period of 4.6 years.

Parallax The tiny shift in a star's apparent position that results as the Earth circles the Sun. The larger the parallax is, the closer the star lies to Earth, so parallax allows astronomers to measure the distances of the nearest stars. Farther stars are too distant to show a detectable parallax.

Parsec A unit of distance equal to 3.261633 light-years.

Pegasian Planet A giant planet that lies near its star. The prototype is 51 Pegasi's planet, which has about 60 percent of the mass of Jupiter and lies so close to the star that it revolves every 4.2 days. Pegasian planets are especially easy for Doppler searchers to detect.

Period See **Orbital Period**.

Phobos The larger of the two Martian moons, Phobos has a mean diameter of 22 kilometers.

Pioneer Pioneer 10 and Pioneer 11 were the first spacecraft to cross the asteroid belt, explore the outer solar system, and venture into interstellar space. Pioneer 10 was launched in 1972 and met Jupiter in 1973. Pioneer 11 was launched in 1973 and met Jupiter in 1974 and Saturn in 1979. See also **Voyager**.

Planet An object that does not emit light but simply reflects the light that strikes it. For an object to be considered a planet, its mass must be at least that of Pluto but not too much more than Jupiter's.

Planetary Nebula The ejected outer atmosphere of a dying red giant star. This gas is set aglow by the star's newly exposed core.

Planetesimal A small asteroidlike body in a young solar system that crashes into other planetesimals and forms a planet.

Planet X The hypothetical tenth planet from the Sun.

Pluto The ninth planet from the Sun. Pluto is the solar system's smallest planet, with a mass only 0.2 percent of the Earth's. Its orbit is highly elliptical (eccentricity 0.25) and inclined (inclination 17.1 degrees). Some do not consider Pluto to be a legitimate planet but instead view it as the largest member of the Kuiper belt. From 1979 to 1999, Pluto skirted inside the orbit of Neptune and temporarily became the eighth planet from the Sun.

Polaris The bright star toward which the northern end of the Earth's rotation axis points. Polaris is a yellow-white F-type star 430 light-years from the Earth. It is in the constellation Ursa Minor, also known as the Little Dipper.

Pollux The nearest giant star to the Earth, Pollux lies 33 light-years away and is the brightest star in the constellation Gemini the Twins. It is an orange star of spectral type K.

Pre-Main-Sequence Star A star so young that it has not yet reached the main sequence. Such a star does not fuse hydrogen-1 into helium, but it does shine, by converting gravitational energy into heat.

Procyon The eighth brightest star in the night sky, Procyon has a spectral type of F and is just beginning to evolve from the main sequence to the subgiant stage. Around the primary star circles a white dwarf. Procyon is 11.4 light-years from the Earth.

Proper Motion The change in a star's position year after year because of its movement through space. The stars with the largest proper motions are Barnard's Star, a nearby red dwarf discovered in 1916, and Kapteyn's Star, a nearby red star discovered in 1897. Their proper motions are 10.3 and 8.7 arcseconds per year, respectively.

Proton The positively charged subatomic particle that exists in every atomic nucleus. The number of protons determines the atomic number and thus the element: all hydrogen atoms (atomic number one) have one proton, all helium atoms (atomic number two) have two protons, all oxygen atoms (atomic number eight) have eight protons, and so on.

Proxima Centauri The nearest and faintest of the three stars in the Alpha Centauri system. See **Alpha Centauri**.

PSR Pulsating Source of Radio. Designation for pulsars.

PSR B0329+54 A pulsar in the constellation Camelopardalis that was reported to have a possible planet in 1979 and again in 1995. This is in dispute.

PSR B1257+12 The site of the first solar system to be discovered outside our own. This millisecond pulsar lies about 1300 light-years away in the constellation Virgo and has at least three planets, similar in many ways to Mercury, Venus, and Earth. The planets akin to Venus and Earth were discovered by Alex Wolszczan and Dale Frail in 1991, and the one similar to Mercury by Wolszczan in 1993.

PSR B1828-11 A pulsar in the constellation Scutum that may have planets. Formerly named PSR 1828-10.

PSR B1829-10 The pulsar that initiated the modern era of pulsar planets. A planet was reported around this pulsar in mid-1991 and dramatically retracted in early 1992. The pulsar lies in the constellation Scutum.

PSR B1919+21 The first pulsar to be found. It was discovered in the constellation Vulpecula by Jocelyn Bell in 1967.

PSR B1957+20 The first black widow pulsar, in which a revived pulsar is tearing apart the star that resuscitated it. See **Black Widow Pulsar**.

Pulsar A rapidly spinning neutron star that like a lighthouse sends a beam of radiation sweeping past Earth every time it spins. The first extrasolar planets were discovered around PSR B1257+12, a pulsar in the constellation Virgo.

Pulsar Planet A planet that revolves around a pulsar.

Quasar The brightest objects in the universe, quasars can emit over a trillion times more light than the Sun. Most quasars lie at enormous distances from Earth, which means they existed long ago, when the universe was young.

Radial Velocity The speed of a star or other celestial object along our line of sight to the object. The radial velocity is therefore revealed by the Doppler shift: if the star is moving toward us, its spectrum is blueshifted, and if the star is moving away from us, its spectrum is redshifted. The greater the Doppler shift one way or the other, the greater the star's radial velocity. Orbiting planets perturb a star's radial velocity and allow astronomers to detect the planets without actually seeing them. The Doppler technique is most sensitive to massive planets near their stars, whereas the astrometric technique is most sensitive to massive planets far from their stars.

Radio Electromagnetic radiation with the longest wavelengths, even longer than those of infrared radiation.

Red Dwarf A main-sequence star of spectral type M, such as Proxima Centauri and Barnard's Star. These stars are faint, cool, and small. Although red dwarfs make up 80 percent of all stars in the Galaxy, none is visible to the naked eye. Because they have little mass, red dwarfs are ideal stars around which to look for planets, because a planet's gravity would tug on the star more than would the same planet around a more massive star like the Sun.

Red Giant The term is used for giant stars with a spectral type of M (red); sometimes it also includes giants with spectral types of K (orange) or G (yellow). Red giants evolve from main-sequence stars that were born with less than eight times the mass of the Sun. Examples include Mira in the constellation Cetus and Gamma Crucis in Crux, the Southern Cross. Someday the Sun will become a red giant. See also **Giant Star**.

Redshift The shift of a spectrum to longer, or redder, wavelengths. There are three different sources of the redshift. One is the *Doppler shift*, which occurs when a star moves away from Earth and stretches its light waves. An orbiting planet causes tiny blueshifts and redshifts in a star's light as the star wobbles toward and away from us.

There are two other causes of a redshift. One is the expansion of space, which stretches the light waves from a distant galaxy while they travel to Earth. This is called a *cosmological redshift*. The final type of redshift results when light climbs out of an object with a strong gravitational field, such as a white dwarf. This is called a *gravitational redshift*.

Regulus A blue B-type main-sequence star in the constellation Leo, Regulus lies 74 light-years from the Earth.

Residual The difference between the predicted and observed values of some quantity. If the residuals in a planet's position are larger than can be explained by observational error, an unseen planet may be pulling the first planet off course. This is how Neptune was predicted to exist before it was actually sighted, because residuals in Uranus's position were far larger than the observational errors allowed.

Revolve To orbit (another object). The Moon revolves around the Earth, the Earth revolves around the Sun, and the Sun revolves around the center of the Galaxy.

Rho[1] Cancri A double star in the constellation Cancer that consists of a yellow G-type main-sequence star and a red dwarf. A Pegasian planet was reported around the yellow star in 1996.

Rho Coronae Borealis A Sunlike star around which astronomers discovered a Pegasian planet in 1997. The star is 57 light-years away.

Rigel The brightest star in the constellation Orion, Rigel is a blue supergiant that lies roughly 900 light-years from the Earth. Someday it will evolve into a red supergiant, like fellow Orion star Betelgeuse, and then explode.

Rotate To spin. The Earth rotates once a day, and the Sun rotates once a month.

Satellite A lesser object that orbits another. In most cases, the term refers to a moon revolving around a planet; sometimes it refers to a galaxy orbiting a larger galaxy. See **Moon**.

Saturn The sixth planet in the solar system, Saturn is best known for its beautiful rings. Like Jupiter, Saturn is a gas giant that consists mostly of hydrogen and helium. When the solar system was young, Jupiter and Saturn cast trillions of comets into interstellar space, leaving few to hit the Earth today and thereby setting the stage for intelligent life.

70 Ophiuchi A double star with two orange K-type dwarfs, 70 Ophiuchi was reported to have a planet in the early 1940s, but later work ruled it out. The system lies 16 light-years from the Earth in the constellation Ophiuchus.

70 Virginis A solar-type star in the constellation Virgo around which a brown dwarf or planet was discovered in late 1995. The star is 59 light-years from Earth.

Sirius The brightest star in the night sky, Sirius is the nearest A-type main-sequence star to the Earth. In orbit around the bright star is the nearest white dwarf to the Earth, Sirius B. The system lies 8.5 light-years away. The two stars revolve around each other every fifty years.

16 Cygni A double yellow star in the constellation Cygnus around which a planet or brown dwarf was discovered in 1996. The newfound object, which orbits 16 Cygni B, has a highly elliptical orbit characteristic of a brown dwarf but a mass only twice that of the planet Jupiter.

61 Cygni A double orange dwarf in the constellation Cygnus that was the first star beyond the Sun to have its parallax measured. The star lies 11.4 light-years from the Earth. In the early 1940s, astrometrists reported a possible planet, but this did not survive further scrutiny.

Solar Mass The amount of mass in the Sun, and the unit in which stellar masses are usually expressed. The Sun is 332,946 times more massive than the Earth and 1048 times more massive than Jupiter.

Solar System The objects that orbit a star: planets, comets, asteroids.

Spectral Type A basic classification of a star by its spectrum, which depends primarily on temperature and color. There are seven main spectral types: O and B, for stars that are hot and blue; A, for stars that are hot and white; F, for stars that are warm and yellow-white; G, for stars that are warm and yellow, like the Sun; K, for stars that are cool and orange; and M, for stars that are cool and red. The OBAFGKM line-up can be memorized via "Oh, Be A Fine Guy, Kiss Me!"

Spectrum The spread of light into a rainbow of colors, which allows astronomers to scrutinize the light and see if it is being Doppler shifted by orbiting planets.

Star A self-luminous celestial body, such as the Sun.

Stellar Parallax See **Parallax**.

Subgiant A star that is making the transition from the main sequence to the red giant stage.

Sun The star that Earth orbits. The Sun is a yellow G-type main-sequence star that powers itself by converting hydrogen into helium at its core. The Sun is much more luminous than the average star, outshining 95 percent of the Milky Way's stars.

Supergiant A large, very luminous star, such as Rigel, Deneb, Antares, and Betelgeuse. Supergiants come in all colors, but the most common are red.

Supernova A huge explosion that destroys a star. Most supernovae arise from supergiants that have run out of fuel; they collapse and explode, leaving behind only a neutron star or black hole. A supernova can also arise from a white dwarf that receives material from another star and becomes unstable, completely annihilating itself and casting its partner into space.

Tachyon A hypothetical particle that travels faster than light.

Tau Boötis A double star in the constellation Boötes that consists of a yellow-white F star and a red dwarf. A Pegasian planet was reported around the F star in 1996.

Tau Ceti The nearest single G-type main-sequence star to the Sun. Tau Ceti is 11.8 light-years from the Earth and was the first star that radio astronomers searched for extraterrestrial intelligence.

Terrestrial Planet A small, rocky planet like Mercury, Venus, Earth, and Mars. The three planets around PSR B1257+12 are probably terrestrial planets.

Tilt See **Axial Tilt**.

Titan The largest moon of Saturn, and the second largest moon in the solar system, after Jupiter's Ganymede. Titan is the only moon in the solar system with a thick atmosphere. This is made mostly of nitrogen, the same gas that makes up most of Earth's atmosphere. Titan has a diameter of 5150 kilometers and revolves around Saturn every 15.95 days.

Titius-Bode Law The planets of the solar system are spaced fairly regularly, and this spacing is encapsulated in the Titius-Bode law, named for Johann Daniel Titius and Johann Elert Bode. Adding 4 to the series 0, 3, 6, 12, . . . produces the new series 4, 7, 10, 16, . . .; these numbers, when divided by 10, then match the distances of the planets from Mercury through Uranus, as well as the asteroids, when those distances are expressed in astronomical units. However, the Titius-Bode law does not work well for Neptune and Pluto.

Transit The passage of an object across the disk of a star or planet. For example, as viewed from Earth, the planets Mercury and Venus occasionally transit the Sun; and someday astronomers may detect extrasolar planets transiting their stars, too.

Triton The largest moon of Neptune, Triton is the smallest of the solar system's "big seven" satellites, having a diameter of 2705 kilometers. It revolves backward around Neptune every 5.88 days. In many ways, Triton resembles Pluto. Triton was once a free object, before Neptune captured it.

Ultraviolet Radiation with a wavelength somewhat shorter than that of visible light. Ozone in Earth's atmosphere shields the planet's surface from this deadly radiation.

Upsilon Andromedae An F-type star around which a Pegasian planet was reported in 1996.

Uranus The seventh planet from the Sun and the first planet to be discovered beyond the Earth, Uranus is a green world that resembles blue Neptune, the next planet out. Both Uranus and Neptune are much smaller than the gas giants, Jupiter and Saturn; sometimes they are called ice giants, because they consist mostly of ices of water, methane, and ammonia, surrounded by hydrogen and helium.

Van Biesbroeck's Star Discovered in 1943 by Georges Van Biesbroeck, this star is a red dwarf that orbits another red dwarf, 19 light-years from the Earth. At the time of its discovery, and for nearly forty years afterward, Van Biesbroeck's Star was the least luminous star known. If put in place of the Sun, it would look about as faint as the full Moon. The star is also called VB 10.

VB 8 An infamous red dwarf in the constellation Ophiuchus around which a false brown dwarf was reported in 1984.

VB 10 Alternate name for Van Biesbroeck's Star. See **Van Biesbroeck's Star**.

Vega The fifth brightest star in the night sky, Vega is an A-type main-sequence star around which the Infrared Astronomical Satellite discovered a ring of dust in 1983, the first time that astronomers had found material orbiting a main-sequence star other than the Sun. Vega lies 25.1 light-years from the Earth in the constellation Lyra the Lyre.

Venus The second planet from the Sun, Venus is an inferno whose thick carbon dioxide atmosphere keeps the planet's surface at a temperature of 860 degrees Fahrenheit. In other ways, though, Venus resembles the Earth, for its mass and size are similar; so too is its carbon dioxide content, but fortunately on Earth most of the carbon dioxide is locked up in rocks.

Vesta The fourth discovered asteroid, Vesta has a diameter of around 500 kilometers, lies 2.36 astronomical units from the Sun, and revolves every 3.6 years.

Voyager Voyager 1 and Voyager 2 explored the outer solar system and are now heading into interstellar space. Both spacecraft left Earth in 1977. Voyager 1 passed Jupiter in 1979, discovering the volcanoes of Io, and met Saturn in 1980, discovering the thick nitrogen atmosphere around Titan; Voyager 2 passed Jupiter in 1979, Saturn in 1981, Uranus in 1986, and Neptune in 1989, discovering geysers on Triton. At their present speed, Voyager 1 and 2 will require some 80,000 years to attain the present distance of Alpha Centauri, although neither craft is heading in that direction. See also **Pioneer**.

Vulcan A purported planet between the Sun and the orbit of Mercury, proposed to explain discrepancies in Mercury's motion. Vulcan does not exist. The discrepancies arose from effects of Einstein's general theory of relativity.

Water A substance (H_2O) that contains the two most abundant reactive elements in the universe—hydrogen and oxygen—and in liquid form is vital to life. As a gas, water contributes to the greenhouse effect on Earth, which keeps the planet warm. As a solid, water was one of the three chief "ices," together with methane and ammonia, that helped the outer solar system give birth to the giant planets.

White Dwarf A dense star that is little larger than the Earth but typically contains 60 percent of the mass of the Sun. A white dwarf is the end for a star, like the Sun, that is born with less than eight times the Sun's mass. After leaving the main sequence, the star evolves into a red giant, then sheds its atmosphere and exposes its hot core, which is a white dwarf. The nearest white dwarfs to the Earth are Sirius B, 8.5 light-years away, and Procyon B, 11.4 light-years away.

Wolf 359 The third nearest star system to the Sun, Wolf 359 is a dim red dwarf in the constellation Leo. At the time of its discovery in 1918, and until the discovery of Van Biesbroeck's Star in 1943, Wolf 359 was the least luminous star known.

Wormhole A shortcut through space that optimists hope will someday allow voyages through the Galaxy.

Xenon Element with atomic number 54, which exists in minute quantities in the Earth's atmosphere. Like helium and neon, xenon is a noble gas, rarely reacting with other atoms.

Year The length of time that a planet takes to revolve around its star. The farther the planet is from the star, the longer is its year: Mercury's year is only 88 days, whereas Pluto's year is 248 Earth years.

Zodiac Because they lie in nearly the same plane, the Sun, the Moon, and the planets appear in a ring of constellations surrounding the Earth. This ring is called the zodiac. The thirteen—yes, thirteen—constellations of the zodiac are Aries, Taurus, Gemini, Cancer, Leo, Virgo, Libra, Scorpius, Ophiuchus, Sagittarius, Capricornus, Aquarius, and Pisces. Next time you meet some "astrologers," tell them you're an Ophiuchus, and see what happens; there may be false planets in your future.

Notes

Introduction: Bruno's Execution

Boulting (1914), Frith (1887), Lindsay (1962), Martin (1921), McIntyre (1903), Michel (1973), Singer (1950), and Yates (1964) all describe Bruno; Singer (1950) translates Bruno's *On the Infinite Universe and Worlds*. The quotation from Guigues appears on page 288 of Michel (1973); Bruno's quotation about the excellence of God appears on page 246 of Singer (1950); the quotation about our being the firmament for others is on page 182 of Michel (1973). The Inquisition's sentence on Bruno appears on pages 175–177 of Singer (1950); Bruno's response appears in many sources, including page 179 of Singer (1950).

Chapter 1. A Good Planet Is Hard to Find

Recent discussions concerning extrasolar planets include Black (1995), Burke (1992), Lissauer (1995), Mammana and McCarthy (1996), and Townes (1996).

Chapter 2. A Living Solar System

Jones (1984) and Beatty and Chaikin (1990) describe the solar system. Brush (1996a,b) recounts the history of theories concerning the solar system's origin; Brush (1978) gives the letter from Chamberlin. Lissauer (1993) describes modern ideas concerning the formation of the solar system.

Chapter 3. The Sun's Distant Outposts

Baumgartner (1987) and Ryden (1990) discuss sunspots' masquerading as planets. Alexander (1965), Bennett (1976, 1982), Gingerich (1984), Grosser (1962), Hoskin (1982), Hoyt (1980), Littmann (1988), Porter (1982), Schaffer

(1981), and Turner (1904) give background on Herschel's discovery of Uranus; Alexander (1965), Forbes (1982), and Grosser (1962) describe prediscovery observations of the planet.

Nieto (1972) and Jaki (1972) describe the development of the Titius-Bode law, which predicted the existence of the asteroids, and Cunningham (1988a,b), Forbes (1971), Grosser (1962), Hoyt (1980), Smith (1982), and Turner (1904) describe the discovery of the first asteroids. Anonymous (1846) relates the discovery of the fifth asteroid.

Grosser (1962) gives the standard history of the discovery of Neptune. The controversy is also described from various viewpoints by Alexander (1965), Chapman (1988), Hoyt (1980), Hubbell and Smith (1992), Littmann (1988), Moore (1988, 1995), Rawlins (1992, 1994), Smart (1947), Smith (1983, 1989), and Turner (1904); in particular, Chapman (1988) attempts the thankless task of rehabilitating Airy, Smith (1989) unearths new information on the discovery, including the search for the planet that was conducted at Paris Observatory, and Rawlins (1992) alleges that the British conspired to steal Neptune from the French.

Fontenrose (1973), Roseveare (1982), and pages 71–74 of Hoyt (1980) describe the hunt for the planet Vulcan. Leverrier (1859) explains why Vulcan might exist; Anonymous (1860a) describes Lescarbault's "observation" of Vulcan and is the source of the quotation praising Lescarbault and Leverrier, while Anonymous (1860b) argues sharply against Vulcan. Young (1878) assured readers of *The New York Times* of Vulcan's likely existence.

The quotation about See—"Personally I have never had such an aversion. . . ."—is from page 34 of Putnam (1994); a slightly different version appears on page 122 of Hoyt (1976).

Hoyt (1980) and Tombaugh and Moore (1980) give excellent accounts of the discovery of Pluto, with the latter book being partly an autobiography by Tombaugh. Grosser (1964) also describes searches for planets beyond Neptune. Levy (1991) is a biography of Tombaugh. Whyte (1980) also discusses Pluto. All of Tombaugh's quotes are from Tombaugh (1996), except for the one concerning See, which is from page 136 of Tombaugh and Moore (1980).

Chapter 4. Planet X

Most quotes are from Harrington (1987, 1990, 1992b), Marsden (1996), Seidelmann (1996), Standish (1996a), and Tombaugh (1996), supplemented by Marsden (1988, 1990), Seidelmann (1987, 1990), and Standish (1990b). Tombaugh (1987) gives additional background.

Brouwer (1955), Duncombe, Klepczynski, and Seidelmann (1968), and Seidelmann, Klepczynski, Duncombe, and Jackson (1971) tried to determine Pluto's mass; Duncombe and Seidelmann (1980) review the history of this problem. Cruikshank, Pilcher, and Morrison (1976) discovered methane ice on Pluto. Christy and Harrington (1978) reported the discovery of Pluto's moon.

Croswell (1988b, 1991b), Kuznik (1993), Littmann (1989), and Tombaugh (1991) discuss Planet X; Kuznik (1993) had the Harrington quote that angered Standish. Seidelmann, Santoro, and Pulkkinen (1985), Seidelmann and Williams (1988), and Seidelmann and Harrington (1988) discuss problems with Uranus and Neptune, and Harrington (1988, 1992a), Gomes (1989), and Powell (1989) propose possible Planet X's. Standish (1990a, 1996b) argues against Planet X, and Standish (1993) employed a new mass for Neptune to remove the need for Planet X. Pierce (1991) describes the changing mass estimates for Neptune. Brady (1972) invoked Planet X to explain problems with Halley's Comet; Foss, Shawe-Taylor, and Whitworth (1972) and Klemola and Harlan (1972) failed to find the planet, and Goldreich and Ward (1972) and Seidelmann, Marsden, and Giclas (1972) argued against its existence. Anderson and Standish (1986) and Anderson et al. (1995) discuss the paths of spacecraft in the outer solar system.

Is Pluto a planet? Levy (1994) says yes; Green, Chambers, and Williams (1994) say no. Yamamoto (1934), Lyttleton (1936), and Kuiper (1956, 1957) proposed that Pluto was an escaped moon of Neptune. Harrington and Van Flandern (1979) invoked Planet X to explain the peculiarities of Neptune's moons and Pluto's orbit. Weissman (1995a) discusses the Kuiper belt, proposed by Edgeworth (1943, 1949) and Kuiper (1951). For more on Edgeworth, see McFarland (1996).

Croswell (1995c) describes Lalande, while Rawlins (1970) and Standish (1994) discuss Lalande's 1795 observations of Neptune. Albers (1979) pointed out that Neptune passed behind Jupiter in 1613, which led Kowal and Drake (1980) to discover Galileo's observations of Neptune. Standish (1981) and Taylor (1985) argue that these do not give Neptune's position accurately.

Tombaugh (1961) described his search for distant planets.

Chapter 5. A Shaky Start

Most quotes are from Gatewood (1996a), Eichhorn (1996), and van de Kamp (1987). Kovalevsky (1990) and Mulholland (1989) describe astrometry. Serio (1990) describes Piazzi's discovery of 61 Cygni's large proper motion, which led Bessel (1838) to detect the star's parallax.

Croswell (1987, 1995b) describes the nearby stars in general, and Croswell (1991a) Alpha Centauri, Croswell (1988a) Barnard's Star, Croswell (1995c) Lalande 21185, and Croswell (1995d) Epsilon Eridani. For more on stellar evolution, see Chapter 5 of Croswell (1995a). Sackmann, Boothroyd, and Kraemer (1993) describe the evolution of the Sun.

Data for the nearby star table come from the following sources. Distances for all the stars are from the new parallax catalogue of van Altena, Lee, and Hoffleit (1995); this is also the source of most other stellar distances in the book. Henry, Kirkpatrick, and Simons (1994) give spectral types for most of the M-type dwarfs, as well as for 61 Cygni; spectral types for Proxima Centauri and Lacaille 9352 are from Henry (1996b); spectral types for the remaining stars are from Hoffleit and Jaschek (1982). For nearly all the red dwarfs as well as 61 Cygni B, Leggett (1992) supplies the apparent magnitudes. Hoffleit and Jaschek (1982) give the apparent magnitudes of Sirius A, Epsilon Indi, 61 Cygni A, Procyon A, and Tau Ceti. The apparent magnitudes of Alpha Centauri A and B are from Bessell (1981), that of Sirius B is from Holberg, Wesemael, and Hubený (1984), that of Luyten 726-8 is from Gliese and Jahreiss (1991), that of Epsilon Eridani is from Soderblom and Däppen (1989), and that of Procyon B is from McCook and Sion (1987). Absolute magnitudes were computed anew from the apparent magnitudes and the new parallaxes.

Bessel (1844) predicted the double nature of Sirius and Procyon. Ashbrook (1964) describes the discovery of the companions to Sirius and Procyon. Gatewood (1976) discusses the astrometric search for planets. Reuyl and Holmberg (1943) reported a planet around 70 Ophiuchi, Strand (1943a,b, 1957) a planet around 61 Cygni, and van de Kamp and Lippincott (1951) and Lippincott (1960) a planet around Lalande 21185. Gatewood (1974) found no planet around Lalande 21185.

Barnard (1916) discovered Barnard's Star; see Sheehan (1995) for a biography of Barnard. The star's mass is derived from relationships given by Henry and McCarthy (1993) and Baraffe and Chabrier (1996). Van de Kamp (1963, 1969a) reported a planet around Barnard's Star; van de Kamp (1969b) presented the possibility that the star had two planets, an idea he subsequently came to favor (van de Kamp 1974, 1975, 1977, 1983a,b, 1986, 1987). The 1977 paper is the source of the John 20:29 quotation, and the 1983b paper the source of the longer quotation. However, Gatewood and Eichhorn (1973) found no planets, nor did Heintz (1976, 1978), Fredrick and Ianna (1985), Harrington (1986), and Harrington and Harrington (1987). Gatewood (1995) gives the latest on the search for planets around the star. Sinnott (1987) reviews the work of Allegheny Observatory. Fredrick (1996) describes van de Kamp's life and work.

Chapter 6. The Vega Phenomenon

Most quotes are from Aumann (1996), Gillett (1996), Smith (1996), and Terrile (1996). Backman and Paresce (1993) review the Vega phenomenon, concentrating on Vega, Fomalhaut, and Beta Pictoris. Aumann et al. (1984) detected Vega's infrared excess in 1983; additional reports appear by Aumann (1985, 1988), Gillett (1986), and Backman and Gillett (1987). Aumann (1987) was pessimistic about using infrared excesses to detect extrasolar planets. Gatewood and de Jonge (1995) measured the precise distance of Vega.

Smith and Terrile (1984) photographed the dust disk around Beta Pictoris. Lagrange-Henri, Vidal-Madjar, and Ferlet (1988) believe comets may occasionally fall into the star. Lanz, Heap, and Hubeny (1995) argue that Beta Pictoris may be a pre-main-sequence star.

Chapter 7. Dark Star

Most quotes are from Kumar (1996), Stevenson (1996), and Burrows (1996). Liebert and Probst (1987) review very low mass stars, and recent reviews of brown dwarfs appear by Stevenson (1991) and Burrows and Liebert (1993); a 1994 conference (Tinney 1995) explored both red and brown dwarfs. Short articles on the subject include Kumar (1990, 1994). Popular-level articles on brown dwarfs appeared by Croswell (1984), Murray (1986), Byrd (1989), Stephens (1995), and Henry (1996a). Articles by Fienberg (1990) and Croswell (1990) are the source of the negative headlines about brown dwarfs mentioned in the text.

Following the apparent discovery of dark companions to 70 Ophiuchi by Reuyl and Holmberg (1943) and 61 Cygni by Strand (1943a,b), Russell (1943, 1944) explored the nature of stars with a few percent of the mass of the Sun. Shapley (1958) also speculated on these dim objects. Kumar (1963a,b) first determined the minimum mass of a main-sequence star and computed the evolution of what Davidson (1975) called infrared dwarfs and Tarter (1975) brown dwarfs. Duquennoy and Mayor (1991) estimate the mean orbital eccentricity of binary stars.

Oort (1932, 1965) and Bahcall (1984a,b) found evidence for dark matter in the Galactic disk, but Kuijken and Gilmore (1989a,b,c) disputed this. Chapter 15 of Croswell (1995a) discusses dark matter in the Galaxy's dark halo, as does Croswell (1996). Innes (1915) discovered Proxima Centauri, Wolf (1918) Wolf 359, and Van Biesbroeck (1944) the star that bears his name.

VB 8 was reported by Van Biesbroeck (1961). McCarthy, Probst, and Low (1985) "found" a brown dwarf around VB 8 that neither Perrier and Mariotti (1987) nor Skrutskie, Forrest, and Shure (1987) nor anyone else ever saw again. The first to report this nonobject were Harrington, Kallarakal, and Dahn (1983); see also Harrington (1985). Kumar (1986) said the object, even if it existed, was not a planet. For the 1985 conference on brown dwarfs, see Kafatos, Harrington, and Maran (1986).

Zuckerman and Becklin (1987) reported a possible brown dwarf around the white dwarf Giclas 29-38, but Graham, Matthews, Neugebauer, and Soifer (1990), Telesco, Joy, and Sisk (1990), and Tokunaga, Becklin, and Zuckerman (1990) argue against this.

Becklin and Zuckerman (1988) and Zuckerman and Becklin (1992) reported a possible brown dwarf orbiting the white dwarf GD 165; see also Kirkpatrick, Henry, and Liebert (1993). Latham et al. (1989) discovered HD

114762's companion, but Cochran, Hatzes, and Hancock (1991) and Hale (1995) argued that we view the star nearly pole-on, so its companion may not be a brown dwarf. Robinson et al. (1990) also put constraints on the orientation of the system. A recent report is by Mazeh, Latham, and Stefanik (1996).

Chapter 8. Pulsar Planets

Quotes are from Lyne (1996), Wolszczan (1996), Frail (1996), and Marsden (1996). Lyne and Graham-Smith (1990) review the nature of pulsars. Baade and Zwicky (1934) said that supernovae may form neutron stars. Hewish et al. (1968) announced the discovery of the first pulsar, in a paper that never says who—Jocelyn Bell—actually made the discovery. Jones (1992) describes Bell's discovery. Richards, Pettengill, Counselman, and Rankin (1970) reported a planet around the Crab pulsar, and Demiański and Prószyński (1979) one around PSR B0329+54 in Camelopardalis.

Bailes, Lyne, and Shemar (1991) reported a planet around PSR B1829-10; Shemar (1991) described the history. The pulsar itself was discovered by Clifton and Lyne (1986). The planet interpretation was challenged by Helfand and Hamilton (1991); Lyne (1991) defended it. Theories for the origin of the planet were put forward by Dawson (1991), Fabian and Podsiadlowski (1991), Lin, Woosley, and Bodenheimer (1991), Nakamura and Piran (1991), Podsiadlowski, Pringle, and Rees (1991), and Postnov and Prokhorov (1992). Lyne and Bailes (1992) retracted the planet. Croswell (1992a) described Lyne's AAS talk.

Wolszczan and Frail (1992) discovered the first true pulsar planets, and the first planets outside the solar system, as reported by Croswell (1991c,d). Wolszczan (1991) had earlier discovered the pulsar itself. The first millisecond pulsar had been reported by Backer et al. (1982). The pulsar planet conference was Phillips, Thorsett, and Kulkarni (1993); see Croswell (1992b) for a report, and Wolszczan (1994a) and Phillips and Thorsett (1994) for technical reviews. Backer (1993) confirmed the oscillations in the pulsar, but Gil, Jessner, and Kramer (1993) and Dolginov and Stepinski (1993) put forth a nonplanetary interpretation. Wolszczan (1994b) confirmed the existence of two of the planets around PSR B1257+12 and reported a third, following a suggestion by Rasio, Nicholson, Shapiro, and Teukolsky (1992) and others.

Siegel (1994) announced the contest results for the names of the planets on National Public Radio. Mazeh and Goldman (1995) described similarities between these planets and the Sun's innermost three. The first black widow pulsar was discovered by Fruchter, Stinebring, and Taylor (1988), a discovery anticipated by Ruderman, Shaham, and Tavani (1989); Lyne et al. (1990) found another black widow pulsar. Livio, Pringle, and Saffer (1992) discuss possible planets around massive white dwarfs. Shabanova (1995) says there is at least one planet around PSR B0329+54. The dismissive quotation concerning pulsar planets is from Black (1995). Bruning (1992) describes the plight of Jupiter seekers in the early 1990s.

Chapter 9. Thanks, Jupiter

Most quotes are from Wetherill (1992, 1996b), Stevenson (1996), Boss (1996b), Weissman (1996), and Gatewood (1996a). Wetherill (1994, 1995) describes the importance of Jupiter and Saturn, as reported by Croswell (1992c). Wetherill (1996a) experimented with moving Jupiter around, to see how this affected the formation of the terrestrial planets, and in Lissauer, Pollack, Wetherill, and Stevenson (1995) he found it difficult to make Jupiters. Owen (1994) gives evidence in favor of the idea that the giant planets started as cores that attracted hydrogen and helium (Mizuno, Nakazawa, and Hayashi 1978; Mizuno 1980). Saumon et al. (1996) provide models of the temperature and luminosity evolution of extrasolar giant planets.

Sekanina (1976) studied the possibility of interstellar comets, as did Stern (1990, 1997) and Kresák (1992). Weissman (1990) describes the Oort cloud of comets, while McGlynn and Chapman (1989), Sen and Rana (1993), and Weissman (1995b) arrive at differing conclusions concerning whether astronomers should have observed any interstellar comets. Laskar, Joutel, and Robutel (1993) describe how the Moon may have stabilized the Earth's axial tilt and thus its climate. Chapter 9 of Croswell (1995a) describes Hoyle's prediction concerning the resonance of carbon.

Campbell, Walker, and Yang (1988) used the Doppler technique and reported two "probable" planets and five "possible" planets that proved false; Walker et al. (1995) found no planets.

Chapter 10. Harvest of Planets

Most quotes are from Mayor (1996), Queloz (1996), Marcy (1996), Butler (1996), Latham (1996), Marsden (1996), Boss (1996b), Kulkarni (1996), Stevenson (1996), Kumar (1996), and Gatewood (1996a,b); Nakajima (1996) and Lissauer (1996) supplied additional information.

Mayor and Queloz (1995, 1996) discovered 51 Pegasi's planet, which was soon confirmed by others, including Marcy et al. (1997). Marcy and Benitz (1989) had earlier searched for brown dwarfs orbiting nearby red dwarfs. Koppel (1995) devoted half an hour to the discovery of 51 Pegasi's planet. Marsden (1995) issued an IAU circular on the planet. For popular accounts of the discovery, see MacRobert and Roth (1996) and Naeye (1996). Guillot et al. (1996) and Lin, Bodenheimer, and Richardson (1996) discuss the odd planet's origin, the latter following up on Lin and Papaloizou (1986). François et al. (1996) measured a precise rotational velocity for 51 Pegasi, thereby helping to confirm that astronomers do not view the star pole-on. But Gray (1997) questioned the planet's existence.

Gliese 229 B was discovered by Nakajima et al. (1995), and the methane spectrum was seen by Oppenheimer, Kulkarni, Matthews, and Nakajima

(1995). For background on the search, see Nakajima, Durrance, Golimowski, and Kulkarni (1994). Tsuji, Ohnaka, and Aoki (1995), Tsuji, Ohnaka, Aoki, and Nakajima (1996), and Fegley and Lodders (1996) describe the expected appearance of brown dwarfs. Additional work on Gliese 229 B was done by Allard, Hauschildt, Baraffe, and Chabrier (1996) and Geballe, Kulkarni, Woodward, and Sloan (1996). Rebolo, Zapatero Osorio, and Martín (1995), Basri, Marcy, and Graham (1996), Martín, Rebolo, and Zapatero-Osorio (1996), and Rebolo et al. (1996) reported brown dwarfs in the Pleiades, based on the lithium test (Rebolo, Martín, and Magazzù 1992; Magazzù, Martín, and Rebolo 1993; Nelson, Rappaport, and Chiang 1993).

Marcy and Butler (1992) and Butler et al. (1996) describe how they attain their Doppler precision. This allowed them to discover the planet around 47 Ursae Majoris (Butler and Marcy 1996), the companion to 70 Virginis (Marcy and Butler 1996), and additional planets (Butler et al. 1997). Perryman et al. (1996) give Hipparcos distances for 51 Pegasi, 47 Ursae Majoris, and 70 Virginis. Gehman, Adams, and Laughlin (1996), Henry et al. (1997), and Baliunas et al. (1997) discuss these and other stars in greater detail. Cochran, Hatzes, Butler, and Marcy (1997) discovered the object around 16 Cygni B, an object whose evolution was explored by Mazeh, Krymolowski, and Rosenfeld (1997) and Holman, Touma, and Tremaine (1997). Noyes et al. (1997) discovered Rho Coronae Borealis's planet. Boss (1996a,c) used mass and orbital eccentricity to distinguish between planets and brown dwarfs; see also Kumar (1967). Lin and Ida (1997) proposed how objects on highly eccentric orbits may have actually originated as planets rather than brown dwarfs. Gatewood (1996c) reported a possible planet around Lalande 21185. Williams, Kasting, and Wade (1997) investigate habitable moons of extrasolar planets.

Chapter 11. Future Worlds

All quotes are from Marcy (1996), Shao (1996), Paczyński (1995), and Angel (1996). Ideas for discovering extrasolar planets appear in Townes (1996), a follow-up to Burke (1992) and Black (1980). Shao and Colavita (1992) describe optical and infrared interferometry; see Davis (1992) and Fischer (1996) for popular-level accounts.

Borucki and Summers (1984), Doyle (1996), Giampapa, Craine, and Hott (1995), and Hale and Doyle (1994) describe the photometric method of planetary detection—searching for transits of planets across a star's disk. Jenkins, Doyle, and Cullers (1996) apply this to eclipsing binaries, especially CM Draconis. Mao and Paczyński (1991) proposed that gravitational microlensing could reveal extrasolar planets, an idea that Gould and Loeb (1992), Bolatto and Falco (1994), Gould and Welch (1996), Bennett and Rhie (1996), and Wambsganss (1997) developed in greater detail. Paczyński (1986) had earlier suggested that gravitational microlensing could reveal the dark halo's composition; for follow-up reports on this, see Croswell (1996) and Paczyński (1996).

Beckers (1993) reviews adaptive optics, an idea that Babcock (1953) suggested; for popular-level accounts, see Fugate and Wild (1994), Wild and Fugate (1994), Hardy (1994), and Peterson (1994). Angel (1994), Nakajima (1994), and Stahl and Sandler (1995) describe using adaptive optics for extrasolar planetary detection; Malbet (1996) also describes adaptive optics.

Lindegren and Perryman (1996) give the scientific goals of a space-borne astrometric interferometer. Angel and Woolf (1996, 1997) and Léger et al. (1996) describe a space-based infrared interferometer that may someday detect Earthlike planets and determine whether they have life; see also Angel, Cheng, and Woolf (1986). Hitchcock and Lovelock (1967), Lovelock (1975), Owen (1980), and Léger, Pirre, and Marceau (1993) discuss the importance of oxygen. Bracewell (1978) suggested a nulling interferometer for extrasolar planetary detection. The atmospheric compositions of Venus, Earth, and Mars are taken from, respectively, von Zahn, Kumar, Niemann, and Prinn (1983), Ahrens (1991), and Owen (1992). Hall et al. (1995) detected tenuous oxygen around Jupiter's moon Europa.

For discussions of extraterrestrial life, see Dick (1996), whose book reviews the history of the idea during the twentieth century, and the book edited by Zuckerman and Hart (1995).

Chapter 12. Into the Cosmos

The books by Mallove and Matloff (1989) and Mauldin (1992) describe interstellar travel in detail, as does Crawford (1990). Purcell (1960) says interstellar travel will never happen; Forward (1996) is more optimistic. Jones (1985) describes the origin of the Fermi paradox. Cesarone, Sergeyevsky, and Kerridge (1984) track the future flights of Pioneers 10 and 11 and Voyagers 1 and 2. Bussard (1960) suggested using interstellar hydrogen to power spacecraft. Matloff (1996) looks at faster-than-light travel. Naeye and Potter (1996) describe the TAU mission beyond Pluto.

Bibliography

Ahrens, C. Donald, 1991. *Meteorology Today*, Fourth Edition (New York: West).

Albers, Steven C., 1979. Mutual Occultations of Planets: 1557 to 2230. *Sky and Telescope*, 57, 220 (March 1979).

Alexander, A. F. O'D., 1965. *The Planet Uranus* (New York: American Elsevier).

Allard, France, Hauschildt, Peter H., Baraffe, Isabelle, and Chabrier, Gilles, 1996. Synthetic Spectra and Mass Determination of the Brown Dwarf Gliese 229B. *Astrophysical Journal Letters*, 465, L123.

Anderson, John D., Lau, Eunice L., Krisher, Timothy P., Dicus, Duane A., Rosenbaum, Doris C., and Teplitz, Vigdor, L., 1995. Improved Bounds on Nonluminous Matter in Solar Orbit. *Astrophysical Journal*, 448, 885.

Anderson, John D., and Standish, E. Myles, Jr., 1986. Dynamical Evidence for Planet X. In *The Galaxy and the Solar System*, edited by Roman Smoluchowski, John N. Bahcall, and Mildred S. Matthews (Tucson: University of Arizona Press), p. 286.

Angel, J. R. P., 1994. Ground-Based Imaging of Extrasolar Planets Using Adaptive Optics. *Nature*, 368, 203.

———, 1996. Interview with Ken Croswell: June 6, 1996.

Angel, J. R. P., Cheng, A. Y. S., and Woolf, N. J., 1986. A Space Telescope for Infrared Spectroscopy of Earth-Like Planets. *Nature*, 322, 341.

Angel, J. Roger P., and Woolf, Neville J., 1996. Searching for Life on Other Planets. *Scientific American*, 274, No. 4 (April 1996), p. 60.

———, 1997. An Imaging Nulling Interferometer to Study Extrasolar Planets. *Astrophysical Journal*, 475, 373.

Anonymous, 1846. Announcements of the Discovery of the New Planet Astraea; with Observations, Elements, &c. *Monthly Notices of the Royal Astronomical Society*, 7, 27.

———, 1860a. A Supposed New Interior Planet. *Monthly Notices of the Royal Astronomical Society*, 20, 98.

———, 1860b. *Monthly Notices of the Royal Astronomical Society*, 20, 265.

Ashbrook, Joseph, 1964. The Companion of Procyon: Some Prediscovery Reports. *Sky and Telescope*, 28, 72 (August 1964).

Aumann, H. H., 1985. IRAS Observations of Matter Around Nearby Stars. *Publications of the Astronomical Society of the Pacific*, 97, 885.

———, 1987. Prospects of Infrared Prospecting for Planets. *Nature*, 328, 208.

———, 1988. Spectral Class Distribution of Circumstellar Material in Main-Sequence Stars. *Astronomical Journal*, 96, 1415.

———, 1996. Interview with Ken Croswell: March 23, 1996.

Aumann, H. H., Gillett, F. C., Beichman, C. A., de Jong, T., Houck, J. R., Low, F. J., Neugebauer, G., Walker, R. G., and Wesselius, P. R., 1984. Discovery of a Shell Around Alpha Lyrae. *Astrophysical Journal Letters*, 278, L23.

Baade, W., and Zwicky, F., 1934. Cosmic Rays from Super-Novae. *Proceedings of the National Academy of Sciences*, 20, 259.

Babcock, H. W., 1953. The Possibility of Compensating Astronomical Seeing. *Publications of the Astronomical Society of the Pacific*, 65, 229.

Backer, D. C., 1993. A Pulsar Timing Tutorial and NRAO Green Bank Observations of PSR 1257+12. In *Planets Around Pulsars*, edited by J. A. Phillips, S. E. Thorsett, and S. R. Kulkarni (San Francisco: Astronomical Society of the Pacific), p. 11.

Backer, D. C., Kulkarni, Shrinivas R., Heiles, Carl, Davis, M. M., and Goss, W. M., 1982. A Millisecond Pulsar. *Nature*, 300, 615.

Backman, D. E., and Gillett, F. C., 1987. Exploiting the Infrared: IRAS Observations of the Main Sequence. In *Cool Stars, Stellar Systems, and the Sun, Fifth Cambridge Workshop*, edited by Jeffrey L. Linsky and Robert E. Stencel (New York: Springer-Verlag), p. 340.

Backman, Dana E., and Paresce, Francesco, 1993. Main-Sequence Stars with Circumstellar Solid Material: The Vega Phenomenon. In *Protostars and Planets III*, edited by Eugene H. Levy and Jonathan I. Lunine (Tucson: University of Arizona Press), p. 1253.

Bahcall, John N., 1984a. The Distribution of Stars Perpendicular to a Galactic Disk. *Astrophysical Journal*, 276, 156.

———, 1984b. Self-Consistent Determinations of the Total Amount of Matter near the Sun. *Astrophysical Journal*, 276, 169.

Bailes, M., Lyne, A. G., and Shemar, S. L., 1991. A Planet Orbiting the Neutron Star PSR1829-10. *Nature*, 352, 311.

Baliunas, Sallie L., Henry, Gregory W., Donahue, Robert A., Fekel, Francis C., and Soon, Willie H., 1997. Properties of Sun-Like Stars with Planets: ρ^1 Cancri, τ Bootis, and υ Andromedae. *Astrophysical Journal Letters*, 474, L119.

Baraffe, Isabelle, and Chabrier, Gilles, 1996. Mass-Spectral Class Relationship for M Dwarfs. *Astrophysical Journal Letters*, 461, L51.

Barnard, E. E., 1916. A Small Star with Large Proper-Motion. *Astronomical Journal*, 29, 181.

Basri, Gibor, Marcy, Geoffrey W., and Graham, James R., 1996. Lithium in

Brown Dwarf Candidates: The Mass and Age of the Faintest Pleiades Stars. *Astrophysical Journal*, 458, 600.

Baumgartner, Frederic J., 1987. Sunspots or Sun's Planets: Jean Tarde and the Sunspot Controversy of the Early Seventeenth Century. *Journal for the History of Astronomy*, 18, 44.

Beatty, J. Kelly, and Chaikin, Andrew (editors), 1990. *The New Solar System*, Third Edition (Cambridge: Cambridge University Press).

Beckers, Jacques M., 1993. Adaptive Optics for Astronomy: Principles, Performance, and Applications. *Annual Review of Astronomy and Astrophysics*, 31, 13.

Becklin, E. E., and Zuckerman, B., 1988. A Low-Temperature Companion to a White Dwarf Star. *Nature*, 336, 656.

Bennett, David P., and Rhie, Sun Hong, 1996. Detecting Earth-Mass Planets with Gravitational Microlensing. *Astrophysical Journal*, 472, 660.

Bennett, J. A., 1976. "On the Power of Penetrating into Space": The Telescopes of William Herschel. *Journal for the History of Astronomy*, 7, 75.

———, 1982. Herschel's Scientific Apprenticeship and the Discovery of Uranus. In *Uranus and the Outer Planets*, edited by Garry Hunt (Cambridge: Cambridge University Press), p. 35.

Bessel, F. W., 1838. A Letter from Professor Bessel to Sir J. Herschel. *Monthly Notices of the Royal Astronomical Society*, 4, 152.

———, 1844. On the Variations of the Proper Motions of *Procyon* and *Sirius*. *Monthly Notices of the Royal Astronomical Society*, 6, 136.

Bessell, M. S., 1981. Alpha Centauri. *Proceedings of the Astronomical Society of Australia*, 4, 212.

Black, David C., 1980. *Project Orion* (Washington, D.C.: NASA).

———, 1995. Completing the Copernican Revolution: The Search for Other Planetary Systems. *Annual Review of Astronomy and Astrophysics*, 33, 359.

Bolatto, Alberto D., and Falco, Emilio E., 1994. The Detectability of Planetary Companions of Compact Galactic Objects from Their Effects on Microlensed Light Curves of Distant Stars. *Astrophysical Journal*, 436, 112.

Borucki, William J., and Summers, Audrey L., 1984. The Photometric Method of Detecting Other Planetary Systems. *Icarus*, 58, 121.

Boss, Alan P., 1996a. Giants and Dwarfs Meet in the Middle. *Nature*, 379, 397.

———, 1996b. Interview with Ken Croswell: May 10, 1996.

———, 1996c. Extrasolar Planets. *Physics Today*, 49, No. 9 (September 1996), p. 32.

Boulting, William, 1914. *Giordano Bruno: His Life, Thought, and Martyrdom* (Freeport, New York: Books for Libraries Press).

Bracewell, R. N., 1978. Detecting Nonsolar Planets by Spinning Infrared Interferometer. *Nature*, 274, 780.

Brady, Joseph L., 1972. The Effect of a Trans-Plutonian Planet on Halley's Comet. *Publications of the Astronomical Society of the Pacific*, 84, 314.

Brouwer, Dirk, 1955. The Motions of the Outer Planets. *Monthly Notices of the Royal Astronomical Society*, 115, 221.

Bruning, David, 1992. Desperately Seeking Jupiters. *Astronomy*, 20, No. 7 (July 1992), p. 36.

Brush, Stephen G., 1978. A Geologist Among Astronomers: The Rise and Fall of the Chamberlin-Moulton Cosmogony, Part 1. *Journal for the History of Astronomy*, 9, 1.

———, 1996a. *Nebulous Earth* (Cambridge: Cambridge University Press).

———, 1996b. *Fruitful Encounters* (Cambridge: Cambridge University Press).

Burke, Bernard F. (chair), 1992. *TOPS: Toward Other Planetary Systems* (Washington, D.C.: NASA).

Burrows, Adam, 1996. Interview with Ken Croswell: April 16, 1996.

Burrows, Adam, and Liebert, James, 1993. The Science of Brown Dwarfs. *Reviews of Modern Physics*, 65, 301.

Bussard, R. W., 1960. Galactic Matter and Interstellar Flight. *Astronautica Acta*, 6, 179.

Butler, R. Paul, 1996. Interview with Ken Croswell: May 10, 1996.

Butler, R. Paul, and Marcy, Geoffrey W., 1996. A Planet Orbiting 47 Ursae Majoris. *Astrophysical Journal Letters*, 464, L153.

Butler, R. Paul, Marcy, Geoffrey W., Williams, Eric, Hauser, Heather, and Shirts, Phil, 1997. Three New "51 Pegasi-Type" Planets. *Astrophysical Journal Letters*, 474, L115.

Butler, R. Paul, Marcy, Geoffrey W., Williams, Eric, McCarthy, Chris, Dosanjh, Preet, and Vogt, Steven S., 1996. Attaining Doppler Precision of 3 m s^{-1}. *Publications of the Astronomical Society of the Pacific*, 108, 500.

Byrd, Deborah, 1989. Do Brown Dwarfs Really Exist? *Astronomy*, 17, No. 4 (April 1989), p. 18.

Campbell, Bruce, Walker, G. A. H., and Yang, S., 1988. A Search for Substellar Companions to Solar-Type Stars. *Astrophysical Journal*, 331, 902.

Cesarone, Robert J., Sergeyevsky, Andrey B., and Kerridge, Stuart J., 1984. Prospects for the Voyager Extra-Planetary and Interstellar Mission. *Journal of the British Interplanetary Society*, 37, 99.

Chapman, Allan, 1988. Private Research and Public Duty: George Biddell Airy and the Search for Neptune. *Journal for the History of Astronomy*, 19, 121.

Christy, James W., and Harrington, Robert S., 1978. The Satellite of Pluto. *Astronomical Journal*, 83, 1005.

Clifton, T. R., and Lyne, A. G., 1986. High-Radio-Frequency Survey for Young and Millisecond Pulsars. *Nature*, 320, 43.

Cochran, William D., Hatzes, Artie P., Butler, R. Paul, and Marcy, Geoffrey W., 1997. The Discovery of a Planetary Companion to 16 Cygni B. *Astrophysical Journal*, in press.

Cochran, William D., Hatzes, Artie P., and Hancock, Terry J., 1991. Constraints on the Companion Object to HD 114762. *Astrophysical Journal Letters*, 380, L35.

Crawford, I. A., 1990. Interstellar Travel: A Review for Astronomers. *Quarterly Journal of the Royal Astronomical Society*, 31, 377.

Croswell, Ken, 1984. Stars Too Small to Burn. *Astronomy*, 12, No. 4 (April 1984), p. 16.

———, 1987. Visit the Nearest Stars. *Astronomy*, 15, No. 1 (January 1987), p. 16.

———, 1988a. Does Barnard's Star Have Planets? *Astronomy*, 16, No. 3 (March 1988), p. 6.

———, 1988b. The Pull of Planet X. *Astronomy*, 16, No. 8 (August 1988), p. 30.

———, 1990. Another Brown Dwarf Bites the Dust. *New Scientist*, 127, No. 1724 (July 7, 1990), p. 28.

———, 1991a. Does Alpha Centauri Have Intelligent Life? *Astronomy*, 19, No. 4 (April 1991), p. 28.

———, 1991b. The Search for Planet X. *Star Date*, 19, No. 5 (September/October 1991), p. 4.

———, 1991c. Trio of Planets Found Orbiting Nearby Pulsar. *New Scientist*, 132, No. 1799 (December 14, 1991), p. 19.

———, 1991d. Pulsar's Planetary Entourage Is Real, Say Astronomers. *New Scientist*, 132, No. 1800 (December 21-28, 1991), p. 11.

———, 1992a. To Kill a Planet. *New Scientist*, 133, No. 1809 (February 22, 1992), p. 48.

———, 1992b. Puzzle of the Pulsar Planets. *New Scientist*, 135, No. 1830 (July 18, 1992), p. 40.

———, 1992c. Why Intelligent Life Needs Giant Planets. *New Scientist*, 136, No. 1844 (October 24, 1992), p. 18.

———, 1995a. *The Alchemy of the Heavens* (New York: Doubleday/Anchor).

———, 1995b. Neighbors of the Sun. *Star Date*, 23, No. 2 (March/April, 1995), p. 4.

———, 1995c. The Story of Lalande 21185. *Sky and Telescope*, 89, No. 6 (June 1995), p. 68.

———, 1995d. Epsilon Eridani: The Once and Future Sun. *Astronomy*, 23, No. 12 (December 1995), p. 46.

———, 1996. The Dark Side of the Galaxy. *Astronomy*, 24, No. 10 (October 1996), p. 40.

Cruikshank, D. P., Pilcher, C. B., and Morrison, D., 1976. Pluto: Evidence for Methane Frost. *Science*, 194, 835.

Cunningham, Clifford J., 1988a. *Introduction to Asteroids* (Richmond, Virginia: Willmann-Bell).

———, 1988b. The Baron and His Celestial Police. *Sky and Telescope*, 75, 271 (March 1988).

Davidson, Kris, 1975. Does the Solar System Include Distant But Discoverable Infrared Dwarfs? *Icarus*, 26, 99.

Davis, John, 1992. The Quest for High Resolution. *Sky and Telescope*, 83, 29 (January 1992).

Dawson, Peter C., 1991. New Possibility for Planet. *Nature*, 352, 763.

Demiański, M., and Prószyński, M., 1979. Does PSR0329+54 Have Companions? *Nature*, 282, 383.

Dick, Steven J., 1996. *The Biological Universe* (Cambridge: Cambridge University Press).

Dolginov, A. Z., and Stepinski, T. F., 1993. On Quasiperiodic Variations of Pulsars' Periods: An Alternative to the Planetary Interpretation of PSR1257+12. In *Planets Around Pulsars*, edited by J. A. Phillips, S. E. Thorsett, and S. R. Kulkarni (San Francisco: Astronomical Society of the Pacific), p. 61.

Doyle, Laurance R., 1996. In the Wink of a Star. *Mercury*, 25, No. 4 (July/August 1996), p. 20.

Duncombe, R. L., Klepczynski, W. J., and Seidelmann, P. K., 1968. Orbit of Neptune and the Mass of Pluto. *Astronomical Journal*, 73, 830.

Duncombe, R. L., and Seidelmann, P. K., 1980. A History of the Determination of Pluto's Mass. *Icarus*, 44, 12.

Duquennoy, A., and Mayor, M., 1991. Multiplicity Among Solar-Type Stars in the Solar Neighbourhood. II. Distribution of the Orbital Elements in an Unbiased Sample. *Astronomy and Astrophysics*, 248, 485.

Edgeworth, K. E., 1943. The Evolution of Our Planetary System. *Journal of the British Astronomical Association*, 53, 181.

———, 1949. The Origin and Evolution of the Solar System. *Monthly Notices of the Royal Astronomical Society*, 109, 600.

Eichhorn, Heinrich, 1996. Interview with Ken Croswell: March 29, 1996.

Fabian, A. C., and Podsiadlowski, Ph., 1991. Binary Precursor for Planet? *Nature*, 353, 801.

Fegley, Bruce, Jr., and Lodders, Katharina, 1996. Atmospheric Chemistry of the Brown Dwarf Gliese 229B: Thermochemical Equilibrium Predictions. *Astrophysical Journal Letters*, 472, L37.

Fienberg, Richard Tresch, 1990. Bad News for Brown Dwarfs. *Sky and Telescope*, 80, 370 (October 1990).

Fischer, Daniel, 1996. Optical Interferometry: Breaking the Barriers. *Sky and Telescope*, 92, No. 5 (November 1996), p. 36.

Fontenrose, Robert, 1973. In Search of Vulcan. *Journal for the History of Astronomy*, 4, 145.

Forbes, Eric G., 1971. Gauss and the Discovery of Ceres. *Journal for the History of Astronomy*, 2, 195.

———, 1982. The Pre-Discovery Observations of Uranus. In *Uranus and the Outer Planets*, edited by Garry Hunt (Cambridge: Cambridge University Press), p. 67.

Forward, Robert L., 1996. Ad Astra! *Journal of the British Interplanetary Society*, 49, 23.

Foss, A. P. O., Shawe-Taylor, J. S., and Whitworth, D. P. D., 1972. Search for a Trans-Plutonian Planet. *Nature*, 239, 266.

Frail, Dale A., 1996. Interview with Ken Croswell: April 26, 1996.

François, P., Spite, M., Gillet, D., Gonzalez, J.-F., and Spite, F., 1996. Accurate

Determination of the Projected Rotational Velocity of 51 Pegasi. *Astronomy and Astrophysics*, 310, L13.

Fredrick, Laurence W., 1996. Peter van de Kamp (1901-1995). *Publications of the Astronomical Society of the Pacific*, 108, 556.

Fredrick, L. W., and Ianna, P. A., 1985. The Barnard's Star Perturbation. *Bulletin of the American Astronomical Society*, 17, 551.

Frith, I., 1887. *Life of Giordano Bruno the Nolan* (London: Trübner and Company).

Fruchter, A. S., Stinebring, D. R., and Taylor, J. H., 1988. A Millisecond Pulsar in an Eclipsing Binary. *Nature*, 333, 237.

Fugate, Robert Q., and Wild, Walter J., 1994. Untwinkling the Stars—Part I. *Sky and Telescope*, 87, No. 5 (May 1994), p. 24.

Gatewood, George, 1974. An Astrometric Study of Lalande 21185. *Astronomical Journal*, 79, 52.

———, 1976. On the Astrometric Detection of Neighboring Planetary Systems. *Icarus*, 27, 1.

———, 1995. A Study of the Astrometric Motion of Barnard's Star. *Astrophysics and Space Science*, 223, 91.

———, 1996a. Interview with Ken Croswell: March 18, 1996.

———, 1996b. Interview with Ken Croswell: May 28, 1996.

———, 1996c. Lalande 21185. *Bulletin of the American Astronomical Society*, 28, 885.

Gatewood, George, and de Jonge, Joost Kiewiet, 1995. MAP-Based Trigonometric Parallaxes of Altair and Vega. *Astrophysical Journal*, 450, 364.

Gatewood, George, and Eichhorn, Heinrich, 1973. An Unsuccessful Search for a Planetary Companion of Barnard's Star (BD +4°3561). *Astronomical Journal*, 78, 769.

Geballe, T. R., Kulkarni, S. R., Woodward, Charles E., and Sloan, G. C., 1996. The Near-Infrared Spectrum of the Brown Dwarf Gliese 229B. *Astrophysical Journal Letters*, 467, L101.

Gehman, Curtis S., Adams, Fred C., and Laughlin, Gregory, 1996. The Prospects for Earth-Like Planets Within Known Extrasolar Planetary Systems. *Publications of the Astronomical Society of the Pacific*, 108, 1018.

Giampapa, Mark S., Craine, Eric R., and Hott, Douglas A., 1995. Comments on the Photometric Method for the Detection of Extrasolar Planets. *Icarus*, 118, 199.

Gil, J. A., Jessner, A., and Kramer, M., 1993. Are There Really Planets Around PSR 1257+12? *Astronomy and Astrophysics*, 271, L17.

Gillett, F. C., 1986. IRAS Observations of Cool Excess Around Main Sequence Stars. In *Light on Dark Matter*, edited by F. P. Israel (Dordrecht: D. Reidel), p. 61.

———, 1996. Interview with Ken Croswell: March 25, 1996.

Gingerich, Owen, 1984. Herschel's 1784 Autobiography. *Sky and Telescope*, 68, 317 (October 1984).

Gliese, W., and Jahreiss, H., 1991. *Catalogue of Nearby Stars*, Third Edition, preliminary version.

Goldreich, Peter, and Ward, William R., 1972. The Case Against Planet X. *Publications of the Astronomical Society of the Pacific*, 84, 737.

Gomes, R. S., 1989. On the Problem of the Search for Planet X Based on Its Perturbation on the Outer Planets. *Icarus*, 80, 334.

Gould, Andrew, and Loeb, Abraham, 1992. Discovering Planetary Systems Through Gravitational Microlenses. *Astrophysical Journal*, 396, 104.

Gould, Andrew, and Welch, Douglas L., 1996. MACHO Proper Motions from Optical/Infrared Photometry. *Astrophysical Journal*, 464, 212.

Graham, James R., Matthews, K., Neugebauer, G., and Soifer, B. T., 1990. The Infrared Excess of G29-38: A Brown Dwarf or Dust? *Astrophysical Journal*, 357, 216.

Gray, David F., 1997. Absence of a Planetary Signature in the Spectra of the Star 51 Pegasi. *Nature*, 385, 795.

Green, Daniel W. E., Chambers, John E., and Williams, Gareth V., 1994. Is Pluto a Major Planet? No. *Sky and Telescope*, 88, No. 2 (August 1994), p. 6.

Grosser, Morton, 1962. *The Discovery of Neptune* (Cambridge: Harvard University Press).

———, 1964. The Search for a Planet Beyond Neptune. *Isis*, 55, 163.

Guillot, T., Burrows, A., Hubbard, W. B., Lunine, J. I., and Saumon, D., 1996. Giant Planets at Small Orbital Distances. *Astrophysical Journal Letters*, 459, L35.

Hale, Alan, 1995. On the Nature of the Companion to HD 114762. *Publications of the Astronomical Society of the Pacific*, 107, 22.

Hale, Alan, and Doyle, Laurance R., 1994. The Photometric Method of Extrasolar Planet Detection Revisited. *Astrophysics and Space Science*, 212, 335.

Hall, D. T., Strobel, D. F., Feldman, P. D., McGrath, M. A., and Weaver, H. A., 1995. Detection of an Oxygen Atmosphere on Jupiter's Moon Europa. *Nature*, 373, 677.

Hardy, John W., 1994. Adaptive Optics. *Scientific American*, 270, No. 6 (June 1994), p. 60.

Harrington, R. S., 1985. The Companion to VB8. *Bulletin of the American Astronomical Society*, 17, 624.

———, 1986. Barnard's Star Once Again. *Bulletin of the American Astronomical Society*, 18, 912.

———, 1987. Interview with Ken Croswell: December 9, 1987.

———, 1988. The Location of Planet X. *Astronomical Journal*, 96, 1476.

———, 1990. Interview with Ken Croswell: September 21, 1990.

———, 1992a. *Observatory*, 112, 87.

———, 1992b. Interview with Ken Croswell: December 18, 1992.

Harrington, Robert S., and Harrington, Betty J., 1987. Barnard's Star: A Status Report on an Intriguing Neighbor. *Mercury*, 16, 77 (May/June 1987).

Harrington, R. S., Kallarakal, V. V., and Dahn, C. C., 1983. Astrometry of the Low-Luminosity Stars VB8 and VB10. *Astronomical Journal*, 88, 1038.

Harrington, R. S., and Van Flandern, T. C., 1979. The Satellites of Neptune and the Origin of Pluto. *Icarus*, 39, 131.

Heintz, W. D., 1976. Systematic Trends in the Motions of Suspected Stellar Companions. *Monthly Notices of the Royal Astronomical Society*, 175, 533.

———, 1978. Reexamination of Suspected Unresolved Binaries. *Astrophysical Journal*, 220, 931.

Helfand, David J., and Hamilton, Thomas T., 1991. A Plasma Cloud, Not a Planet? *Nature*, 352, 481.

Henry, Gregory W., Baliunas, Sallie L., Donahue, Robert A., Soon, Willie H., and Saar, Steven H., 1997. Properties of Sun-Like Stars with Planets: 51 Pegasi, 47 Ursae Majoris, 70 Virginis, and HD 114762. *Astrophysical Journal*, 474, 503.

Henry, Todd J., 1996a. Brown Dwarfs Revealed—At Last! *Sky and Telescope*, 91, No. 4 (April 1996), p. 24.

———, 1996b. Private communication.

Henry, Todd J., Kirkpatrick, J. Davy, and Simons, Douglas A., 1994. The Solar Neighborhood. I. Standard Spectral Types (K5-M8) for Northern Dwarfs Within Eight Parsecs. *Astronomical Journal*, 108, 1437.

Henry, Todd J., and McCarthy, Donald W. Jr., 1993. The Mass-Luminosity Relation for Stars of Mass 1.0 to 0.08 M⊙. *Astronomical Journal*, 106, 773.

Hewish, A., Bell, S. J., Pilkington, J. D. H., Scott, P. F., and Collins, R. A., 1968. Observation of a Rapidly Pulsating Radio Source. *Nature*, 217, 709.

Hitchcock, Dian R., and Lovelock, James E., 1967. Life Detection by Atmospheric Analysis. *Icarus*, 7, 149.

Hoffleit, Dorrit, and Jaschek, Carlos, 1982. *The Bright Star Catalogue*, Fourth Revised Edition (New Haven: Yale University Observatory).

Holberg, J. B., Wesemael, F., and Hubený, I., 1984. The Far-Ultraviolet Energy Distribution of Sirius B from Voyager 2. *Astrophysical Journal*, 280, 679.

Holman, Matthew, Touma, Jihad, and Tremaine, Scott, 1997. Chaotic Variations in the Eccentricity of the Planet Orbiting 16 Cygni B. *Nature*, 386, 254.

Hoskin, Michael, 1982. Herschel and the Construction of the Heavens. In *Uranus and the Outer Planets*, edited by Garry Hunt (Cambridge: Cambridge University Press), p. 55.

Hoyt, William Graves, 1976. *Lowell and Mars* (Tucson: University of Arizona Press).

———, 1980. *Planets X and Pluto* (Tucson: University of Arizona Press).

Hubbell, John G., and Smith, Robert W., 1992. Neptune in America: Negotiating a Discovery. *Journal for the History of Astronomy*, 23, 261.

Innes, R., 1915. A Faint Star of Large Proper Motion. *Union Observatory Circular*, No. 30.

Jaki, Stanley L., 1972. The Early History of the Titius-Bode Law. *American Journal of Physics*, 40, 1014.

Jenkins, Jon M., Doyle, Laurance R., and Cullers, D. K., 1996. A Matched Filter Method for Ground-Based Sub-Noise Detection of Terrestrial Extrasolar Planets in Eclipsing Binaries: Application to CM Draconis. *Icarus*, 119, 244.

Jones, Barrie William, 1984. *The Solar System* (Oxford: Pergamon Press).

Jones, Eric M., 1985. Where Is Everybody? *Physics Today*, 38, No. 8 (August 1985), p. 11.

Jones, Glyn, 1992. When Stardom Beckoned. *New Scientist*, 135, No. 1830 (July 18, 1992), p. 36.

Kafatos, Minas C., Harrington, Robert S., and Maran, Stephen P. (editors), 1986. *Astrophysics of Brown Dwarfs* (Cambridge: Cambridge University Press).

Kirkpatrick, J. Davy, Henry, Todd J., and Liebert, James, 1993. The Unique Spectrum of the Brown Dwarf Candidate GD 165B and Comparison to the Spectra of Other Low-Luminosity Objects. *Astrophysical Journal*, 406, 701.

Klemola, A. R., and Harlan, E. A., 1972. Search for Brady's Hypothetical Trans-Plutonian Planet. *Publications of the Astronomical Society of the Pacific*, 84, 736.

Koppel, Ted, 1995. New Worlds. *Nightline* (New York: ABC News), October 19, 1995.

Kovalevsky, Jean, 1990. Astrometry from Earth and Space. *Sky and Telescope*, 79, 493 (May 1990).

Kowal, Charles T., and Drake, Stillman, 1980. Galileo's Observations of Neptune. *Nature*, 287, 311.

Kresák, Ľubor, 1992. Are There Any Comets Coming from Interstellar Space? *Astronomy and Astrophysics*, 259, 682.

Kuijken, Konrad, and Gilmore, Gerard, 1989a. The Mass Distribution in the Galactic Disc—I. A Technique to Determine the Integral Surface Mass Density of the Disc near the Sun. *Monthly Notices of the Royal Astronomical Society*, 239, 571.

———, 1989b. The Mass Distribution in the Galactic Disc—II. Determination of the Surface Mass Density of the Galactic Disc near the Sun. *Monthly Notices of the Royal Astronomical Society*, 239, 605.

———, 1989c. The Mass Distribution in the Galactic Disc—III. The Local Volume Mass Density. *Monthly Notices of the Royal Astronomical Society*, 239, 651.

Kuiper, Gerard P., 1951. On the Origin of the Solar System. In *Astrophysics*, edited by J. A. Hynek (New York: McGraw-Hill), p. 357.

———, 1956. The Formation of the Planets, Part III. *Journal of the Royal Astronomical Society of Canada*, 50, 158.

———, 1957. Further Studies on the Origin of Pluto. *Astrophysical Journal*, 125, 287.

Kulkarni, Shrinivas, 1996. Interview with Ken Croswell: May 13, 1996.

Kumar, Shiv S., 1963a. The Structure of Stars of Very Low Mass. *Astrophysical Journal*, 137, 1121.

———, 1963b. The Helmholtz-Kelvin Time Scale for Stars of Very Low Mass. *Astrophysical Journal*, 137, 1126.

———, 1967. On Planets and Black Dwarfs. *Icarus*, 6, 136.

———, 1986. Stellar Companions. *Physics Today*, 39, No. 3 (March 1986), p. 154.

———, 1990. The Nature of the Luminous and Dark Objects of Very Low Mass. *Comments on Astrophysics*, 15, 55.

———, 1994. Very Low Mass Stars, Black Dwarfs and Planets. *Astrophysics and Space Science*, 212, 57.

———, 1996. Interview with Ken Croswell: April 15, 1996.

Kuznik, Frank, 1993. Is Something Out There? *Air and Space/Smithsonian*, 8, No. 1 (April/May 1993), p. 46.

Lagrange-Henri, A. M., Vidal-Madjar, A., and Ferlet, R., 1988. The ß Pictoris Circumstellar Disk. VI. Evidence for Material Falling on to the Star. *Astronomy and Astrophysics*, 190, 275.

Lanz, Thierry, Heap, Sara R., and Hubeny, Ivan, 1995. HST/GHRS Observations of the ß Pictoris System: Basic Parameters and the Age of the System. *Astrophysical Journal Letters*, 447, L41.

Laskar, J., Joutel, F., and Robutel, P., 1993. Stabilization of the Earth's Obliquity by the Moon. *Nature*, 361, 615.

Latham, David W., 1996. Private communications: August 11, 1996; October 4, 1996.

Latham, David W., Mazeh, Tsevi, Stefanik, Robert P., Mayor, Michel, and Burki, Gilbert, 1989. The Unseen Companion of HD114762: A Probable Brown Dwarf. *Nature*, 339, 38.

Léger, A., Mariotti, J. M., Mennesson, B., Ollivier, M., Puget, J. L., Rouan, D., and Schneider, J., 1996. Could We Search for Primitive Life on Extrasolar Planets in the Near Future? The DARWIN Project. *Icarus*, 123, 249.

Léger, A., Pirre, M., and Marceau, F. J., 1993. Search for Primitive Life on a Distant Planet: Relevance of O_2 and O_3 Detections. *Astronomy and Astrophysics*, 277, 309.

Leggett, S. K., 1992. Infrared Colors of Low-Mass Stars. *Astrophysical Journal Supplement Series*, 82, 351.

Leverrier, U., 1859. Suspected Existence of a Zone of Asteroids Revolving Between Mercury and the Sun. *Monthly Notices of the Royal Astronomical Society*, 20, 24.

Levy, David H., 1991. *Clyde Tombaugh* (Tucson: University of Arizona Press).

———, 1994. Is Pluto a Major Planet? Yes. *Sky and Telescope*, 88, No. 2 (August 1994), p. 8.

Liebert, James, and Probst, Ronald G., 1987. Very Low Mass Stars. *Annual Review of Astronomy and Astrophysics*, 25, 473.

Lin, D. N. C., Bodenheimer, P., and Richardson, D. C., 1996. Orbital Migration of the Planetary Companion of 51 Pegasi to Its Present Location. *Nature*, 380, 606.

Lin, D. N. C., and Ida, Shigeru, 1997. On the Origin of Massive Eccentric Planets. *Astrophysical Journal*, 477, 781.

Lin, D. N. C., and Papaloizou, John, 1986. On the Tidal Interaction Between Protoplanets and the Protoplanetary Disk. III. Orbital Migration of Protoplanets. *Astrophysical Journal*, 309, 846.

Lin, D. N. C., Woosley, S. E., and Bodenheimer, P. H., 1991. Formation of a Planet Orbiting Pulsar 1829-10 from the Debris of a Supernova Explosion. *Nature*, 353, 827.

Lindegren, L., and Perryman, M. A. C., 1996. GAIA: Global Astrometric Interferometer for Astrophysics. *Astronomy and Astrophysics Supplement Series*, 116, 579.

Lindsay, Jack, 1962. *Cause, Principle, and Unity: Five Dialogues by Giordano Bruno* (Castle Hedingham, Essex: Daimon Press).

Lippincott, Sarah Lee, 1960. Astrometric Analysis of Lalande 21185. *Astronomical Journal*, 65, 445.

Lissauer, Jack J., 1993. Planet Formation. *Annual Review of Astronomy and Astrophysics*, 31, 129.

———, 1995. Urey Prize Lecture: On the Diversity of Plausible Planetary Systems. *Icarus*, 114, 217.

———, 1996. Interview with Ken Croswell: May 15, 1996.

Lissauer, Jack J., Pollack, James B., Wetherill, George W., and Stevenson, David J., 1995. Formation of the Neptune System. In *Neptune and Triton*, edited by Dale P. Cruikshank (Tucson: University of Arizona Press), p. 37.

Littmann, Mark, 1988. *Planets Beyond* (New York: Wiley).

———, 1989. Where Is Planet X? *Sky and Telescope*, 78, 596 (December 1989).

Livio, M., Pringle, J. E., and Saffer, R. A., 1992. Planets Around Massive White Dwarfs. *Monthly Notices of the Royal Astronomical Society*, 257, 15P.

Lovelock, J. E., 1975. Thermodynamics and the Recognition of Alien Biospheres. *Proceedings of the Royal Society of London B*, 189, 167.

Lyne, Andrew, 1991. A Planet, Not a Plasma Cloud? *Nature*, 352, 573.

———, 1996. Interview with Ken Croswell: April 25, 1996.

Lyne, A. G., and Bailes, M., 1992. No Planet Orbiting PSR1829-10. *Nature*, 355, 213.

Lyne, A. G., and Graham-Smith, F., 1990. *Pulsar Astronomy* (Cambridge: Cambridge University Press).

Lyne, A. G., Manchester, R. N., D'Amico, N., Staveley-Smith, L., Johnston, S., Lim, J., Fruchter, A. S., Goss, W. M., and Frail, D., 1990. An Eclipsing Millisecond Pulsar in the Globular Cluster Terzan 5. *Nature*, 347, 650.

Lyttleton, Raymond A., 1936. On the Possible Results of an Encounter of Pluto with the Neptunian System. *Monthly Notices of the Royal Astronomical Society*, 97, 108.

MacRobert, Alan M., and Roth, Joshua, 1996. The Planet of 51 Pegasi. *Sky and Telescope*, 91, No. 1 (January 1996), p. 38.

Magazzù, Antonio, Martín, Eduardo L., and Rebolo, Rafael, 1993. A Spectroscopic Test for Substellar Objects. *Astrophysical Journal Letters*, 404, L17.

Malbet, F., 1996. High Angular Resolution Coronography for Adaptive Optics. *Astronomy and Astrophysics Supplement Series*, 115, 161.

Mallove, Eugene F., and Matloff, Gregory L., 1989. *The Starflight Handbook* (New York: Wiley).

Mammana, Dennis L., and McCarthy, Donald, 1996. *Other Suns. Other Worlds?* (New York: St. Martin's Press).

Mao, Shude, and Paczyński, Bohdan, 1991. Gravitational Microlensing by Double Stars and Planetary Systems. *Astrophysical Journal Letters*, 374, L37.

Marcy, Geoffrey W., 1996. Interview with Ken Croswell: May 15, 1996; May 16, 1996.

Marcy, Geoffrey W., and Benitz, Karsten J., 1989. A Search for Substellar Companions to Low-Mass Stars. *Astrophysical Journal*, 344, 441.

Marcy, Geoffrey W., and Butler, R. Paul, 1992. Precision Radial Velocities with an Iodine Absorption Cell. *Publications of the Astronomical Society of the Pacific*, 104, 270.

———, 1996. A Planetary Companion to 70 Virginis. *Astrophysical Journal Letters*, 464, L147.

Marcy, Geoffrey W., Butler, R. Paul, Williams, Eric, Bildsten, Lars, Graham, James R., Ghez, Andrea M., and Jernigan, J. Garrett, 1997. The Planet Around 51 Pegasi. *Astrophysical Journal*, in press.

Marsden, Brian G., 1988. Interview with Ken Croswell: March 19, 1988.

———, 1990. Interview with Ken Croswell: September 21, 1990.

———, 1995. 51 Pegasi. *International Astronomical Union Circular*, No. 6251.

———, 1996. Interview with Ken Croswell: March 1, 1996; March 4, 1996.

Martín, Eduardo L., Rebolo, Rafael, and Zapatero-Osorio, María Rosa, 1996. Spectroscopy of New Substellar Candidates in the Pleiades: Toward a Spectral Sequence for Young Brown Dwarfs. *Astrophysical Journal*, 469, 706.

Martin, Eva, 1921. *Giordano Bruno: Mystic and Martyr* (London: William Rider and Son).

Matloff, Gregory L., 1996. Wormholes and Hyper-Drives. *Mercury*, 25, No. 4 (July/August 1996), p. 10.

Mauldin, John H., 1992. *Prospects for Interstellar Travel* (San Diego: American Astronautical Society).

Mayor, Michel, 1996. Interview with Ken Croswell: May 20, 1996.

Mayor, Michel, and Queloz, Didier, 1995. A Jupiter-Mass Companion to a Solar-Type Star. *Nature*, 378, 355.

———, 1996. A Search for Substellar Companions to Solar-Type Stars via Precise Doppler Measurements: A First Jupiter Mass Companion Detected. In *Cool Stars, Stellar Systems, and the Sun, Ninth Cambridge Workshop*, edited by Roberto Pallavicini and Andrea K. Dupree (San Francisco: Astronomical Society of the Pacific), p. 35.

Mazeh, Tsevi, and Goldman, Itzhak, 1995. Similarities Between the Inner Solar System and the Planetary System of PSR B1257+12. *Publications of the Astronomical Society of the Pacific*, 107, 250.

Mazeh, Tsevi, Krymolowski, Yuval, and Rosenfeld, Gady, 1997. The High Eccentricity of the Planet Orbiting 16 Cygni B. *Astrophysical Journal Letters*, 477, L103.

Mazeh, Tsevi, Latham, David W., and Stefanik, Robert P., 1996. Spectroscopic Orbits for Three Binaries with Low-Mass Companions and the Distribution of Secondary Masses near the Substellar Limit. *Astrophysical Journal*, 466, 415.

McCarthy, D. W. Jr., Probst, Ronald G., and Low, F. J., 1985. Infrared Detection of a Close Cool Companion to Van Biesbroeck 8. *Astrophysical Journal Letters*, 290, L9.

McCook, George P., and Sion, Edward M., 1987. A Catalog of Spectroscopically Identified White Dwarfs. *Astrophysical Journal Supplement Series*, 65, 603.

McFarland, John, 1996. Kenneth Essex Edgeworth—Victorian Polymath and Founder of the Kuiper Belt? *Vistas in Astronomy*, 40, 343.

McGlynn, Thomas A., and Chapman, Robert D., 1989. On the Nondetection of Extrasolar Comets. *Astrophysical Journal Letters*, 346, L105.

McIntyre, J. Lewis, 1903. *Giordano Bruno* (New York: Macmillan).

Michel, Paul Henri, 1973. *The Cosmology of Giordano Bruno*, translated by R. E. W. Maddison (Ithaca: Cornell University Press).

Mizuno, Hiroshi, 1980. Formation of the Giant Planets. *Progress of Theoretical Physics*, 64, 544.

Mizuno, Hiroshi, Nakazawa, Kiyoshi, and Hayashi, Chushiro, 1978. Instability of a Gaseous Envelope Surrounding a Planetary Core and Formation of Giant Planets. *Progress of Theoretical Physics*, 60, 699.

Moore, Patrick, 1988. *The Planet Neptune* (Chichester, England: Ellis Horwood).

———, 1995. The Discoveries of Neptune and Triton. In *Neptune and Triton*, edited by Dale P. Cruikshank (Tucson: University of Arizona Press), p. 15.

Mulholland, Derral, 1989. Mapping the Sky. *Mosaic*, 20, No. 1 (Spring 1989), p. 34.

Murray, Mary, 1986. The Search for Unlit Suns. *Science News*, 130, 282 (November 1, 1986).

Naeye, Robert, 1996. Is This Planet for Real? *Astronomy*, 24, No. 3 (March 1996), p. 34.

Naeye, Robert, and Potter, Seth, 1996. Journey to the Outer Limits. *Astronomy*, 24, No. 8 (August 1996), p. 36.

Nakajima, Tadashi, 1994. Planet Detectability by an Adaptive Optics Stellar Coronagraph. *Astrophysical Journal*, 425, 348.

———, 1996. Interview with Ken Croswell: May 9, 1996.

Nakajima, T., Durrance, S. T., Golimowski, D. A., and Kulkarni, S. R., 1994. A Coronagraphic Search for Brown Dwarfs Around Nearby Stars. *Astrophysical Journal*, 428, 797.

Nakajima, T., Oppenheimer, B. R., Kulkarni, S. R., Golimowski, D. A., Matthews, K., and Durrance, S. T., 1995. Discovery of a Cool Brown Dwarf. *Nature*, 378, 463.

Nakamura, Takashi, and Piran, Tsvi, 1991. The Origin of the Planet Around PSR 1829-10. *Astrophysical Journal Letters*, 382, L81.

Nelson, L. A., Rappaport, S., and Chiang, E., 1993. On the Li and Be Tests for Brown Dwarfs. *Astrophysical Journal*, 413, 364.

Nieto, Michael Martin, 1972. *The Titius-Bode Law of Planetary Distances* (New York: Pergamon).

Noyes, Robert W., Jha, Saurabh, Korzennik, Sylvain G., Krockenberger, Martin, Nisenson, Peter, Brown, Timothy M., Kennelly, Edward J., and Horner, Scott D., 1997. A Planet Orbiting the Star Rho Coronae Borealis. *Astrophysical Journal Letters*, in press.

Oort, J. H., 1932. The Force Exerted by the Stellar System in the Direction

Perpendicular to the Galactic Plane and Some Related Problems. *Bulletin of the Astronomical Institutes of the Netherlands*, 6, 249.

———, 1965. Stellar Dynamics. In *Galactic Structure*, edited by Adriaan Blaauw and Maarten Schmidt (Chicago: University of Chicago Press), p. 455.

Oppenheimer, B. R., Kulkarni, S. R., Matthews, K., and Nakajima, T., 1995. Infrared Spectrum of the Cool Brown Dwarf Gl 229B. *Science*, 270, 1478.

Owen, Tobias, 1980. The Search for Early Forms of Life in Other Planetary Systems: Future Possibilities Afforded by Spectroscopic Techniques. In *Strategies for the Search for Life in the Universe*, edited by Michael D. Papagiannis (Dordrecht: D. Reidel), p. 177.

———, 1992. The Composition and Early History of the Atmosphere of Mars. In *Mars*, edited by Hugh H. Kieffer, Bruce M. Jakosky, Conway W. Snyder, and Mildred S. Matthews (Tucson: University of Arizona Press), p. 818.

———, 1994. The Search for Other Planets: Clues from the Solar System. *Astrophysics and Space Science*, 212, 1.

Paczyński, Bohdan, 1986. Gravitational Microlensing by the Galactic Halo. *Astrophysical Journal*, 304, 1.

———, 1995. Interview with Ken Croswell: November 13, 1995.

———, 1996. Gravitational Microlensing in the Local Group. *Annual Review of Astronomy and Astrophysics*, 34, 419.

Perrier, C., and Mariotti, J.-M., 1987. On the Binary Nature of Van Biesbroeck 8. *Astrophysical Journal Letters*, 312, L27.

Perryman, M. A. C., Lindegren, L., Arenou, F., Bastian, U., Bernstein, H. H., van Leeuwen, F., Schrijver, H., Bernacca, P. L., Evans, D. W., Falin, J. L., Froeschlé, M., Grenon, M., Hering, R., Høg, E., Kovalevsky, J., Mignard, F., Murray, C. A., Penston, M. J., Petersen, C. S., Le Poole, R. S., Söderhjelm, S., and Turon, C., 1996. Hipparcos Distances and Mass Limits for the Planetary Candidates: 47 Ursae Majoris, 70 Virginis, and 51 Pegasi. *Astronomy and Astrophysics*, 310, L21.

Peterson, Ivars, 1994. Guiding Light. *Science News*, 146, 56 (July 23, 1994).

Phillips, J. A., and Thorsett, S. E., 1994. Planets Around Pulsars: A Review. *Astrophysics and Space Science*, 212, 91.

Phillips, J. A., Thorsett, S. E., and Kulkarni, S. R. (editors), 1993. *Planets Around Pulsars* (San Francisco: Astronomical Society of the Pacific).

Pierce, David A., 1991. The Mass of Neptune. *Icarus*, 94, 413.

Podsiadlowski, Ph., Pringle, J. E., and Rees, M. J., 1991. The Origin of the Planet Orbiting PSR1829-10. *Nature*, 352, 783.

Porter, Roy, 1982. William Herschel, Bath, and the Philosophical Society. In *Uranus and the Outer Planets*, edited by Garry Hunt (Cambridge: Cambridge University Press), p. 23.

Postnov, K. A., and Prokhorov, M. E., 1992. Planet Survival? *Astronomy and Astrophysics*, 258, L17.

Powell, Conley, 1989. A Mathematical Search for Planet X. *Journal of the British Interplanetary Society*, 42, 327.

Purcell, Edward, 1960. Radioastronomy and Communication Through

Space. *Brookhaven National Laboratory Lectures in Science: Vistas in Research*, 1, 3-13 (USAEC Report BNL-658). Also in *Interstellar Communication*, edited by A. G. W. Cameron (New York: W. A. Benjamin, 1963), p. 121.

Putnam, William Lowell, 1994. *The Explorers of Mars Hill* (West Kennebunk, Maine: Phoenix).

Queloz, Didier, 1996. Interview with Ken Croswell: May 10, 1996.

Rasio, F. A., Nicholson, P. D., Shapiro, S. L., and Teukolsky, S. A., 1992. An Observational Test for the Existence of a Planetary System Orbiting PSR1257+12. *Nature*, 355, 325.

Rawlins, Dennis, 1970. The Great Unexplained Residual in the Orbit of Neptune. *Astronomical Journal*, 75, 856.

———, 1992. The Neptune Conspiracy: British Astronomy's Post-Discovery Discovery. *Dio*, 2, 115.

———, 1994. The "Theft" of the Neptune Papers, or: Does the Astronomer Royal Merit an Amnesty? *Dio*, 4, 92.

Rebolo, R., Martín, E. L., Basri, G., Marcy, G. W., and Zapatero-Osorio, M. R., 1996. Brown Dwarfs in the Pleiades Cluster Confirmed by the Lithium Test. *Astrophysical Journal Letters*, 469, L53.

Rebolo, Rafael, Martín, Eduardo L., and Magazzù, Antonio, 1992. Spectroscopy of a Brown Dwarf Candidate in the α Persei Open Cluster. *Astrophysical Journal Letters*, 389, L83.

Rebolo, R., Zapatero Osorio, M. R., and Martín, E. L., 1995. Discovery of a Brown Dwarf in the Pleiades Star Cluster. *Nature*, 377, 129.

Reuyl, Dirk, and Holmberg, Erik, 1943. On the Existence of a Third Component in the System 70 Ophiuchi. *Astrophysical Journal*, 97, 41.

Richards, D. W., Pettengill, G. H., Counselman III, C. C., and Rankin, J. M., 1970. Quasi-Sinusoidal Oscillation in Arrival Times of Pulses from NP 0532. *Astrophysical Journal*, 160, L1.

Robinson, Edward L., Cochran, Anita L., Cochran, William D., Shafter, Allen W., and Zhang, Er-Ho, 1990. A Search for Eclipses of HD 114762 by a Low-Mass Companion. *Astronomical Journal*, 99, 672.

Roseveare, N. T., 1982. *Mercury's Perihelion from Le Verrier to Einstein* (Oxford: Clarendon Press).

Ruderman, M., Shaham, J., and Tavani, M., 1989. Accretion Turnoff and Rapid Evaporation of Very Light Secondaries in Low-Mass X-Ray Binaries. *Astrophysical Journal*, 336, 507.

Russell, Henry Norris, 1943. Physical Characteristics of Stellar Companions of Small Mass. *Publications of the Astronomical Society of the Pacific*, 55, 79.

———, 1944. Notes on White Dwarfs and Small Companions. *Astronomical Journal*, 51, 13.

Ryden, Barbara, 1990. Spots on the Sun. *Star Date*, 18, No. 1 (January/February 1990), p. 19.

Sackmann, I.-Juliana, Boothroyd, Arnold I., and Kraemer, Kathleen E., 1993. Our Sun. III. Present and Future. *Astrophysical Journal*, 418, 457.

Saumon, D., Hubbard, W. B., Burrows, A., Guillot, T., Lunine, J. I., and Chabrier, G., 1996. A Theory of Extrasolar Giant Planets. *Astrophysical Journal*, 460, 993.

Schaffer, Simon, 1981. Uranus and the Establishment of Herschel's Astronomy. *Journal for the History of Astronomy*, 12, 11.

Seidelmann, P. Kenneth, 1987. Interview with Ken Croswell: December 11, 1987.

———, 1990. Interview with Ken Croswell: September 21, 1990.

———, 1996. Interview with Ken Croswell: February 29, 1996.

Seidelmann, P. K., and Harrington, R. S., 1988. Planet X—the Current Status. *Celestial Mechanics*, 43, 55.

Seidelmann, P. K., Klepczynski, W. J., Duncombe, R. L., and Jackson, E. S., 1971. Determination of the Mass of Pluto. *Astronomical Journal*, 76, 488.

Seidelmann, P. K., Marsden, B. G., and Giclas, H. L., 1972. Note on Brady's Hypothetical Trans-Plutonian Planet. *Publications of the Astronomical Society of the Pacific*, 84, 858.

Seidelmann, P. K., Santoro, E. J., and Pulkkinen, K. F., 1985. Systematic Differences Between Planetary Observations and Ephemerides. In *Dynamical Astronomy*, edited by Victor G. Szebehely and Béla Balázs (Austin: University of Texas Press), p. 55.

Seidelmann, P. K., and Williams, C. A., 1988. Discussion of Current Status of Planet X. *Celestial Mechanics*, 43, 409.

Sekanina, Zdenek, 1976. A Probability of Encounter with Interstellar Comets and the Likelihood of Their Existence. *Icarus*, 27, 123.

Sen, A. K., and Rana, N. C., 1993. On the Missing Interstellar Comets. *Astronomy and Astrophysics*, 275, 298.

Serio, Giorgia Foderà, 1990. Giuseppe Piazzi and the Discovery of the Proper Motion of 61 Cygni. *Journal for the History of Astronomy*, 21, 275.

Shabanova, Tatiana V., 1995. Evidence for a Planet Around the Pulsar PSR B0329+54. *Astrophysical Journal*, 453, 779.

Shao, Michael, 1996. Interview with Ken Croswell: June 4, 1996.

Shao, M., and Colavita, M. M., 1992. Long-Baseline Optical and Infrared Stellar Interferometry. *Annual Review of Astronomy and Astrophysics*, 30, 457.

Shapley, Harlow, 1958. *Of Stars and Men* (Boston: Beacon Press).

Sheehan, William, 1995. *The Immortal Fire Within* (Cambridge: Cambridge University Press).

Shemar, Setnam, 1991. How We Found the First Planet Beyond the Solar System. *Astronomy Now*, 5, No. 10 (October 1991), p. 10.

Siegel, Robert, 1994. Contest Results to Name 3 New Planets. On *All Things Considered* (Washington, D.C.: National Public Radio), April 29, 1994.

Singer, Dorothea Waley, 1950. *Giordano Bruno: His Life and Thought* (New York: Henry Schuman).

Sinnott, Roger W., 1987. The Wandering Stars of Allegheny. *Sky and Telescope*, 74, 360 (October 1987).

Skrutskie, M. F., Forrest, W. J., and Shure, M. A., 1987. Direct Infrared Imaging of VB 8. *Astrophysical Journal Letters*, 312, L55.

Smart, W. M., 1947. John Couch Adams and the Discovery of Neptune. *Occasional Notes of the Royal Astronomical Society*, 2, 33.

Smith, Bradford A., 1996. Interview with Ken Croswell: March 27, 1996.

Smith, Bradford A., and Terrile, Richard J., 1984. A Circumstellar Disk Around ß Pictoris. *Science*, 226, 1421.

Smith, Robert W., 1982. The Impact on Astronomy of the Discovery of Uranus. In *Uranus and the Outer Planets*, edited by Garry Hunt (Cambridge: Cambridge University Press), p. 81.

———, 1983. William Lassell and the Discovery of Neptune. *Journal for the History of Astronomy*, 14, 30.

———, 1989. The Cambridge Network in Action: The Discovery of Neptune. *Isis*, 80, 395.

Soderblom, David R., and Däppen, Werner, 1989. Modeling Epsilon Eridani and Its Oscillations. *Astrophysical Journal*, 342, 945.

Stahl, Steven M., and Sandler, David G., 1995. Optimization and Performance of Adaptive Optics for Imaging Extrasolar Planets. *Astrophysical Journal Letters*, 454, L153.

Standish, E. Myles, Jr., 1981. Galileo's Observation of Neptune. *Nature*, 290, 164.

———, 1990a. Planet X: Where Aren't You? *Sky and Telescope*, 79, 581 (June 1990).

———, 1990b. Interview with Ken Croswell: September 21, 1990.

———, 1993. Planet X: No Dynamical Evidence in the Optical Observations. *Astronomical Journal*, 105, 2000.

———, 1994. Meridian Circle Observations of the Planets. In *Galactic and Solar System Optical Astrometry*, edited by L. V. Morrison and G. F. Gilmore (Cambridge: Cambridge University Press), p. 253.

———, 1996a. Interview with Ken Croswell: February 29, 1996.

———, 1996b. Pluto and Planets X. In *Completing the Inventory of the Solar System*, edited by Terrence W. Rettig and Joseph M. Hahn (San Francisco: Astronomical Society of the Pacific), p. 163.

Stephens, Sally, 1995. Needles in the Cosmic Haystack. *Astronomy*, 23, No. 9 (September 1995), p. 50.

Stern, S. Alan, 1990. On the Number Density of Interstellar Comets as a Constraint on the Formation Rate of Planetary Systems. *Publications of the Astronomical Society of the Pacific*, 102, 793.

———, 1997. Seeking Rogue Comets. *Astronomy*, 25, No. 2 (February 1997), p. 46.

Stevenson, David J., 1991. The Search for Brown Dwarfs. *Annual Review of Astronomy and Astrophysics*, 29, 163.

———, 1996. Interview with Ken Croswell: April 16, 1996.

Strand, K. Aa., 1943a. 61 Cygni as a Triple System. *Publications of the Astronomical Society of the Pacific*, 55, 29.

———, 1943b. The Orbital Motion of 61 Cygni. *Proceedings of the American Philosophical Society*, 86, 364.

———, 1957. The Orbital Motion of 61 Cygni. *Astronomical Journal*, 62, 35.

Tarter, Jill C., 1975. *The Interaction of Gas and Galaxies Within Galaxy Clusters.* Ph.D. thesis, University of California at Berkeley.

Taylor, Gordon E., 1985. The Observations of Neptune by Galileo. *Journal of the British Astronomical Association,* 95, 116.

Telesco, C. M., Joy, M., and Sisk, C., 1990. Observations of G29-38 at 10 Microns. *Astrophysical Journal Letters,* 358, L17.

Terrile, Richard J., 1996. Interview with Ken Croswell: March 28, 1996.

Tinney, Christopher G. (editor), 1995. *The Bottom of the Main Sequence—and Beyond* (New York: Springer-Verlag).

Tokunaga, A. T., Becklin, E. E., and Zuckerman, B., 1990. The Infrared Spectrum of G29-38. *Astrophysical Journal Letters,* 358, L21.

Tombaugh, Clyde W., 1961. The Trans-Neptunian Planet Search. In *Planets and Satellites,* edited by Gerard P. Kuiper and Barbara M. Middlehurst (Chicago: University of Chicago Press), p. 12.

———, 1987. Interview with Ken Croswell: December 9, 1987.

———, 1991. Plates, Pluto, and Planets X. *Sky and Telescope,* 81, 360 (April 1991).

———, 1996. Interview with Ken Croswell: February 18, 1996.

Tombaugh, Clyde W., and Moore, Patrick, 1980. *Out of the Darkness: The Planet Pluto* (Harrisburg, Pennsylvania: Stackpole Books).

Townes, Charles (chair), 1996. *Exploration of Neighboring Planetary Systems (ExNPS): Mission and Technology Road Map* (Pasadena: Jet Propulsion Laboratory).

Tsuji, Takashi, Ohnaka, Keiichi, and Aoki, Wako, 1995. Spectra and Colours of Brown Dwarfs: Predictions by Model Atmospheres. In *The Bottom of the Main Sequence—and Beyond,* edited by Christopher G. Tinney (New York: Springer-Verlag), p. 45.

Tsuji, T., Ohnaka, K., Aoki, W., and Nakajima, T., 1996. Evolution of Dusty Photospheres Through Red to Brown Dwarfs: How Dust Forms in Very Low Mass Objects. *Astronomy and Astrophysics,* 308, L29.

Turner, Herbert Hall, 1904. *Astronomical Discovery* (London: Edward Arnold).

van Altena, William F., Lee, John Truen-liang, and Hoffleit, E. Dorrit, 1995. *The General Catalogue of Trigonometric Stellar Parallaxes,* Fourth Edition (New Haven: Yale University Observatory).

Van Biesbroeck, G., 1944. The Star of Lowest Known Luminosity. *Astronomical Journal,* 51, 61.

———, 1961. A Search for Stars of Low Luminosity. *Astronomical Journal,* 66, 528.

van de Kamp, Peter, 1963. Astrometric Study of Barnard's Star from Plates Taken with the 24-Inch Sproul Refractor. *Astronomical Journal,* 68, 515.

———, 1969a. Parallax, Proper Motion, Acceleration, and Orbital Motion of Barnard's Star. *Astronomical Journal,* 74, 238.

———, 1969b. Alternate Dynamical Analysis of Barnard's Star. *Astronomical Journal,* 74, 757.

———, 1974. A Study of Barnard's Star. *Bulletin of the American Astronomical Society,* 6, 306.

———, 1975. Unseen Astrometric Companions of Stars. *Annual Review of Astronomy and Astrophysics*, 13, 295.

———, 1977. Barnard's Star 1916-1976: A Sexagintennial Report. *Vistas in Astronomy*, 20, 501.

———, 1983a. The Planetary System of Barnard's Star. *Vistas in Astronomy*, 26, 141.

———, 1983b. The Fainter End of the Main Sequence. In *The Nearby Stars and the Stellar Luminosity Function, IAU Colloquium No. 76*, edited by A. G. Davis Philip and Arthur R. Upgren (Schenectady, New York: L. Davis Press), p. 15.

———, 1986. Astrometric Study of Barnard's Star. *Space Science Reviews*, 43, 312.

———, 1987. Interview with Ken Croswell: May 26, 1987.

van de Kamp, Peter, and Lippincott, Sarah Lee, 1951. Astrometric Study of Lalande 21185. *Astronomical Journal*, 56, 49.

von Zahn, U., Kumar, S., Niemann, H., and Prinn, R., 1983. Composition of the Venus Atmosphere. In *Venus*, edited by D. M. Hunten, L. Colin, T. M. Donahue, and V. I. Moroz (Tucson: University of Arizona Press), p. 299.

Walker, Gordon A. H., Walker, Andrew R., Irwin, Alan W., Larson, Ana M., Yang, Stephenson L. S., and Richardson, Derek C., 1995. A Search for Jupiter-Mass Companions to Nearby Stars. *Icarus*, 116, 359.

Wambsganss, Joachim, 1997. Discovering Galactic Planets by Gravitational Microlensing: Magnification Patterns and Light Curves. *Monthly Notices of the Royal Astronomical Society*, 284, 172.

Weissman, Paul R., 1990. The Oort Cloud. *Nature*, 344, 825.

———, 1995a. The Kuiper Belt. *Annual Review of Astronomy and Astrophysics*, 33, 327.

———, 1995b. Where Are the Interstellar Comets? *Bulletin of the American Astronomical Society*, 27, 1123.

———, 1996. Interview with Ken Croswell: May 14, 1996.

Wetherill, George W., 1992. Interview with Ken Croswell: September 29, 1992.

———, 1994. Possible Consequences of Absence of "Jupiters" in Planetary Systems. *Astrophysics and Space Science*, 212, 23.

———, 1995. How Special Is Jupiter? *Nature*, 373, 470.

———, 1996a. The Formation and Habitability of Extra-Solar Planets. *Icarus*, 119, 219.

———, 1996b. Interview with Ken Croswell: May 10, 1996.

Whyte, A. J., 1980. *The Planet Pluto* (New York: Pergamon Press).

Wild, Walter J., and Fugate, Robert Q., 1994. Untwinkling the Stars—Part II. *Sky and Telescope*, 87, No. 6 (June 1994), p. 20.

Williams, Darren M., Kasting, James F., and Wade, Richard A., 1997. Habitable Moons Around Extrasolar Giant Planets. *Nature*, 385, 234.

Wolf, M., 1918. Zwei Sterne mit Großer Eigenbewegung in Leo. *Astronomische Nachrichten*, 206, 237.

Wolszczan, A., 1991. A Nearby 37.9-Ms Radio Pulsar in a Relativistic Binary System. *Nature*, 350, 688.

———, 1994a. Toward Planets Around Neutron Stars. *Astrophysics and Space Science*, 212, 67.

———, 1994b. Confirmation of Earth-Mass Planets Orbiting the Millisecond Pulsar PSR B1257+12. *Science*, 264, 538.

———, 1996. Interview with Ken Croswell: May 2, 1996.

Wolszczan, A., and Frail, D. A., 1992. A Planetary System Around the Millisecond Pulsar PSR1257+12. *Nature*, 355, 145.

Yamamoto, Issei, 1934. Prof. Yamamoto's Suggestion on the Origin of Pluto. *Bulletin of the Kwasan Observatory*, No. 288.

Yates, Frances A., 1964. *Giordano Bruno and the Hermetic Tradition* (London: Routledge and Kegan Paul).

Young, C. A., 1878. Vulcan and the Corona. *New York Times*, August 16, 1878, p. 5.

Zuckerman, B., and Becklin, E. E., 1987. Excess Infrared Radiation from a White Dwarf—an Orbiting Brown Dwarf? *Nature*, 330, 138.

———, 1992. Companions to White Dwarfs: Very Low-Mass Stars and the Brown Dwarf Candidate GD 165B. *Astrophysical Journal*, 386, 260.

Zuckerman, Ben, and Hart, Michael H. (editors), 1995. *Extraterrestrials: Where Are They?*, Second Edition (Cambridge: Cambridge University Press).

Index

References to figures are *italicized.*